# Sustainable Forestry Management and Wood Production in a Global Economy

*Sustainable Forestry Management and Wood Production in a Global Economy* has been co-published simultaneously as *Journal of Sustainable Forestry*, Volume 24, Number 1 and Numbers 2/3 2007.

## Monographic Separates from the *Journal of Sustainable Forestry*

For additional information on these and other Haworth Press titles, including descriptions, tables of contents, reviews, and prices, use the QuickSearch catalog at http://www.HaworthPress.com.

***Sustainable Forestry Management and Wood Production in a Global Economy***, Robert L. Deal, Rachel White, and Gary L. Benson (Vol. 24, No. 1 & No. 2/3, 2007). *An examination of sustainable forest management from a global view using diverse examples of important forestry issues.*

***Plantations and Protected Areas in Sustainable Forestry***, William C. Price, Naureen Rana, and V. Alaric Sample (Vol. 21, No. 4, 2005). *This updated collection of papers from the 2001 two-day symposium at Grey Towers National Historic Site sponsored by the Pinchot Institute explores crucial forest management issues in the United States and elsewhere.*

***Sustainable Forestry in Southern Sweden: The SUFOR Research Project***, edited by Kristina Blennow and Mats Niklasson (Vol. 21, No. 2/3, 2005). *"The authors are all very competent and well-known members of their respective disciplines. The text reads well and will be OF INTEREST TO FOREST MANAGERS AND SCIENTISTS ALIKE." (Klaus von Gadow, PhD, Professor of Forest Management, Institute of Forest Assessment and Forest Growth, Georg-August-University Göttingen, Germany)*

***Environmental Services of Agroforestry Systems***, edited by Florencia Montagnini (Vol. 21, No. 1, 2005). *"TIMELY AND APPROPRIATE. . . . All those interested in agroforestry and the ongoing environmental debate will find this book quite useful." (P. K. Ramachandran Nair, DrSc, PhD, Distinguished Professor, Agroforestry & International Forestry Director, Center for Subtropical Agroforestry, University of Florida)*

***Illegal Logging in the Tropics: Strategies for Cutting Crime***, Ramsay M. Ravenel, Ilmi M. E. Granoff, and Carrie Magee (Vol. 19, No. 1/2/3, 2004). *"This book goes far beyond the usual ailing against illegal logging, to explore with utter realism what to do about it." (William Ascher, PhD, Donald C. McKenna Professor of Government and Economics, Claremont McKenna College)*

***Transboundary Protected Areas: The Viability of Regional Conservation Strategies***, edited by Uromi Manage Goodale, Marc J. Stern, Cheryl Margoluis, Ashley G. Lanfer, and Matthew Fladeland (Vol. 17, No. 1/2, 2003). *Top researchers share their expertise on conservation and sustainability in protected regions that extend across national borders.*

***War and Tropical Forests: Conservation in Areas of Armed Conflict***, edited by Steven V. Price (Vol. 16, No. 3/4, 2003). *"AN EXEMPLARY COLLECTION . . . HIGHLY RELEVANT for academic, policy, and activist audiences, and a state-of-the-art account of what environmental governance might mean in areas of armed conflict. . . . The first systematic effort to explore the profound implications for policy and practice of forest conservation operating in the context of war and civil strife. Comparative and interdisciplinary in scope and approach, this book powerfully shows how conservation practice can become militarized, how conservation efforts can become an important part of peacemaking, how local communities must be active in such settings, and how global market and political forces can fuel violently exploitative resource extraction." (Michael Watts, PhD, Director and Chancellor's Professor, Institute of International Studies, University of California, Berkeley)*

***Non-Timber Forest Products: Medicinal Herbs, Fungi, Edible Fruits and Nuts, and Other Natural Products from the Forest***, edited by Marla R. Emery, Rebecca J. McLain (Vol. 13, No. 3/4, 2001). *Focuses on NTFP use, research, and policy concerns in the United States. Discusses historical and contemporary NTFP use, ongoing research on NTFPs, and socio-political considerations for NTFP management.*

***Understanding Community-Based Forest Ecosystem Management***, edited by Gerald J. Gray, Maia J. Enzer, and Jonathan Kusel (Vol. 12, No. 3/4 & Vol. 13, No. 1/2, 2001). *Here is a state-of-the-art reference and information source for scientists, community groups and their leaders, resource managers, and ecosystem management practitioners. Healthy ecosystems and community well-being go hand in hand, and the interdependence between the two is the focal point of community-based ecosystem management. The information you'll find in* ***Understanding***

***Community-Based Forest Ecosystem Management*** *will be invaluable in your effort to manage and maintain the ecosystems in your community.* ***Understanding Community-Based Forest Ecosystem Management*** *examines the emergence of community-based ecosystem management (CBEM) in the United States. This comprehensive book blends diverse perspectives, enabling you to draw on the experience and expertise of forest-based practitioners, researchers, and leaders in community-based efforts in the ecosystem management situations that you deal with in your community.*

***Climate Change and Forest Management in the Western Hemisphere***, edited by Mohammed H. I. Dore (Vol. 12, No. 1/2, 2001). *This valuable book examines integrated forest management in the Americas, covering important global issues including global climate change and the conservation of biodiversity. Here you will find case studies from representative forests in North, Central, and South America. The book also explores the role of the Brazilian rainforest in the global carbon cycle and implications for sustainable use of rainforests, as well as the carbon cycle and the valuation of forests for carbon sequestration.*

***Mapping Wildfire Hazards and Risks***, edited by R. Neil Sampson, R. Dwight Atkinson, and Joe W. Lewis (Vol. 11, No. 1/2, 2000). *Based on the October 1996 workshop at Pingree Park in Colorado,* ***Mapping Wildfire Hazards and Risks*** *is a compilation of the ideas of federal and state agencies, universities, and non-governmental organizations on how to rank and prioritize forested watershed areas that are in need of prescribed fire. This book explains the vital importance of fire for the health and sustainability of a watershed forest and how the past acceptance of fire suspension has consequently led to increased fuel loadings in these landscapes that may lead to more severe future wildfires. Complete with geographic maps, charts, diagrams, and a list of locations where there is the greatest risk of future wildfires,* ***Mapping Wildfire Hazards and Risks*** *will assist you in deciding how to set priorities for land treatment that might reduce the risk of land damage.*

***Frontiers of Forest Biology: Proceedings of the 1998 Joint Meeting of the North American Forest Biology Workshop and the Western Forest Genetics Association***, edited by Alan K. Mitchell, Pasi Puttonen, Michael Stoehr, and Barbara J. Hawkins (Vol. 10, No. 1/2 & 3/4, 2000). *Based on the 1998 Joint Meeting of the North American Forest Biology Workshop and the Western Forest Genetics Association, Frontiers of Forest Biology addresses changing priorities in forest resource management. You will explore how the emphasis of forest research has shifted from productivity-based goals to goals related to sustainable development of forest resources. This important book contains fascinating research studies, complete with tables and diagrams, on topics such as biodiversity research and the productivity of commercial species that seek criteria and indicators of ecological integrity.*

*"There is clear emphasis on the genetics, genecology, and physiology of trees, particularly temperate trees. . . . These proceedings are also testimony to what does or should distinguish forest biology from other sciences: a focus on intra- and inter-specific interactions between forest organisms and their environment, over scales of both time and place." (Robert D. Guy, PhD, Associate Professor, Department of Forest Sciences, University of British Columbia, Vancouver, Canada)*

***Contested Issues of Ecosystem Management***, edited by Piermaria Corona and Boris Zeide (Vol. 9, No. 1/2, 1999). *Provides park rangers, forestry students and personnel with a unique discussion of the premise, goals, and concepts of ecosystem management. You will discover the need for you to maintain and enhance the quality of the environment on a global scale while meeting the current and future needs of an increasing human population. This unique book includes ways to tackle the fundamental causes of environmental degradation so you will be able to respond to the problem and not merely the symptoms.*

***Protecting Watershed Areas: Case of the Panama Canal***, edited by Mark S. Ashton, Jennifer L. O'Hara, and Robert D. Hauff (Vol. 8, No. 3/4, 1999). *"This book makes a valuable contribution to the literature on conservation and development in the neo-tropics. . . . These writings provide a fresh yet realistic account of the Panama landscape." (Raymond P. Guries, Professor of Forestry, Department of Forestry, University of Wisconsin at Madison, Wisconsin)*

***Sustainable Forests: Global Challenges and Local Solutions***, edited by O. Thomas Bouman and David G. Brand (Vol. 4, No. 3/4 & Vol. 5, No. 1/2, 1997). *"Presents visions and hopes and the challenges and frustrations in utilization of our forests to meet the economical and social needs of communities, without irreversibly damaging the renewal capacities of the world's forests." (Dvoralai Wulfsohn, PhD, PEng, Associate Professor, Department of Agricultural and Bioresource Engineering, University of Saskatchewan)*

ALL HAWORTH FOOD & AGRICULTURAL
PRODUCTS PRESS BOOKS AND JOURNALS
ARE PRINTED ON CERTIFIED ACID-FREE PAPER

# Sustainable Forestry Management and Wood Production in a Global Economy

Robert L. Deal
Rachel White
Gary L. Benson
Editors

*Sustainable Forestry Management and Wood Production in a Global Economy* has been co-published simultaneously as *Journal of Sustainable Forestry*, Volume 24, Number 1 and Numbers 2/3 2007.

Haworth Food & Agricultural Products Press™
An Imprint of The Haworth Press, Inc.

Published by

Haworth Food & Agricultural Products Press™, 10 Alice Street, Binghamton, NY 13904-1580 USA

Haworth Food & Agricultural Products Press™ is an imprint of The Haworth Press, Inc., 10 Alice Street, Binghamton, NY 13904-1580 USA.

*Sustainable Forestry Management and Wood Production in a Global Economy* has been co-published simultaneously as *Journal of Sustainable Forestry*, Volume 24, Number 1 and Numbers 2/3 2007.



**Library of Congress Cataloging-in-Publication Data**

Sustainable forestry management and wood production in a global economy / Robert L. Deal, Rachel White, Gary L. Benson, editors.

p. cm.

"Sustainable forestry management and wood production in a global economy has been co-published simultaneously as Journal of sustainable forestry, volume 24, number 1 and numbers 2/3."

Includes bibliographic references and index.

ISBN-13: 978-1-56022-165-4 (hard cover : alk. paper)

ISBN-10: 1-56022-165-8 (hard cover : alk. paper)

1. Sustainable forestry. 2. Forest management. 3. Wood products. I. Deal, Robert L. (Robert Leslie) II. White, Rachel. III. Benson, Gary L. IV. Journal of sustainable forestry.

SD387.S87S853 2007

333.75–dc22 2006031755

This section provides you with a list of major indexing & abstracting services and other tools for bibliographic access. That is to say, each service began covering this periodical during the the year noted in the right column. Most Websites which are listed below have indicated that they will either post, disseminate, compile, archive, cite or alert their own Website users with research-based content from this work. (This list is as current as the copyright date of this publication.)

Abstracting, Website/Indexing Coverage . . . . . . . . . Year When Coverage Began

- ***(CAB ABSTRACTS, CABI) <http://www.cabi.org>*** . . . . . . . . . . . . . . . *
- ***(IBR) International Bibliography of Book Reviews on the Humanities and Social Sciences (Thomson) <http://www.saur.de>*** . . . . . . . . . . . . . . . . . . . . . . . . . . . . . . **2006**
- ***(IBZ) International Bibliography of Periodical Literature on the Humanities and Social Sciences (Thomson) <http://www.saur.de>*** . . . . . . . . . . . . . . . . . . . . . . . . . . . . . . **2002**
- *****Academic Search Premier (EBSCO)** <http://search.ebscohost.com>*** . . . . . . . . . . . . . . . . . . . . . . . . . **2006**
- *****MasterFILE Premier (EBSCO)** <http://search.ebscohost.com>*** . . . . . . . . . . . . . . . . . . . . . . . . . **2006**
- ***Abstract Bulletin of Paper Science & Technology (Elsevier) <http://www.elsevier.com>*** . . . . . . . . . . . . . . . . . . . . . . . . . . . **1993**
- ***Abstracts in Anthropology <http://www.baywood.com/Journals/PreviewJournals.asp?Id=0001-3455>*** . . . . . . . . . . . . . . . . . . . . . **2006**
- ***Academic Source Premier (EBSCO) <http://search.ebscohost.com>*** . . . . . . . . . . . . . . . . . . . . . . . . . **2007**
- ***AgBiotech News & Information (CAB ABSTRACTS, CABI ) <http://www.cabi.org>*** . . . . . . . . . . . . . . . . . . . . . . . . . . . . . . **2006**
- ***Agricultural Engineering Abstracts (CAB ABSTRACTS, CABI) <http://www.cabi.org>*** . . . . . . . . . . . . . . . . . . . . . . . . . . **2006**

(continued)

- *AGRIS/CARIS <http://FAO-Agris-Caris@fao.org>* . . . . . . . . . . . . **1993**
- *Agroforestry Abstracts (CAB ABSTRACTS, CABI) <http://www.cabi.org>* . . . . . . . . . . . . *
- *Animal Breeding Abstracts (CAB ABSTRACTS, CABI ) <http://www.cabi.org>* . . . . . . . . . . . . **2006**
- *Biology Digest (in print & online) <http://www.infotoday.com>* . . **2000**
- *British Library Inside (The British Library) <http://www.bl.uk/services/current/inside.html>* . . . . . . . . . . . . **2006**
- *Cambridge Scientific Abstracts <http://www.csa.com>* . . . . . . . . . . **2004**
- *Compendex (Elsevier) <http://www.ei.org/eicorp/compendex.html>* . . . . . . . . . . . . **2006**
- *Crop Physiology Abstracts (CAB ABSTRACTS, CABI) <http://www.cabi.org>* . . . . . . . . . . . . *
- *Current Abstracts (EBSCO) <http://search.ebscohost.com>* . . . . . **2007**
- *Current Citations Express (EBSCO) <http://search.ebscohost.com>* . . . . . . . . . . . . **2007**
- *EBSCOhost Electronic Journals Service (EJS) <http://search.ebscohost.com>* . . . . . . . . . . . . **2001**
- *Electronic Collections Online (OCLC) <http://www.oclc.org/electroniccollections/>* . . . . . . . . . . . . **2006**
- *Elsevier Eflow-D <http://www.elsevier.com>* . . . . . . . . . . . . **2006**
- *Elsevier Scopus <http://www.info.scopus.com>* . . . . . . . . . . . . **2005**
- *EMBiology (Elsevier) <http://www.elsevier.com>* . . . . . . . . . . . . **2006**
- *Engineering Information (PAGE ONE) (Elsevier)* . . . . . . . . . . . . **1999**
- *Environment Abstracts (LexisNexis) <http://www.cispubs.com>* . . **1993**
- *Environment Complete (EBSCO) <http://search.ebscohost.com>* . . **2002**
- *Environment Index (EBSCO) <http://search.ebscohost.com>* . . . . **2007**
- *Environmental Engineering Abstracts (Cambridge Scientific Abstracts) <http://csa.com>* . . . . . . . . . . . . **2006**
- *Environmental Sciences and Pollution Management (Cambridge Scientific Abstracts) <http://www.csa.com>* . . . . . . **1995**
- *Forest Products Abstracts (CAB ABSTRACTS, CABI) <http://www.cabi.org/>* . . . . . . . . . . . . *
- *Forestry Abstracts (CAB ABSTRACTS, CABI) <http://www.cabi.org>* . . . . . . . . . . . . **1993**

(continued)

- *Garden, Horticulture, Landscape Index (EBSCO) <http://search.ebscohost.com>* . . . . . . . . . . . . . . . . . . . . . . . . . . . . **2006**
- *GEOBASE (Elsevier) <http://www.elsevier.com>* . . . . . . . . . . . . . . **2000**
- *GLOBAL HEALTH (CAB ABSTRACTS, CABI) <http://www.cabi.org>* . . . . . . . . . . . . . . . . . . . . . . . . . . . . . . **2006**
- *Google <http://www.google.com>* . . . . . . . . . . . . . . . . . . . . . . . . . . . . . **2004**
- *Google Scholar <http://scholar.google.com>* . . . . . . . . . . . . . . . . **2004**
- *Grasslands and Forage Abstracts (CAB ABSTRACTS, CABI) <http://www.cabi.org>* . . . . . . . . . . . . . . . . . . . . . . . . . . . . . . **2006**
- *Haworth Document Delivery Center <http://www.HaworthPress.com/journals/dds.asp>* . . . . . . . . . . **1993**
- *Horticultural Abstracts (CAB ABSTRACTS, CABI) <http://www.cabi.org>* . . . . . . . . . . . . . . . . . . . . . . . . . . . . . . . . . . *
- *Index Guide to College Journals* . . . . . . . . . . . . . . . . . . . . . . . . . **1999**
- *IndexCopernicus <http://www.indexcopernicus.com>* . . . . . . . . . . **2006**
- *Internationale Bibliographie der geistes- und sozialwissenschaftlichen Zeitschriftenliteratur . . . See IBZ . . . . . . . <http://www.saur.de>* . . . . . . . . . . . . . . . . . . . **2002**
- *Irrigation and Drainage Abstracts (CAB ABSTRACTS, CABI) <http://www.cabi.org>* . . . . . . . . . . . . . . . . . . . . . . . . . . . . . . **2006**
- *JournalSeek <http://www.journalseek.net>* . . . . . . . . . . . . . . . . . . . **2006**
- *Leisure, Recreation & Tourism Abstracts (CAB ABSTRACTS, CABI) <http://www.cabi.org>* . . . . . . . . . . . . . . . . . . . . . . . . . **2006**
- *Links@Ovid (via CrossRef targeted DOI links) <http://www.ovid.com>* . . . . . . . . . . . . . . . . . . . . . . . . . . . . . . . **2005**
- *Maize Abstracts (CAB ABSTRACTS, CABI) <http://www.cabi.org>* . . . . . . . . . . . . . . . . . . . . . . . . . . . . . . **2006**
- *National Academy of Agricultural Sciences (NAAS) <http://www.naas-india.org>* . . . . . . . . . . . . . . . . . . . . . . . . . **2006**
- *NewJour (Electronic Journals & Newsletters) <http://gort.ucsd.edu/newjour/>* . . . . . . . . . . . . . . . . . . . . . . . . **2006**
- *Nutrition Abstracts & Reviews, Series B: Livestock Feeds & Feeding (CAB ABSTRACTS, CABI) <http://www.cabi.org>* . . . **2006**
- *OCLC ArticleFirst <http://www.oclc.org/services/databases/>* . . . . **2006**
- *Ornamental Horticulture (CAB ABSTRACTS, CABI) <http://www.cabi.org>* . . . . . . . . . . . . . . . . . . . . . . . . . . . . . . **2006**
- *Ovid Linksolver (OpenURL link resolver via CrossRef targeted DOI links) <http://www.linksolver.com>* . . . . . . . . . . . . . . . . . . . **2005**

(continued)

- *PaperChem (Elsevier) <http://www.elsevier.com>* . . . . . . . . . . . . . 2006
- *Plant Breeding Abstracts (CAB ABSTRACTS, CABI) <http://www.cabi.org>* . . . . . . . . . . . . . . . . . . . . *
- *Plant Genetic Resources Abstracts (CAB ABSTRACTS, CABI) <http://www.cabi.org>* . . . . . . . . . . . . . . . . . . . . *
- *Plant Growth Regulator Abstracts (CAB ABSTRACTS, CABI) <http://www.cabi.org>* . . . . . . . . . . . . . . . . . . 2006
- *Pollution Abstracts (Cambridge Scientific Abstracts) <http://csa.com>* . . . . . . . . . . . . . . . . . . 2006
- *Referativnyi Zhurnal (Abstracts Journal of the All-Russian Institute of Scientific and Technical Information-in Russian) <http://www.viniti.ru>* . . . . . . . . . . . . . . . . . . 1993
- *Review of Agricultural Entomology (CAB ABSTRACTS, CABI) <http://www.cabi.org>* . . . . . . . . . . . . . . . . . . 2006
- *Review of Aromatic & Medicinal Plants (CAB ABSTRACTS, CABI) <http://www.cabi.org>* . . . . . . . . . . . . . . . . . . 2006
- *Review of Plant Pathology (CAB ABSTRACTS, CABI) <http://www.cabi.org>* . . . . . . . . . . . . . . . . . . . . *
- *Rice Abstracts (CAB ABSTRACTS, CABI) <http://www.cabi.org>* . . . . . . . . . . . . . . . . . . 2006
- *Scopus (See instead Elsevier Scopus) <http://www.info.scopus.com>* . . . . . . . . . . . . . . . . . . 2005
- *Seed Abstracts (CAB ABSTRACTS, CABI) <http://www.cabi.org>* . . . . *
- *Soils & Fertilizers Abstracts (CAB ABSTRACTS, CABI) <http://www.cabi.org>* . . . . . . . . . . . . . . . . . . . . *
- *Soybean Abstracts (CAB ABSTRACTS, CABI) <http://www.cabi.org>* . . . . . . . . . . . . . . . . . . 2006
- *Sustainability Science Abstracts (Cambridge Scientific Abstracts) <http://csa.com>* . . . . . . . . . . . . . . . . . . 2006
- *TOC Premier (EBSCO) <http://search.ebscohost.com>* . . . . . . . . . 2007
- *TROPAG & RURAL <http://www.kit.nl>* . . . . . . . . . . . . . . . . . . 1993
- *Tropical Diseases Bulletin (CAB ABSTRACTS, CABI) <http://www.cabi.org>* . . . . . . . . . . . . . . . . . . 2006
- *Water Resources Abstracts (Cambridge Scientific Abstracts) <http://csa.com>* . . . . . . . . . . . . . . . . . . 2006
- *Weed Abstracts (CAB ABSTRACTS, CABI) <http://www.cabi.org>* . . . . . . . . . . . . . . . . . . 2006
- *Wheat, Barley & Triticale Abstracts (CAB ABSTRACTS, CABI) <http://www.cabi.org>* . . . . . . . . . . . . . . . . . . 2006

(continued)

- ***Wildlife & Ecology Studies Worldwide (NISC USA) <http://www.nisc.com>*** . . . . . . . . . . . . . . . . . . . . . . . . . . . . . . . . **1993**
- ***World Agricultural Economics & Rural Sociology Abstracts (CAB ABSTRACTS, CABI) <http://www.cabi.org>*** . . . . . . . . . . . . . *****
- ***zetoc (The British Library) <http://www.bl.uk>*** . . . . . . . . . . . . . . . . **2004**

***Exact start date to come**

## *Bibliographic Access*

- ***Magazines for Libraries (Katz)***
- ***MediaFinder <http://www.mediafinder.com/>***
- ***Ulrich's Periodicals Directory: The Global Source for Periodicals Information Since 1932 <http://www.bowkerlink.com>***

*Special Bibliographic Notes related to special journal issues (separates) and indexing/abstracting:*

- indexing/abstracting services in this list will also cover material in any "separate" that is co-published simultaneously with Haworth's special thematic journal issue or DocuSerial. Indexing/abstracting usually covers material at the article/chapter level.
- monographic co-editions are intended for either non-subscribers or libraries which intend to purchase a second copy for their circulating collections.
- monographic co-editions are reported to all jobbers/wholesalers/approval plans. The source journal is listed as the "series" to assist the prevention of duplicate purchasing in the same manner utilized for books-in-series.
- to facilitate user/access services all indexing/abstracting services are encouraged to utilize the co-indexing entry note indicated at the bottom of the first page of each article/chapter/contribution.
- this is intended to assist a library user of any reference tool (whether print, electronic, online, or CD-ROM) to locate the monographic version if the library has purchased this version but not a subscription to the source journal.

As part of Haworth's continuing committment to better serve our library patrons, we are proud to be working with the following electronic services:

*AGGREGATOR SERVICES*

- EBSCOhost
- Ingenta
- J-Gate
- Minerva
- OCLC FirstSearch
- Oxmill
- SwetsWIse

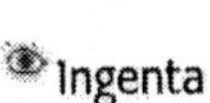

*LINK RESOLVER SERVICES*

- 1Cate (Openly Informatics)
- CrossRef
- Gold Rush (Coalliance)
- LinkOut (PubMed)
- LINKplus (Atypon)
- LinkSolver (Ovid)
- LinkSource with A-to-Z (EBSCO)
- Resource Linker (Ulrich)
- SerialsSolutions (ProQuest)
- SFX (Ex Libris)
- Sirsi Resolver (SirsiDynix)
- Tour (TDnet)
- Vlink (Extensity, *formerly Geac*)
- WebBridge (Innovative Interfaces)

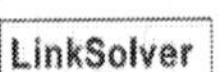

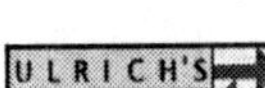

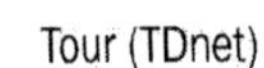
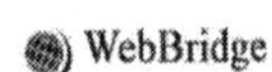

# Sustainable Forestry Management and Wood Production in a Global Economy

## CONTENTS

Preface xix

INTEGRATING WOOD PRODUCTION WITHIN SUSTAINABLE FOREST MANAGEMENT

Integrating Concerns About Wood Production and Sustainable Forest Management in the United States 1
*Richard W. Haynes*

Integrating Wood Production Within Sustainable Forest Management: An Australian Viewpoint 19
*Ian Ferguson*

Silviculture of Scottish Forests at a Time of Change 41
*W. L. Mason*

Changes in Wood Product Proportions in the Douglas-Fir Region with Respect to Size, Age, and Time 59
*Robert A. Monserud*
*Xiaoping Zhou*

Measuring Sustainable Forest Management in Tierra del Fuego, Argentina 85
*Esteban Carabelli*
*Hugh Bigsby*
*Ross Cullen*
*Pablo Peri*

Increasing and Sustaining Productivity of Tropical Eucalypt Plantations Over Multiple Rotations 109
*K. V. Sankaran*
*D. S. Mendham*
*K. C. Chacko*
*R. C. Pandalai*
*T. S. Grove*
*A. M. O'Connell*

Structural and Biometric Characterization of *Nothofagus betuloides* Production Forests in the Magellan Region, Chile 123
*Gustavo Cruz Madariaga*
*Raúl Caprile Navarro*
*Álvaro Promis Baeza*
*Gustavo Cabello Verrugio*

EMERGING ISSUES FOR SUSTAINABLE FOREST MANAGEMENT

A Working Definition of Sustainable Forestry and Means of Achieving It at Different Spatial Scales 141
*Chadwick Dearing Oliver*
*Robert L. Deal*

Environmental Change and the Sustainability of European Forests 165
*Peter Freer-Smith*

Barriers to Sustainable Forestry in Central America and Promising Initiatives to Overcome Them 189
*Glenn E. Galloway*
*Dietmar Stoian*

A United States View on Changes in Land Use and Land Values Affecting Sustainable Forest Management 209
*Ralph J. Alig*

Landowner-Driven Sustainable Forest Management and Value-Added Processing: A Case Study in Massachusetts, USA 229
*David T. Damery*

Monitoring Sustainable Forest Management in the Pacific Rim Region 245
*Gordon M. Hickey*
*Craig R. Nitschke*

A European Network in Support of Sustainable Forest Management 279
*Folke Andersson*
*Anders Mårell*

Index 295

## ABOUT THE EDITORS

**Robert L. Deal** is a research silviculturist and team leader for the Sustainable Wood Production Initiative (SWPI), Pacific Northwest Research Station, USDA Forest Service, Portland, Oregon, USA. He has 25 years of professional forester and research experience working throughout the Pacific Northwest and Alaska. As team leader of the SWPI he manages a research team of PNW scientists and university cooperators who conduct research on economic, ecological and social issues related to sustainable wood production in the Pacific Northwest. He received a BS in biology from Evergreen State College, an MS in silviculture from the University of Washington and a PhD in forest resources from Oregon State University. Deal's research interests include stand development, regeneration, silvicultural systems including alternatives to clearcutting, and practices to enhance compatible forest management. He has spent much of his career conducting research in Sitka spruce/western hemlock forests of southeast Alaska and is internationally recognized for his expertise in this forest type. His current research focuses on emerging issues for sustainable forest management and synthesizing information on the barriers and opportunities for sustainable wood production in the Pacific Northwest. He is active in the Society of American Foresters (SAF), is a certified forester (C.F.) and serves on the SAF National Science Committee. He is Editor for the *Western Journal of Applied Forestry*.

**Rachel White** is a science writer and editor for the Focused Science Delivery Program in the Pacific Northwest Research Station, USDA Forest Service, Portland, Oregon, USA. Formerly a field biologist with the U.S. Fish and Wildlife Service, she is the author of *Shapers of the Great Debate on Conservation*, was a contributing author on *American Environmental Leaders*, and a contributing author and editorial assistant on *Birds of Oregon*.

**Gary L. Benson** is Staff Ecologist on the Science Delivery Team, Focused Science Delivery Program, Pacific Northwest (PNW) Research Station, USDA Forest Service, Portland, Oregon, USA. He has 32 years of professional ecology and forestry research experience working throughout the PNW and California. As a Science Delivery team member, he facilitates the exchange of mature science information and management needs between scientists, managers, policy-makers and those who influence policy in the PNW area. He received a BA in biology from San Jose State College, an MA in biology from the State University of New York, Plattsburgh, and a PhD in Botany (Plant Ecology) from Washington State University. Benson's areas of interest include landscape ecology, wildlife ecology, climate change, and long-term climate dynamics. He has spent parts of his career developing remote sensing applications for natural resources in Oregon and developing ecological guides for Sierran National Forests in northern California. He has worked on modeling and describing wildlife habitat needs in northern California, as well as vegetative habitat needs for the northern Spotted Owl in the PNW. He has worked in research and application programs involving ecosystem management in the PNW, as well as research oversight of the Northwest Forest Plan for the PNW Research Station. He is active in the Ecological Society of America (ESA), and the Society of American Foresters (SAF).

# Preface

During the International Union of Forest Research Organizations [IUFRO] World Congress, August 8-13, 2005, in Brisbane, Australia, two large technical sessions were held on a variety of topics related to sustainable forest management. These sessions were entitled "integrating wood production within sustainable forest management" and "emerging issues for sustainable forest management." The primary goal of these sessions was to bring together international researchers who were interested in producing wood while maintaining other forest resource values, and also to discuss some of the emerging issues for sustainable forest management.

The session on integrating wood production within sustainable forest management included 7 oral presentations and 12 posters from researchers in Europe, North America, South America and Asia. It included two state of the art presentations from North America and Australia. Richard Haynes (USDA Forest Service) gave a paper on wood production and sustainable forest management in the United States which discussed the influence of forest products markets and management decisions on sustainable forestry. Ian Ferguson (professor emeritus, University of Melbourne) discussed the development of conservation reserves and changes in forestry policy in Australia and the different tradeoffs associated with particular forest management policies. Mason provided an overview of some of the changes and challenges for forestry in Scotland with the establishment of even-aged plantations of fast growing conifer species, most of which are not native to the British Isles. Monserud and Zhou discussed different wood prod-

[Haworth co-indexing entry note]: "Preface." Deal, Robert L., Rachel White, and Gary L. Benson. Co-published simultaneously in *Journal of Sustainable Forestry* (Haworth Food & Agricultural Products Press, an imprint of The Haworth Press, Inc.) Vol. 24, No. 1, 2007, pp. xxiii-xxv; and: *Sustainable Forestry Management and Wood Production in a Global Economy* (ed: Robert L. Deal, Rachel White, and Gary L. Benson) Haworth Food & Agricultural Products Press, an imprint of The Haworth Press, Inc., 2007, pp. xix-xxi. Single or multiple copies of this article are available for a fee from The Haworth Document Delivery Service [1-800-HAWORTH, 9:00 a.m. - 5:00 p.m. (EST). E-mail address: docdelivery@haworthpress.com].

Available online at http://jsf.haworthpress.com

ucts obtained from trees in the Douglas-fir region of the northwestern Unites States as the region shifts from high quality old-growth trees to young growth plantations. Carabelli et al. provided strategies for sustainable forest management in Argentina including the development of indicators of sustainable management of lenga forests. Sankaran et al. discussed management options to increase the productivity of eucalypt plantations in India. Finally, Cruz Madariaga et al. provided structural and biometric informaiton for sustainable forest management of *Nothofagus* forests in Chile.

The session on emerging issues for sustainable forest management included 7 oral presentations and 10 posters from researchers in Europe, North America, Central America, Australia and Asia on a suite of emerging issues related to sustainable forest management. It included two state of the art presentations and perspectives from North America and Europe. Chadwick Oliver (Yale University, USA) and Robert Deal gave a fascinating paper on sustainable forestry at different scales ranging from local to international. Peter Freer-Smith (UK Research) discussed the sustainability of European forests using examples from a variety of recent research projects. Galloway and Stoian discussed illegal logging, and the economic challenges of forest management in Central America and some initiatives to address these problems. Alig highlighted changes in forest land use and land values on sustainable forest management in the United States. Damery discussed the success of forest cooperatives in Massachusetts, USA, and their potential use for sustainable forest management in other regions. Hickey and Nitschke assessed similarities and differences in sustainable forest management criteria and indicators and their role for sustainable forestry for the Pacific Rim region. Finally, Andersson and Mårell discussed policy issues related to criteria and indicators for sustainable forest management in Europe.

Following the IUFRO World Congress, researchers were invited to submit papers as part of this collection. Sixteen papers were submitted and 14 were accepted after peer-review including all of the state of the art presentations. Although we were not able to include papers from all the meeting presenters, the papers that are included in this volume capture many of the key findings from the session and provide a broad international perspective on some of the emerging issues for sustainable forest management. In summary, the papers published here cover a wide range of topics on issues important for sustainable wood production throughout the world. We thank all those involved in planning, presenting, or contributing in some way to the meeting. We hope that this volume will provide new and useful information to its readers.

*Robert L. Deal*
*Rachel White*
*Gary L. Benson*

# *INTEGRATING WOOD PRODUCTION WITHIN SUSTAINABLE FOREST MANAGEMENT*

# Integrating Concerns About Wood Production and Sustainable Forest Management in the United States

Richard W. Haynes

**SUMMARY.** The implementation of Sustainable Forest Management (SFM) in the United States is strongly influenced by U.S. forest products markets and the numerous management decisions made by individual landowners and managers. These decisions are influenced by a mix of market incentives and regulatory actions reducing predictability in as-

Richard W. Haynes is Research Forester, U.S. Department of Agriculture, Forest Service, Pacific Northwest Research Station, P.O. Box 3890, Portland, OR 97208 (E-mail: rhaynes@fs.fed.us).

[Haworth co-indexing entry note]: "Integrating Concerns About Wood Production and Sustainable Forest Management in the United States." Haynes, Richard W. Co-published simultaneously in *Journal of Sustainable Forestry* (Haworth Food & Agricultural Products Press, an imprint of The Haworth Press, Inc.) Vol. 24, No. 1, 2007, pp. 1-18; and: *Sustainable Forestry Management and Wood Production in a Global Economy* (ed: Robert L. Deal, Rachel White, and Gary L. Benson) Haworth Food & Agricultural Products Press, an imprint of The Haworth Press, Inc., 2007, pp. 1-18. Single or multiple copies of this article are available for a fee from The Haworth Document Delivery Service [1-800-HAWORTH, 9:00 a.m. - 5:00 p.m. (EST). E-mail address: docdelivery@haworthpress.com].

Available online at http://jsf.haworthpress.com
doi:10.1300/J091v24n01_01

sessing progress toward SFM and causing angst for some proponents of SFM because prices might provide insufficient incentive for what they believe are necessary forest practices. At the same time, conservation proponents are advocating forest management regimes that lead to reduced financial returns. These dual concerns are leading to new alliances and new approaches for forest-based conservation and management. doi:10.1300/J091v24n01_01 *[Article copies available for a fee from The Haworth Document Delivery Service: 1-800-HAWORTH. E-mail address: <docdelivery@haworthpress.com> Website: <http://www. HaworthPress.com>.]*

**KEYWORDS.** Forest management, conservation, sustainability

## *INTRODUCTION*

The U.S. forestland base is 298.3 million ha (737 million acres) that includes 197.9 million ha (489 million acres) of timberland defined by criteria of productivity and availability for harvest (see Smith et al., 2001 for a detailed description of the U.S. timber resource). The largest share of timberland (73%) is privately owned by a variety of owners including forest industry, farmers, and numerous private individuals.

Over the past century, persistent rising stumpage prices have provided adequate returns to maintain or expand forests and forest management. This improved forest management and protection has created a vast commons of forest or ecosystem services and goods, many of which are free to those who choose to enjoy them. By the end of the 20th century, the presence of these forested commons led to societal expectations for relatively well managed and productive forests capable of producing a wide array of goods and services. Coincidentally, institutions motivated by concerns about forests and sustainable development have and are continuing to evolve to assist in governing this forest estate.

But the recent weakening of expected returns for forest management raises concerns about whether North American forests can meet societal expectations for both making progress toward sustainable forest management (see Haynes, 2004) and the provision of ecosystem services expected by increasingly urbanized populations. At the same time, questions are emerging about the effectiveness of the institutions that have evolved to promote forest management and to control (or modify) human behavior to conserve non-market ecosystem services (see Brechin et al., 2002).

We also need to be clear about what we mean when we use the term 'forest management.' Here it is used to describe the process of managing a stand, collection of stands, or a forest to meet the objectives of the landowners. In the case of private forestland owners, particularly those interested in financial returns (timber is considered a capital asset and part of an individual's portfolio of investments), their objectives often center on the production of marketable goods such as timber, hunting rights, and selected nontimber forest products such as floral greens in an environmentally sound fashion. Public forestland owners typically have broader management objectives including the production of both market and nonmarket goods.

The purpose of this paper is to discuss the changing basis for forest-based conservation and management in the coming century. There are several issues to consider. First, there are the dual arguments that proponents of forest management relied on in the past. What is the specter for changes in these arguments? Related to these, are two management issues that will challenge landowners.

The first issue is the role that long-term stumpage price increases have played in generally enabling the public to have both relatively cheap timber products and a variety of nonmarket goods that have been provided free to the public. With weakening stumpage prices and continued loss of forestland to development, concern is growing that the public may have to provide financial incentives to maintain the expected level of environmental goods and services.

There are two related issues to the question of incentives that will challenge policy makers. First, what are the sizes of the incentives that might be necessary for landowners to remain financially neutral to changes in management for public goods? Second, why do some landowners continue to invest in forestry in spite of what might seem like weak price expectations?

A second set of issues includes the emerging challenges of how to govern the forested ecosystem when forest management takes place across relatively large geographic spaces composed of many different landowners with a myriad of land management objectives. Here governance refers to the various models of decision making and power sharing. While not covered here, the issue of governance also includes various justice issues such as who has the right to participate and how to include people who are typically underrepresented.

## *WHAT IS THE CONTRIBUTION OF RISING PRICES?*

Past prices and markets for various types and sizes of timber have influenced the goals for land management and the evolution and application of various management regimes. As shown in Figures 1 and 2, these forest product prices are often characterized as highly volatile but are increasing faster than the overall rate of inflation. In both the Douglas-fir region (western Oregon and Washington) and in the South, for those landowners sensitive to financial returns, rising prices for timber products have led to the adoption of relatively systematic forest management regimes consisting of practices that tend to speed development of well-stocked forests (see Haynes et al., 2003 for a history of increasing forest management in the Pacific Northwest). At the same time, rotation ages for these managed forests have dropped, leading to more frequent entries and a greater proportion of the forest in relatively young stands (stands 20 years or less). This increases the financial returns to those (individuals, companies, and various publics) who own timberlands. Current and expected trends show private timberland owners continuing to invest in forest management, subject to increasing regulations of various forest practices (see Haynes, 2003). Public lands, on the other hand, are expected to be managed for diverse goals–many not involving the marketplace–reflecting increased recognition of the benefits of many nontimber forest goods and services.

The data in Table 1 show three periods with separate price trends: until 1990 for each of the data sets, the last decade, and projected price trends for the next 50 years (see Haynes, 2003 for an expanded discussion of the timber situation and projections for the U.S. forest sector). Since World War II, the long-term trends reflect slightly faster growth in consumption relative to timber supplies, which has led to increased imports to help meet demand.

From an economic perspective, the general history of rising prices (as shown in Figure 2) suggests that timber is relatively scarce and, all else being equal, that there should be changes in various market factors to alleviate the price increases. This behavior is expected as these are relatively efficient (free) markets comprising numerous producers and consumers making decisions based on understanding of available information. The history of rising prices should also encourage consumers to substitute nonwood material in some uses such as residential construction (e.g., steel studs for framing) if the real prices of the substitute are less (or more stable) than wood prices. These rising stumpage prices have encouraged increased efficiency and diversity in the mix of forest products as producers looked for ways to increase returns. Rising prices should also encourage landowners and managers to increase the intensity or extent of land management to produce more timber (to the point where timber prices become stable).

FIGURE 1. Softwood sawtimber stumpage prices in 1982 dollars per thousand board feet (MBF) Scribner log rule and 1982 dollars per cubic meter ($M^3$), by major U.S. region, 1950-2050.

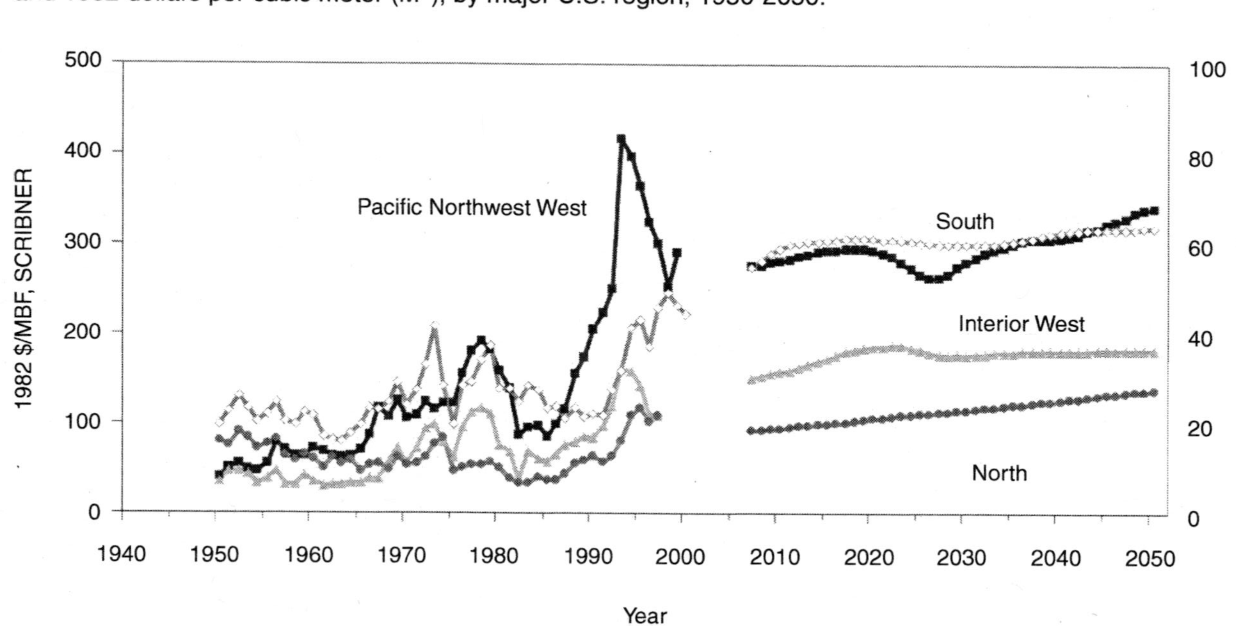


Sources: Douglas-fir, ponderosa pine, and southern pine sawtimber prices are updated from series published in *The Outlook for Timber in the United States* (USDA, 1973 Appendix 5, Table 2). Northeast prices are updated from data published by Sendak (1994). Projections for 2010 to 2050 are from Haynes (2003).

FIGURE 2. Stumpage prices by species, 1910-2002.

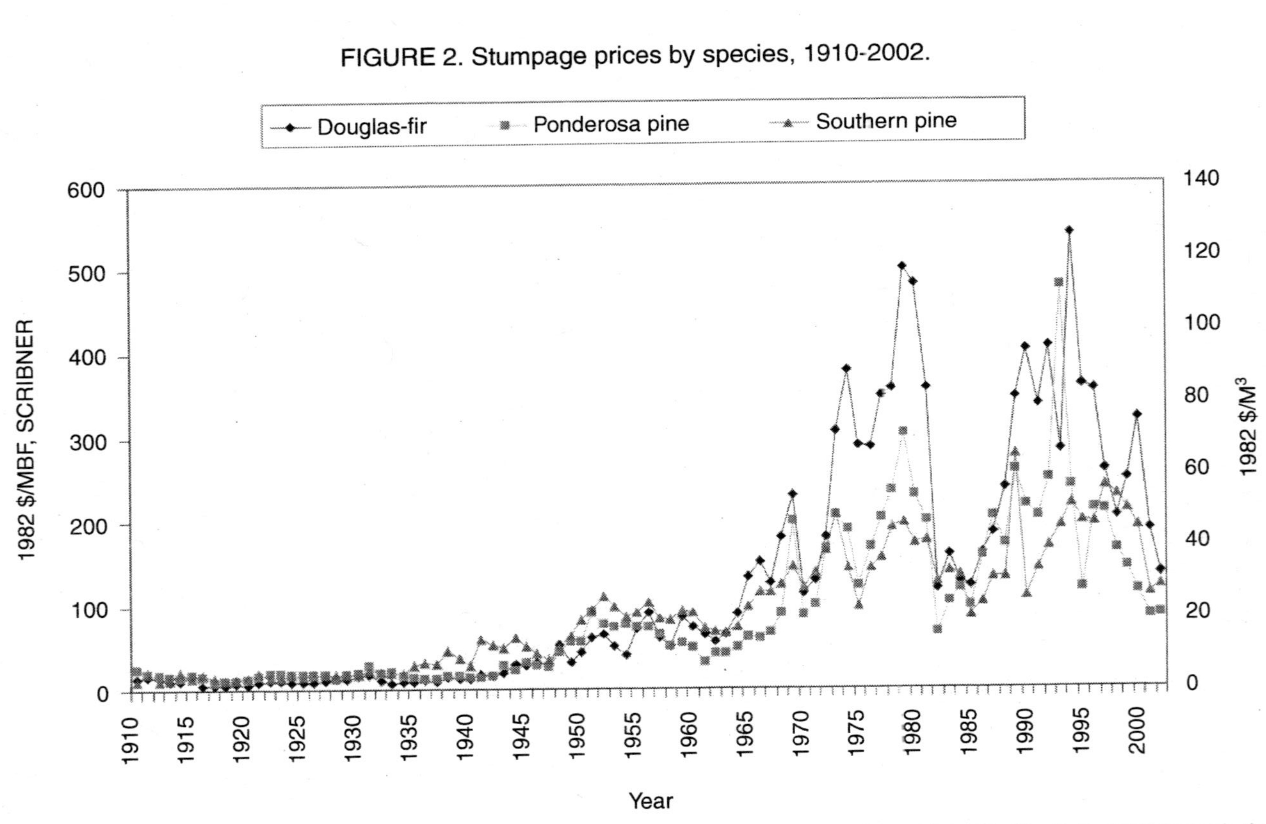


Sources: Douglas-fir, ponderosa pine, and southern pine sawtimber prices are updated from series published in *The Outlook for Timber in the United States* (USDA, 1973 Appendix 5, Table 2).

TABLE 1. Rates of stumpage price appreciation.

A. Traditional softwood sawtimber prices

| Years | Douglas-fir sawtimber | Ponderosa pine sawtimber | Southern pine sawtmber |
|---|---|---|---|
| | | *Percent* | |
| 1910-2002 | 5.1 | 3.6 | 3.4 |
| 1910-1990 | 5.6 | 4.0 | 3.8 |
| 1991-2002 | −6.9 | −8.8 | 0 |
| 2000-2050 | .2 | N/A | .6 |

B. Northeast prices

| Years | Hardwood sawtimber | Hardwood pulpwood | Softwood sawtimber | Softwood pulpwood |
|---|---|---|---|---|
| | | | *Percent* | |
| 1961-2002 | 4.6 | .5 | 1.3 | .7 |
| 1961-1990 | 4.5 | .9 | .8 | 0 |
| 1991-2002 | 3.8 | 0 | 3.9 | 0 |
| 2000-2050 | .2 | 0 | .4 | .7 |

Sources: Douglas-fir, ponderosa pine, and southern pine sawtimber prices are updated from series published in *The Outlook for Timber in the United States* (USDA, 1973 Appendix 5, Table 2). Northeast prices are updated from data published by Sendak (1994). Projections for 2000 to 2050 are from Haynes (2003).

The general slowing of the price trends in the 1990s reflects the growing importance that managed forests (with higher volumes of wood per hectare) are playing in providing timber supplies in the United States (see Table 20 in Haynes, 2003). The projected price trends (also shown in Figure 1) reflect this increased role of managed forests that are located on a relatively small part of the forest land base (that is, we anticipate a large increase in both harvest and inventory volumes on a relatively constant area base-see Haynes, 2003 for details). Such forests offer a relatively certain supply of timber at lower logging costs. The projected price trends reflect continued growth in U.S. consumption (roughly 0.9% per year), increases in imports (U.S. net trade continues to favor imports relative to exports) and increases in U.S. forest resources (both total inventories and net growth increase).

In the United States, we frequently debate the role that prices play in private timberland management. In periods of weak stumpage markets (e.g., for some hardwoods or for small softwood trees in the West) there is concern that landowners and managers will not implement certain forest practices such as thinning to reduce fire risk. In the case of private timberlands, relatively low returns to forestry may also lead to changes in land use as landowners seek higher returns by converting to agriculture or residential developments (see USDA FS, 1988; Kline and Alig, 2001; Ahn et al., 2002; Haynes, 2003, for additional details). At the same time, the lack of perceived effectiveness of markets in increasing quantities of timber supplies (or other forest outputs) in periods of rising prices and perceived scarcity have often prompted the establishment of public and private programs designed to improve forest management with the intention of slowing the expected rises in timber prices (see Cubbage and Haynes, 1988 for a summary).

The recent flat to decreasing stumpage prices (see Table 1) will lower expectations for long-term returns for forest management. They will also raise, in some manager's minds, questions about the incentives for sustainable forest management. From their perspective, the issue is how to produce a range of goods and services that also meets their financial expectations. This concern for financial returns has long been recognized as the heart of the forest management question (see Baker, 1950; Davis, 1966; Davis and Johnson, 1987; Davis et al., 2001). The concerns about financial returns pose a challenge to advocates for sustainable forest management who may find themselves championing selected management practices with doubtful financial returns. Further complicating these challenges are the diverse array of management objectives of numerous landowners and managers. The importance of financial returns also means that many of the benefits of forest management need to accrue to those who pay the costs. Some landowners with more financially oriented management objectives may lack the incentive to implement costly practices that have mostly broad societal benefits.

## *FORESTS AS A SET OF COMMONS*

The public has come to expect that forest management contributes to maintaining a broad set of environmental values including timber, wildlife habitat, aesthetics, biological diversity, water flows, ecological integrity, and recreation. At the broad scale, the public acts as if forested ecosystems are a set of 'commons' capable of producing both marketed and nonmarket goods and services. In this context, forest management decisions are viewed as involving broader environmental problems dealing with complex tradeoffs (or compatibility) among a broad set of environmental values. In the United States, many of these goods and services are thought to be fairly available to anyone. For example, consider how clean air and water, scenic integrity, and wildlife habitat are taken for granted by much of the public.

This interest in forested 'commons' is a consequence of the shift to managing broader ecosystems. Increasingly, this shift has been motivated by ecological as well as socioeconomic considerations. The ecological arguments rest on the need to place management actions within a spatial and temporal hierarchy that considers the broad context for actions at a specific place as well as an understanding of how ecological processes work at that site (these are often scaled up from site-specific information). The socioeconomic considerations include the role of civic engagement about the importance of ecosystem goods and services to overall societal well-being. These considerations have also manifested themselves in the discussions of sustainable development (and sustainable forest management) where the goal is to promote economic prosperity that is socially just and environmentally sound.

Considering forested ecosystems as a set of commons whose goods and services are fairly available to anyone raises three issues. These are concerns about common property, institutions to guide actions, and how property rights are assigned.

First, there are the traditional economic arguments about how common property is abused rather than protected when no property rights exist. Hardin (1968) laid out these 'open access' property issues in his classic article the *Tragedy of the Commons*.[1] The essence of his argument and the one that has become part of economic dogma was that if no one held property rights to various goods and services, then there was no incentive to manage the resource to sustain production but the incentive was to capture as much of the value as quickly as possible before others seized the various goods and services.

Second, champions of greater civic engagement argue for more attention to be paid to the goods supplied by the 'commons' (see the 22 January, 2005 *Economist* for an article on the movement for corporate social responsibility). They advocate for expanded institutions and greater governance as a means to provide guidelines for private land management. Often these champions are motivated by the actions of private landowners who act as if they are more interested in making money or sustaining the values of their asset portfolios than in the provision of a broad array of ecosystem goods and services.

Third, there are also questions about how and who could assign property rights to sustain various environmental services and goods. Some of the contemporary arguments (such as those for corporate social responsibility) focus assigning property rights to various stakeholder groups who have traditionally been marginalized in market based approaches to resource allocation and management. In this case, the issue of governance is more about how firms, agencies, and organizations conduct themselves in relation to stakeholders than about actual shared power.

Interest in this last issue is rapidly expanding among champions for greater civic involvement in managing ecosystems who argue that many ecosystem goods and services benefit all of society and for the most part (except for the small set where property rights have been established and assigned) are nominally free to users. The management of these goods involves active engagement of stakeholder groups and other regulatory costs to insure a balance between the marketed goods and services, and those provided altruistically. Many of the public see this as establishing the "license to operate" for those individuals involved in land management and, in essence, dividing forest property rights into those that can be traded in some form and those that cannot. Among these champions are advocates for systemic reforms of the market system involving increased regulatory approaches that many land managers consider encroaching on their decision making.

There are two management needs to clarify if we are to consider managing forested ecosystems as a set of 'commons.' First, there is the specification of the set of goods and services produced from forested ecosystems. These include goods and services like timber, wildlife habitat, aesthetics, biological diversity, carbon storage, water flows, ecological integrity, and recreation. In addition to the list of products, we also need to be able to provide measures of them and their associated values so we can determine the values associated with ecosystem management. Second, the increased level of civic engagement raises questions about the role and effectiveness of governance, especially in those cases where the social prices/values are perceived to be greater than what might be market determined. Classic cases of this are wilderness values or habitat conservation values where markets are thought to undersupply goods or services with positive externalities. In these cases civic interaction exercised through a mix of governmental and non-governmental organizations seeks to advance social well being.

In an economic sense, one of the roles of expanded governance is a need to assign property rights to what had been formally open access public goods and another is to intervene in markets where the specter of market failures seem high. Related to this last point is the need to consider different power-sharing perspectives in allocation decisions. There is a growing recognition that broad-scale conservation goals can be achieved if we can include added values associated with forest management for a range of goods and services, especially on private forest lands (Best, 2002). A key first step is to assign property rights to some of the goods for which we can develop notions of values. The next step is to assist in developing markets and trading. By providing or improving returns to landowners we will improve the conservation and management of those goods or services. This approach is workable for only those goods or services for which property rights can be assigned. Other goods or services such as clean air or water for habitat are considered public consumption goods and remain a governmental function.

## *TWO PRIVATE LAND MANAGEMENT CHALLENGES*

The increased civic interest in the broad set of ecosystem goods and services raises two possible policy issues. First, there are questions about what it would cost the public to make landowners financially neutral to producing environmental goods and services. That is, how much would landowners have to be compensated to broaden their management objectives to produce both their primary management objectives as well as a set of goods (and services) of interest to the broader public? The second issue is the general tendency to make judgments about management intentions from price signals alone. This tendency was illustrated in the earlier discussion about markets being a barrier to sustainable forest management.

In the first case, we see various proposals in contemporary forestry literature about the need to consider compensation of private landowners for contributing to greater conservation goals (such as protection of biodiversity). Best (2002) introduced these notions in her discussion of challenges to private forest conservation. This raises questions about how much compensation would be needed to make landowners indifferent to broadening management objectives. For example, in the Douglas-fir region there have been proposals, described as high-quality forestry, to lengthen rotations to provide both high-quality habitat and timber (for a discussion of the specific high-quality regimes see Weigand, 1994; Barbour et al., 2003). This raises the question of how much the public would need to pay landowners to extend the length of their rotation assuming that a greater array of public goods could be produced.

In the Douglas-fir region where rotations on private timberland are 40 to 50 years (see Haynes et al., 2003 for more details), there are proposals to double such rotation lengths. In this case, the present value at a real interest rate of 4% (expected returns) for a 40-year rotation is $14,321 or an annualized value of $25/ha (see Haynes, 2005 for details). The present value per hectare of two consecutive 40-year rotations is $14,941 or $27/year. The present value per hectare of a single 80-year rotation is $8,956 or $16 per year. To make a landowner indifferent to rotation length over this 80-year period, a payment of $11/ha/year will be required. Although this may not seem like a great sum (and may not be needed in all cases), there are 4.36 million ha of private timberland in Oregon; if all private timberland owners require compensation, this involves 48 million dollars per year in payments.

The second issue is the propensity of landowners to intensify land management even when stumpage prices are expected to remain relatively stable. As discussed in an earlier section, there has been a tendency among foresters to assume that rising prices were key in stimulating higher levels of investment in forest management. But the calculation of financial returns (or the soil expectation value[2]) to intensifying management includes both price and quantity harvested. In this context Figure 3 shows that as long as quantity increases as prices decrease, financial returns remain the same for the landowner (the data are from Haynes, 2005). The slope of the relation in Figure 3A is −15.625 suggesting that land owners will remain indifferent to financial returns as long as there is a gain from increased management of 4.53 $m^3$ (1,000 board feet [mbf]) per reduction of $3.45/$m^3$ (or $15.62/mbf) of stumpage prices. As indicated by the data used to develop Figure 3, landowners with a high propensity to invest would be indifferent to their returns as long as increases in management intensity would lead to the same or greater returns.

Another way to consider the implications of Figure 3 is to look at the volume increases among different management regimes. In the example used before, the yields were computed from using an average yield of 25 $m^3$/ha at age 50, but in the Douglas-fir region there are several management regimes generally used. Custodial management involves only site preparation and planting sufficient to meet State forest practices act requirements. A second regime involves more site preparation and a more intensive plantation regime but little else until harvest at stand age 40 to 50 years. A third type of regime involves site preparation, planting, precommercial thinning (or some form of stocking control), and thinning (depending on final age of harvest). This last regime was popular among landowners managing for the export market where a premium was placed on larger logs that had little taper and few limbs. At age 50, the yields computed from empirical yield functions developed from remeasured plot data were 24.5, 29.9, and 35.8 $m^3$/ha for these three regimes. This range in values represents the entire price range illustrated in Figure 3A or B. There are also cost differences among the

FIGURE 3. The tradeoff between stumpage prices and volume per hectare (or acre) while maintaining the same per hectare (or acre) expected returns for a 50-year rotation.

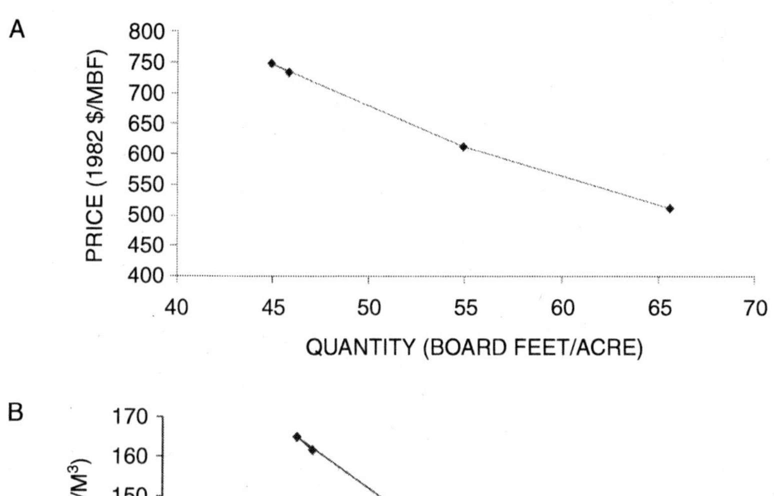


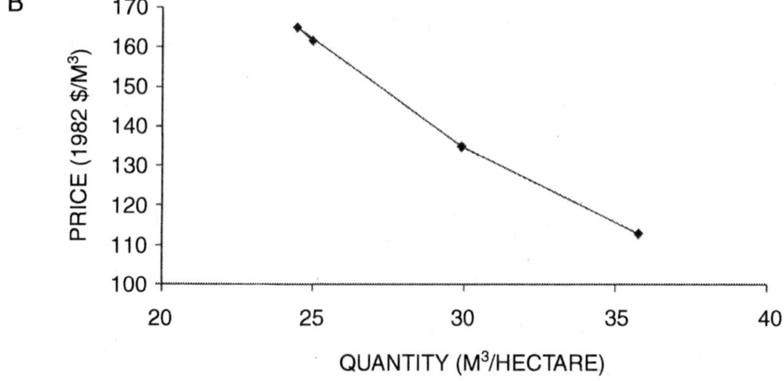


Source: Developed from data published in Haynes (2005).

regimes; regime 1 does not contain precommercial thinning costs, which increases net returns by 5 to 6% (all else held equal). The costs for the two other regimes are assumed to be the same for a 50-year rotation (the third regime contains a thinning, which adds to overall returns).

## *CHANGING THE PROPENSITY TO REGULATE*

The growing public recognition of the array of goods and services provided by forests will change the nature of governance for forest management actions. In this context, governance is defined as excising authority over actions and has evolved in the United States from being market based to being a mix of market and regulatory functions (see Haynes et al., 2003 for an expanded discussion of how this has evolved in the Pacific Northwest). For federal forestlands, forest planning has been developed to implement forest management. It includes very formal processes, broad management objectives, and increased stakeholder participation. Management on private forestlands is determined by a mix of market and regulatory functions. A different set of regulations (e.g., State forest practice acts) influence both the design and applications of forest management practices on private lands.

For the most part, these regulations reflect public concerns about forestlands or forest conditions. These growing public concerns have long been a determinant of forest policies, and since the early 1990s, forest policy has increasingly been internationalized (see the discussion in Haynes, 2003: pp. 173-179) in contexts of economic globalization and sustainable development. Currently much of the international debate deals with suggestions about the need to supplement market-determined actions with processes to attain equilibrium among interests advocating environmental protection, employment that contributes to economic prosperity, public access, and social justice (see Andersson et al., 2004 for a variety of perspectives on these issues).

The adoption of broad-scale land management plans such as the Northwest Forest Plan (USDA and USDI 1994) for federal lands in the Pacific Northwest is a step in this evolution of shifting societal expectations for forest management. It involves the development of institutions to supplement the existing mix of market and regulatory processes already present in the region. These institutions include a mix of formal and informal groups and organizations whose roles are evolving and which now provide a forum for discussing emerging problems beyond just plan implementation. At the same time, some of these groups and organizations are engaged in the development of collaborative efforts in the subdivision of forest property rights. The Northwest Forest Plan included 10 formal adaptive management areas, several of which illustrated that successful collaboration depended on early engagement of stakeholders in the assessment part of the planning process and on more fully involving stakeholders with the goal of gaining social acceptability for designed treatments (Stankey et al., 2003). In selected cases, engagement with informal groups led to partnerships that were able to accomplish specific actions in a collaborative fashion.

## *CONCLUSION*

Changing expectations of returns from forest management raise concerns about whether North American forests can meet societal expectations for making progress towards both sustainable forest management and the provision of the suite of ecosystem services expected by increasingly urbanized populations. At the same time, questions are emerging about the effectiveness of the institutions that have evolved to both promote forest management and to control (or modify) human behavior to conserve nonmarket ecosystem services.

The projections of relatively constant real prices (Figure 1) for the next five decades will provide an incentive for those landowners with a strong propensity to manage (industrial landowners, large private landowners, timberland investment organizations, public timberlands) but by themselves will not engender much enthusiasm among the majority of landowners who display a lower propensity to manage.

What does this mean in the context of sustainable forest management? Simply put, it means that the majority of timberland will be lightly managed while a small minority of hectares will be heavily (or actively) managed (on relatively short rotations). The net effect is that prices, although not a barrier, will also not provide a strong incentive to improve land management or to expand the production of ecosystem services. This raises a dilemma among the advocates for improved forest management when they find themselves supporting either more regulation to insure progress toward sustainable forest management across a broader area of forestland or developing alliances with the broader conservation community who are starting to consider financial incentives for the production of various ecosystem services.

This predicament is troublesome to many traditional forest managers who tend to see regulations as impeding their management discretion and conservation advocates as challenging long-held views about the "proper goals" for land management. But there is an emerging alliance between forest owners and conservation advocates motivated by a shared desire to protect open spaces and the associated goods and services (that many consider as public goods) from dwindling away as human populations increase. A frequently cited strategy to protect open space is to increase the economic attractiveness of current land uses by compensating private landowners for their contributions to broad scale conservation goals. The early examples (see Best and Wayburn, 2001) have largely involved private funding as part of environmental impact mitigation, but for larger scales it will require public funding. This could proceed in a number of ways such as various conservation programs, but to reach the scale needed, it will eventually involve public compensation of private owners for their contributions to conservation goals. These latter approaches will depend on some key steps, but one early step will be assigning property rights to some of the former public goods derived from the forested land. Clear property rights enable those who own the resources to gain the income that can be derived from the production of goods and services. Another early step is development of markets and price signals necessary to sustain the production of a range of goods and services. Markets will evolve only for the tradable subset of goods and services derived from forested land; there will always be public consumption goods produced jointly that must be protected as trading develops.

## NOTES

1. The distinction between 'open access' and 'common property rights' is important because much has been made of behavior under open access, but incorrectly identifying it as behavior relating to common property. Common property rights (sometimes called communal property rights) are those applying formally or informally to a particular group of people and use of a particular set of assessts.

2. The present value of all future net returns, the soil expectation value or SEV, from a simple plant-and-harvest management regime can be expressed as:

$$SEV = \frac{Price * Volume\ at\ rotation - Planting\ cost}{(1 + Interest\ rate)^{Rotation} - 1}$$

## REFERENCES

Ahn, S., A.J. Plantinga and R.J. Alig. 2002. Determinants and projections of land use in the South Central United States. *Southern Journal of Applied Forestry* 26(2):78-84.

Andersson, F., Y. Birot and R. Päivinen. (Eds.). 2004. Towards the sustainable use of Europe's forests–forest ecosystem and landscape research: scientific challenges and opportunities. EFI Proceedings No. 49. European Forest Institute, Joensuu, Finland.

Baker, F.S. 1950. The principles of silviculture. McGraw-Hill, New York.

Barbour, R.J., D.D. Marshall and E.C. Lowell. 2003. Managing for wood quality. In: pp. 299-336. Chapter 11. Monserud, R.A., R.W. Haynes and A.C. Johnson. (Eds.) Compatible forest management. Kluwer Academic Publishers, Dordrecht, The Netherlands.

Best, C. 2002. America's private forests. Challenges for conservation. *Journal of Forestry* 100(3):14-17.

Best, C. and L.A. Wayburn. 2001. America's private forests status and stewardship. Island Press, Washington, DC.

Brechin, S.R., P.R. Wilshusen, C.L. Fortwangler and P.C.West. 2002. Beyond the square wheel: toward a more comprehensive understanding of biodiversity conservation as social and political process. *Society and Natural Resources* 15:41-64.

Cubbage, F. and R. Haynes. 1988. Evaluation of the effectiveness of market responses to timber scarcity problems. Marketing Res. Rep. 1149. U.S. Department of Agriculture, Forest Service, Washington, DC.

Davis, L., K.N. Johnson, P. Bettinger and T. Howard. 2001. Forest management. 4th ed. McGraw Hill, New York.

Davis, L.S. 1966. Forest management: regulation and valuation. 2nd ed. McGraw-Hill, New York.

Davis, L.S. and K.N. Johnson. 1987. Forest management. 3rd ed. McGraw-Hill, New York.

Hardin, G. 1968. The tragedy of the commons. *Science* 162:1243-1248.

Haynes, R.W. (Tech. coord.) 2003. An analysis of the timber situation in the United States: 1952 to 2050. A technical document supporting the 2000 USDA Forest Service RPA assessment. Gen. Tech. Rep. PNW-GTR-560. U.S. Department of Agriculture, Forest Service, Pacific Northwest Research Station, Portland, OR.

Haynes, R.W. 2004. Do markets provide barriers or incentives for sustainable forest management: the US experience. 2004 SAF Convention Proceedings. Edmonton, Alberta, Canada. Society of American Foresters, Bethesda, MD, USA.

Haynes, R.W. 2005. Economic feasibility of longer management regimes in the Douglas-fir region. Res. Note. PNW-RN-547. U.S. Department of Agriculture, Forest Service, Pacific Northwest Research Station, Portland, OR.

Haynes, R.W., D.M. Adams and J.R. Mills. 2003. Contemporary management regimes in the Pacific Northwest: balancing biophysical and economic concerns. In: pp. 267-296. Chapter 10. Monserud, R.A., R.W. Haynes and A.C. Johnson. (Eds.) Compatible forest management. Kluwer Academic Publishers, Dordrecht, The Netherlands.

Kline, J. and R. Alig. 2001. A spatial model of land use change for Western Oregon and Western Washington. In support of the 2000 RPA Assessment. Res. Pap. PNW-RP-528. U.S. Department of Agriculture, Forest Service, Pacific Northwest Research Station, Portland, OR.

Sendak, P.E. 1994. Northeastern regional timber stumpage prices: 1961-91. Res. Pap. NE-683. U.S. Department of Agriculture, Forest Service, Northeastern Forest Experiment Station. Radnor, PA.

Smith, W.B., J.S. Visage, D.R. Darr and R.M. Sheffield. 2001. Forest resources of the United States, 1997. Gen. Tech. Rep. NC-219. U.S. Department of Agriculture, Forest Service, Northeastern Forest Experiment Station. St. Paul, MN.

Stankey, G.H., B.T. Bormann, C. Ryan, B. Shindler, V. Sturtevant, R.N. Clark and C. Philpot. 2003. Adaptive management and the Northwest Forest Plan: Rhetoric and reality. *Journal of Forestry* 101(1):40-46.

The Economist. 2005. A survey of corporate social responsibility. January 22:3-22.

U.S. Department of Agriculture, Forest Service [USDA FS]. 1973. The outlook for timber in the United States. Forest Resource Report 20. Washington, DC.

U.S. Department of Agriculture, Forest Service [USDA FS]. 1988. The South's fourth forest: alternatives for the future. Forest Resource Report 24. Washington, DC.

U.S. Department of Agriculture, Forest Service; U.S. Department of the Interior, Bureau of Land Management [USDA and USDI]. 1994. Record of decision for amendments to Forest Service and Bureau of Land Management planning documents within the range of the northern spotted owl. Standards and guidelines for management of habitat for late-successional and old-growth forest related species within the range of the northern spotted owl. [Place of publication unknown]. 74 p.

Weigand, J.F. 1994. First cuts at long-rotation forestry: projecting timber supply and value for the Olympic National Forest. In: pp. 177-202. Weigand, J.F., R.W. Haynes and J.L. Mikowski. (Comps.). High quality forestry workshop: the idea of long rotations. Proceedings. CINTRAFOR SP 15. College of Forest Resources, University of Washington, Seattle, WA.

doi:10.1300/J091v24n01_01

# Integrating Wood Production Within Sustainable Forest Management: An Australian Viewpoint

Ian Ferguson

**SUMMARY.** Australia is a federation in which the Constitution vested forest management to the states. A joint Commonwealth-state process was developed to implement a National Forest Policy, involving joint oversight of resource assessment and planning for each of eleven regions. Codes of forest practice and planning systems were greatly improved by incorporating available research on nonwood as well as wood uses. The 39% increase in the area of conservation reserves system necessitated major withdrawals from wood production. Despite the increases in reserves, conservation groups and allied interests were generally dissatisfied with the outcomes. Furthermore, resource security for the industry was not realised. Pre-election promises made by incoming governments, both state and Commonwealth, involved breaking the spirit, if not the letter of the agreements. Nevertheless, lower order processes of third-party audits and joint reviews of practices and planning

---

Ian Ferguson is Professor Emeritus at the School of Forest and Ecosystem Science, University of Melbourne, Parkville, Victoria 3010, Australia (E-mail: iansf@unimelb.edu.au).

The author acknowledges helpful comments from Richard Haynes and an anonymous reviewer and the views of and debates with numerous colleagues involved in the Regional Forest Agreement process.

[Haworth co-indexing entry note]: "Integrating Wood Production Within Sustainable Forest Management: An Australian Viewpoint." Ferguson, Ian. Co-published simultaneously in *Journal of Sustainable Forestry* (Haworth Food & Agricultural Products Press, an imprint of The Haworth Press, Inc.) Vol. 24, No. 1, 2007, pp. 19-40; and: *Sustainable Forestry Management and Wood Production in a Global Economy* (ed: Robert L. Deal, Rachel White, and Gary L. Benson) Haworth Food & Agricultural Products Press, an imprint of The Haworth Press, Inc., 2007, pp. 19-40. Single or multiple copies of this article are available for a fee from The Haworth Document Delivery Service [1-800-HAWORTH, 9:00 a.m. - 5:00 p.m. (EST). E-mail address: docdelivery@haworthpress.com].

Available online at http://jsf.haworthpress.com

doi:10.1300/J091v24n01_02

should be continued to improve sustainability and the information made available in forms accessible to the community. The contribution of science, however, is limited by the inability to adequately evaluate the tradeoffs between uses. The higher level processes should identify the major options and leave the political process, be that state or Commonwealth, to resolve the choice between them. doi:10.1300/J091v24n01_02


**KEYWORDS.** Sustainable forest management, forest policy, regional agreements, conservation reserves, wood production, codes of practice, planning systems

## *INTRODUCTION*

This paper reviews the ways in which sustainable management of forests in Australia is being implemented through the regional forest agreement process and specifically how wood production is being integrated into that process in native (*syn*, natural) forests. Being a federation, the relationships and arrangements between the Commonwealth and state governments are critical to this process. These rest on a joint Commonwealth-State National Forest Policy Statement that has been fleshed out at considerable expense in the form of regional forest agreements, to carry out a comprehensive review of the uses, planning, and management of native forests. Before considering the policies and their implementation, however, a brief summary of Australian native forests is needed to place these policies in context.

Almost all the so-called "tall forests" (capable of attaining over 30 metres [m] in total height) that are productive in the sense of commercial wood production are to be found around the rim of the continent (about one-third of the land area) where rainfall exceeds 500 millimetres per year (mm/yr). These forests are characterised by a diversity of natural ecosystems and species in which the genera *Eucalyptus* and *Corymbia* (formerly part of genus *Eucalyptus*) play a dominant role. This is the most densely populated part of the continent. Much of the population (80%) resides in major cities such as Sydney, Melbourne, and other state capitals, scattered around the coast. Within the tall forest zone, there are some 50 million hectares (ha) of forests that are potentially productive for timber and related products. This is divided approximately 23% state-owned conservation reserves, 22% state-owned multiple-use forests (including wood production), 40% private ownership, and 15% state-owned but privately leased forest (generally pastoral use).

## *NATIONAL FOREST POLICY STATEMENT*

Australia is a federation initially formed in 1901, now comprising 6 states and 2 territories. Under the Constitution, land management, including forest management, was to be administered by the states. More recent administrative arrangements of the Commonwealth government extend this devolution to the territories. Until the 1970s, this division of responsibilities was relatively simple and clear, but as the conservation movement evolved, the responsibilities regarding forests have become much less clear. Major issues inevitably attracted the attention of the Commonwealth Parliament, increasingly so as the media became more national in orientation.

Over the past 30 years, the Commonwealth government has increasingly become involved in forest issues, often taking a position in opposition to the views of the state government concerned, especially when of the opposite political persuasion. Although the Commonwealth had no direct controls over forest management under the Constitution, it soon developed several forms of indirect control. By the early 1990s, it had become apparent that a joint Commonwealth-state forest policy framework was needed. A joint policy statement was negotiated between the Commonwealth and the states in 1992 and after further negotiation finally signed by the last state in 1995 (Commonwealth of Australia, 1992, 1995). The statement rests on three main principles as the basis for sustainable forest management:

- Maintain ecological processes.
- Maintain biological diversity.
- Manage for the full range of environmental, economic, and social benefits.

The statement is an unusually comprehensive and far-sighted document, especially given the complexity of the Commonwealth-state relationships. Two provisions in this statement deserve special mention because of the changes they were to institute:

- Jointly agreed and legally binding codes of forest practice were to guide forest management where wood production or other commercial extractive uses were involved.
- Comprehensive and joint regional assessments were to be instituted in developing a national reserve system.

While the process to implement the National Forest Policy Statement was being developed, a dispute between Commonwealth Ministers over the issuing of a woodchip export license led to a chaotic national protest in 1994 and the establishment of the regional forest agreement process under the control of the Department of Prime Minister and Cabinet (McDonald, 1999; Zammit, 1999; Hollander, 2004). These events accelerated the realisation of the various governments that continuing political gamesmanship between the Commonwealth and state levels was counterproductive to rational resolution of the issues.

## *REGIONAL FOREST AGREEMENTS*

Davey (2005) provided a comprehensive description of the regional forest agreement process. Initially some eleven regions (see Figure 1) were delineated. Victoria had 5, New South Wales 3 (one of which was effectively split for most purposes), Queensland, Tasmania, and Western Australia, one each. The total forest area covered by existing agreements is 23.2 million ha, representing the overwhelming majority of Australian tall forests other than the north Queensland rainforest in conservation reserves and southeast Queensland region (see later). About 7 million ha of the area under agreements is now in state-owned multiple-use forests, 10 million ha is in state-owned conservation reserves, and the remainder is privately owned, based on interpolation of National Forest Inventory (2003) data.

The agreements focussed mainly on publicly owned native forests that were directly owned by the state concerned and managed by either a state forest agency (if wood production and extractive uses were permitted) or a park or watershed agency; or, in the case of Western Australia, an integrated agency. The regions predominantly cover the "tall forests" because these are the forests in which the wood production versus conservation debate has been greatest (Zammit, 1999). Some small areas of privately owned native forest were purchased or otherwise protected for the special ecosystems they could add to the national conservation system.

FIGURE 1. Tall forest regions of Australia with regional forest agreements.

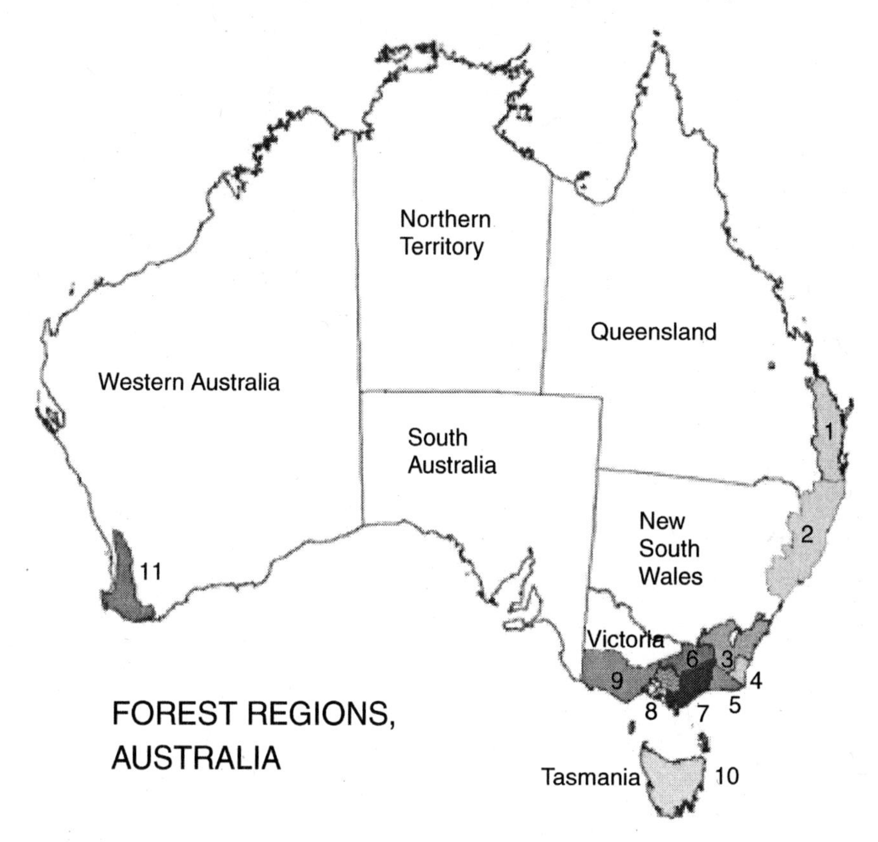


The regional forest agreement process commenced in 1997 and sought to achieve two main objectives that extended or amplified those set out in the National Forest Policy statement, each being an attempt to assuage one of the two sides in what had become a highly polarised debate:

- Establish a comprehensive, adequate, and representative national reserve system.
- Provide greater certainty regarding the native forest resource available for wood production by integrating industry and conservation policy, encouraging downstream processing of the native forest resource, and encouraging the export of unique Australian wood products.

Following advice by a scientific advisory committee and negotiations by an intergovernmental committee (JANIS, 1997), the national reserve process was framed around the broad objective of achieving representation of 15% of the pre-1750 areas in native forest types as mapped at a 1:100,000 scale. Special provisions were also made to reserve at least 60% of remaining old-growth forest (100% in the case of rare species of types), and other special provisions were made for rare and threatened ecosystems or heterogeneous forest types.

The creation of the national conservation reserve system meant that some of the timber resource in public ownership was withdrawn from that use. In return, the 20-year agreements offered the hope of greater security of timber supply from the remaining areas of publicly owned native forest. Commonwealth export controls, that had been a primary battleground for political influence by conservation and industry interests, were also abolished for those regions under agreements. The agreements were made subject to 5-yearly reviews but the administrative arrangements were such that major alterations in tenure were unlikely (McDonald, 1999).

The regional forest agreements gave little attention to plantations because taxation and related changes to encourage plantations had been largely put in place soon after the National Forest Policy statement was drafted. An ambitious "Plantations 2020" policy was drafted to treble the area of plantations (circa 1.0 million ha in 1990) by the year 2020 (see Gerrand et al., 2003). Implicitly, this also alleviated any pressure on native forests to cater for future increases in demand.

## *OUTCOMES AND FUTURE CHALLENGES*

Not all the proposed regional forest agreements were consummated, even though that was the intention. A comprehensive resource assessment was completed for southeast Queensland, but the agreement did not proceed to completion (Brown, 2001). Instead, the forest industry negotiated an arrangement with the conservation movement and state government whereby harvesting of publicly owned native forest could continue until 2025, albeit at a somewhat lesser rate than previously, after which it was to cease entirely and timber supply was then to be taken up by fast-grown plantations. Although many saw this as a Faustian arrangement relative to the concept of sustainable forest management, substantial reductions in sustainable yield would have been required under sustainable forest management and would have necessitated an immediate major relocation and re-equipping of the industry. This arrangement highlights the precarious nature of a joint Commonwealth-state agreement, with the Commonwealth essentially being bypassed, notwithstanding the National Forest Policy agreement.

The Regional Forest Agreement process had many critics–Dargavel (1998a, 1998b), Horwitz and Calver (1998), Kirkpatrick (1998), Mercer (2000), Conacher and Conacher (2000), and Musselwhite and Herath (2004)–to name but a few. They tended to focus on the defects, sometimes without giving due recognition to the realised or potential accomplishments through a process of continual improvement through monitoring and third-party audits. Hence a systematic review of the major outcomes seems appropriate here, together with an assessment of future challenges.

### *Conservation Objectives*

The creation of a national conservation reserve system that is based on consistent and tenure-neutral criteria and contains at least 15 per cent of the areas of the forest types extant prior to 1750 (where those areas still existed) was an outstanding achievement. The development of the criteria (JANIS, 1997) for this "comprehensive, adequate and representative" system of reserves was itself a major feat in achieving agreement between leading forest managers, scientists, and nongovernment organisations, at least for a time. Associated legislative changes also rationalised most of the (often inexplicable) disparities between the environmental and heritage legislation of the Commonwealth and the individual States, as it related to forests.

Critics of the process argued that the extent of reservation was not sufficient, but that is an attempt to shift the goals agreed between the Commonwealth and states. There are legitimate criticisms that can be levelled at the comprehensiveness hinging largely on the 1:100,000 scale for mapping of forest types, the boundaries of the regions, and the consequent recognition and treatment of heterogeneity within that scale. However, it is difficult to see how a system based on a finer geographic scale could have been completed in any realistic initial time scale, given the process.

The process resulted in an immediate increase of 2.5 million ha (i.e., 39%) in the area of reserves according to Mobbs (2003). Election-prompted events after this have expanded the area of reserves still further. This has left the states to carry the financial burden of the additional areas to be managed. Despite the fact that this is a *national* conservation reserve system devised jointly by the Commonwealth and state government concerned, the Commonwealth does not contribute financially to its ongoing management, other than through the periodic review processes of the regional forest agreements and those obligations it has to World Heritage areas under the international treaty. In the longer term, the lesser-known conservation reserves and national parks are likely to suffer (Hoggett, 2005) because they generate little if any revenue or public interest.

With some notable exceptions, there has generally been a reluctance or indifference to explicitly identifying objectives, specifying goals, and monitoring outcomes for conservation reserves (e.g. Public Land Use Commission, 1995a, 1995b). Typically, these objectives assumed a direct translation of all of the objectives of the national park legislation to that reserve, rather than identifying (1) what conservation or other attributes the particular area was best capable of supplying, (2) strategies to manage these, and (3) addressing the inevitable tradeoffs between multiple uses. In formal reviews of the Ecologically Sustainable Forest Management process by expert committees in at least three States, the need for better integration of the management of reserve and off-reserve forests was noted. Ecologically sustainable forest management applies just as much to conservation and other reserves as it does to forests in which harvesting and extractive uses are permitted, and the integration of reserve and off-reserve management is fundamental to sustainability.

### *Wood Production Objectives*

Resource security for wood production in native forests, the other principal policy objective of the regional forest agreements, could not be achieved by a land-use allocation process alone: it required that wood production be integrated within a system that was seen to lead to sustainable forest management.

McKinnell et al. (1991) provided descriptions of silvicultural practices in the various major native forest types in contention in these states, as they existed prior to the Regional Forest Agreement process. For those publicly-owned native forests outside the national conservation reserve system, the subsequent introduction or refinement of *codes of practice* regulating operations and of formal processes of planning, management, monitoring, and review of activities and uses at the regional, landscape, and stand levels, together with public participation in review of them, have greatly improved progress toward sustainable management (e.g., Ferguson et al., 1997).

The following discussion will cover *Codes of practice*, *Plans of management*, *Sustainable yield*, *Pricing of wood and funding of nonwood uses*, *Tradeoffs with other values*, *Social and economic values*, *Public participation*, and *Certification*.

### *Codes of Practice*

Formal Codes of Practice were first introduced in Tasmania in 1985 and then in Victoria in 1988, followed by Western Australia and other states (McCormack, 1996). The regional forest agreements subjected them to detailed review and improvement to provide a regulatory framework for field operations, including fire management. In many respects, the Codes of Practice simply represent an evolution of past regulatory prescriptions, as Lee and Abbott (2004) described in their historical review for Western Australia. But the codes (e.g., Wilkinson, 1999) now incorporate the most recent scientific results, the precautionary principle, postharvest amelioration requirements, provisions for monitoring and reporting, and penalties for noncompliance. The fundamental change is that these Codes are publicly available, being generally accessible on the Web, and are subject to a public review process periodically (generally at least every 5 years). They also provide a basis for informal monitoring and reporting of performance by environmental nongovernment organisations, which have been extremely active in monitoring those areas of particular conservation interest.

The Codes of Practice are implemented by a preharvest inventory of a proposed coupe (syn. stand) that aims to identify the boundaries of the area available for harvesting, with due regard to prescribed stream buffers, wildlife corridors, habitat trees, vulnerable and endangered species, landscape values, and other constraints. This inventory is generally guided by prior aerial photographic interpretation of forest types, and their crown cover and height. The pattern of reasonably homogeneous communities so defined can be quite detailed, the minimum area for delineation as a distinct community generally being about 1 ha but occasionally reaching thousands of hectares. The preharvest inventory phase deals with a variety of scales in space and time. The complexity involved prompted the adoption of the International Organization for Standardisation (ISO) 14001 Environmental Management System framework, together with the use of geographic information systems. This has seen a notable advance in integrating the management of the various uses across time and space.

In carrying out this work, the forester has the authority to exclude additional areas beyond the minima prescribed to give better effect to conserving the ecosystems involved, such as to rationalise an otherwise irregular boundary or to encompass local topographic variation. In other cases, there may be a need to aggregate habitat trees in clumps or to link them with other retained buffers or corridors for protection from wind or fire, where burning forms part of the regeneration or later fuel reduction process.

On the other hand, there is virtually no scope to vary the prescriptions in the other direction. Although this latter provision may seem a sensible safeguard, it is not always so. Given the complexity of the tradeoffs between uses, some scope is needed for the exercise of intelligent judgment in either direction, subject to proper documentation of reasons. However, environmental nongovernment organisations, acting informally as environmental watchdogs, commonly pillory public agencies over any apparent breaches of the code. Thus the community is only likely to grant flexibility if the person exercising it can be held accountable, and that requires the use of more sophisticated field recording and reference systems–a prospect that is now likely to materialize via hand-held and geo-referenced field computing devices.

The agreement process represented a major achievement in systematically considering and incorporating the results of research into codes of practice and identifying gaps to be pursued. Although research findings have greatly assisted the refinement of prescriptions to limit damage to nonwood uses and values, and sometimes even to identify tradeoffs between them (Ferguson, 2005), much remains to be done.

Some aspects of research and policy findings remain poorly implemented. The most important of these concerns regeneration systems used in wood production operations. Policy and ecological reviews dating at least from Ferguson (1985) to Dovers (2003) and Lindenmayer (2003), to cite a few of many, have stressed the desirability of tailoring the choice to the particular ecosystem, rather than adopting a uniform system across the entire forest type. With some notable exceptions (e.g., Burrows et al., 2002), that is what is still being done in some regions. Commonly in the temperate forests of Australia, a clearfelling system has been adopted, clearfelling being followed by burning to provide a receptive seedbed, and artificial seeding. Typically, a fixed rotation of 80 to 120 years is prescribed. Although such a regeneration system may be the most cost-effective, and can thus be justified on some of the area, it is not always the most appropriate from the viewpoint of other uses and social acceptability (Williams et al., 2005). Scope exists for cross-site tradeoffs if rotation lengths are reduced somewhat on the better sites and lengthened on poorer sites. That rotation lengths and regeneration systems are not being varied as much as is desirable probably reflects the institutional characteristics regarding oversight by conservation groups noted earlier rather than any lack of knowledge on the part of the agencies.

*Plans of Management*

The span of scales in time and space is further extended in developing plans of management, which form a prescribed part of the planning process. The scope of these plans differs widely in different states. In Western Australia, the legislation requires a single plan for the entire southwest region. For Tasmania, the plans for multiple use forests are based on seven administrative districts. For Victoria and New South Wales, they are based on individual large regions, although there have been several changes in this scale over the last two decades, generally to larger units. These differences and changes reflect the peculiarities of the forests and of the institutional structures.

Although much of the process and rationale of planning may be summarized in a single document, the reality is that a plan of management rests on a diverse and complex bundle of documents, records, and databases. Hence computerized information and retrieval systems are used to store, update, or audit plans of management and to integrate management across uses and spatial and time scales.

The rotation lengths prescribed for sustainable yield far exceed any purely financial investment decision, being from 80 years to 200 years, depending on species. Despite the seemingly low impact that this has on the area of forest harvested annually, the public at large react adversely to the resulting visual impact of clearfelling in publicly owned native forests (Ford et al., 2005) indicating the need to develop lower visual impact silviculture for these sites.

Structural goals (Bradshaw and Rayner, 1997a, 1997b, Bradshaw, 2002) have been used in Western Australia to integrate the management of reserves and the so-called "off-reserves" on which harvesting and other extractive uses are permitted. They represent a direct planning approach to maintaining a full suite of age and forest structures in terms of canopy and tree size distributions, rather than using sustainable yield as the sole mechanism. Thus while 200- and 100-year nominal rotations were scheduled for the two major forest types on about 85 to 90% of the respective areas, the remainder had longer rotations to maintain a sufficient supply of older structures.

Research relating to or stemming from the agreements has improved the standards of inventory, the scheduling of yield, and the calculation of sustainable yield in most regions. Risk management has also been given greater attention in recent management plans. Notable recent examples include research on yield scheduling in New South Wales (Turner et al., 2002); disease management in Western Australia (e.g., Strelein et al., 2004), alternative silvicultural systems in Tasmania (Hickey and Brown, 2003), and fauna management in Gippsland, Victoria (e.g., Alexander et al., 2002).

### *Sustainable Yield*

The sustainable yield is an especially important control variable for wood production. To use Western Australia as an example, some 600 strata are recognised in the *E. marginata* (jarrah) state forests, based on tenure, region, rainfall, height, history of cutting, current structure, and whether or not mined. Within each such stratum, there is further subdivision into six silvicultural classes (called cohorts)–thinned, gap creation, shelterwood, temporary exclusion areas, selective logging, and dieback. This complexity would be difficult to use in practice without access to geographic information systems and computerized planning systems.

Areas of local reservations and other exclusions in which timber production is not permitted in multiple-use forests affect sustainable yield. The extent of the so-called informal reserves that are associated with buffer strips, wildlife corridors, and other local exclusions from wood production is substantial. Although reductions in sustainable yield were anticipated to follow the introduction of more stringent Codes of Practice, the impacts have proven larger than expected. Comparison of actual net areas with "prescribed" areas in Victoria summarized by Vanclay and Turner (2001) suggests reductions of 15 to 24% relative to the total areas of forest that could potentially support economic harvesting. Allowing also for forest components that are incapable of supporting economic wood production, the proportions of the total area in which harvesting is allowable varies from 60 to 70% (National Forest Inventory, 2003). Some of the reductions can be detected with aerial photography but a number cannot, so that post facto comparisons need to be monitored to provide guidance on the losses to be expected for future harvesting operations.

Harvest scheduling itself is presently largely built around the industrial issues of supplying relatively constant periodic amounts of wood to support processing plants. Most of the sustainable yield systems work on the basis of nondeclining flow of sawlogs over extraordinarily long planning horizons. Given the uncertainties and changing technologies and markets in relation to wood use, this basis is difficult to justify in terms other than political inertia, and it limits the flexibility of planning wood use. A shift to the use of a more realistic planning horizon of say, 50 years, and age-class and other structural goals to be achieved at the end of that horizon would provide a more transparent and accountable basis for planning (Ferguson et al., 2003).

*Pricing of Wood and Funding of Nonwood Uses*

In most cases, the State agency responsible for management of native forests for wood production and other commercial purposes has been corporatised or at least commercialised and separated from the regulatory agency. In part, this reflects the joint Commonwealth-State National Competition Policy (Hilmer et al., 1993), in which all commit to managing publicly owned commercial entities under provisions that remove subsidies (other than those allowed under the General Agreement on Tariffs and Trade) and make them operate on equal terms to the private sector. However, the focus on commercial outcomes poses marked uncertainties as to the commitment of the new state agencies associated with wood production to fund the management of and research on noncommercial uses and services and their integration with the reserve system.

*Tradeoffs with Other Values*

In addition to the progress made through the dedication of additional conservation and wilderness reserves, the regional forest agreements made substantial progress in recognising, albeit often somewhat arbitrarily, the trade-offs between wood production and nonwood uses. If the production possibility locus for joint production of wood and a nonwood use is only moderately competitive (Ferguson, 1996), as the agreements suggest is often the case, tradeoffs provide greater gains to society over those from either use alone. Where it is supplementary, which may occasionally be the case, there may be gains to one or the other use as well as gains to society.

However, the agreements were less than successful in addressing the issues of indigenous and cultural heritage (Centre for Social Research, 1997; Dargavel, 1998a; Lennon, 1998), although they at least made a start. The Commonwealth has recently funded the preparation of an indigenous heritage forest strategy, and this may accelerate progress.

### *Social and Economic Values*

Some innovative research in social values and indicators was carried out (Coakes, 1998), Nevertheless, Dargavel (1995, 1998a, 1998b) and McDonald (1999) have criticised the failure to consult those employed in the industry and regions concerned, although this may also reflect the limited way in which the media reported such issues (Collins, 1999).

However, research concerning economic values (e.g., the direct opportunity costs of changes of land use) was largely sterile, being directed at the regional economic effects that had already been implicitly accepted by the process. There was virtually no attempt to explicitly address the detail of measuring wood and nonwood values in a way that might have illuminated the appropriate trade-offs between them. This is not entirely surprising, given the technical difficulties and imprecision of measuring many of the nonwood values and services (Ferguson, 1998), but it highlights a key weakness in the notion of a scientifically based allocation of forest uses.

### *Public Participation*

The regional forest agreement process placed considerable emphasis on public participation. All expert committee and consultant reports were publicly available. Much of the scientific community with environmental interests distrusts the forest-related agencies, so future work by independent consultants and scientists should be subject to peer review to improve the scientific standing of the outcomes (see Horwitz and Calver, 1998) or, alternatively, independent panels of experts used. Some of the previous expert committees held various forms of public consultation, ranging from workshops to submissions and later interviews with respondents. Nevertheless, the general public reaction was that there were insufficient opportunities, time, and resources for adequate public involvement. Some conservation groups stood outside the process for political reasons, but others participated to varying degrees (McDonald, 1999; Stewart and Jones, 2003).

Public participation was well handled in Tasmania by the then Public Land Use Commission (now replaced by a special tribunal), an independent body established to conduct public land use inquiries. This was in contrast to Western Australia, where Mercer (2000), Conacher and Conacher (2000) and Stewart and Jones (2003) claimed that the forestry agency, which was closely associated with the arrangements, was not genuinely interested in public participation.

In general, public participation was the only initiative involving bottom-up inputs–most of the process was dominated by top-down inputs. Given the inability to make scientific distinctions in the valuations and tradeoffs, this was probably a major weakness in the design of the process.

### *Certification*

The regional forest agreements have promoted interest in forest certification, partly because much of the necessary preparation was accomplished in the process. Public and private forest owners are increasingly moving to adopt either the Australian Forestry Standard or Forest Stewardship Council certification systems, the latter being largely restricted to plantations because of the political issues relating to the utilisation of native forests. Hence the Australian and state governments, industry groups, and state forest agencies established an Australian Forestry Standard (2003) through processes prescribed by Standards Australia, which represents Australia in the International Standards Organisation. This provides a system of certification applicable to all forest types, whether native forests and plantations, and across all ownerships and land tenures, and is capable of independent third-party verification and certification. Certification is not compulsory, and the choice between the two schemes is voluntary.

## *ONE TENURE-ONE USE VS. SUSTAINABLE MANAGEMENT*

The debate between two principal sets of stakeholders, the conservation groups on the one side and the industry and unions on the other, had become extremely polarised at the time the regional forest agreement process commenced. The national park tenure became associated exclusively with conservation and the state forest tenure with logging. In the "one tenure-one use" argument (Dovers, 2003) that characterised that debate, any gains for one side meant losses for the other. The regional forest agreement process used two principal strategies that were intended to assuage conservation and industry interests respectively. The first element of the strategy was concerned with a National Conservation Reserve System, and the second concerned resource security for the forest products industry. Although assiduously cast in the operational mould of ecologically sustainable forest management, this tended to perpetuate a "one tenure-one use" image in the media reporting, albeit with a wider definition of tenures and uses.

The agreements generally achieved the goals set out for the establishment of a National Conservation Reserve System but did not achieve resource security for the forest industry in any of the states involved. Each of the four states and the Commonwealth has breached the spirit, if not the letter, of the agreements in the years since their first agreement was signed, principally as a result of pre-election promises made by the leaders of political parties that were subsequently honoured when taking up government. The breakdowns highlight the inability of the Commonwealth to hold individual states, and vice versa, to a joint agreement of this kind when there are no effective sanctions available under the Constitution.

Toyne (1994, 2002) has argued that the Commonwealth could use its powers if it so wished. However, successive Commonwealth governments have been reluctant to do so because of the divisions this could create in the federation. Within the Constitution, state governments that have a majority in their parliaments can make or unmake legislation giving effect to policies under their control. This so-called "sovereign risk" also pertains to the Commonwealth. For example, the conservation of the remaining Tasmanian old-growth forests not in reserves again became a major (some would say, defining) issue in the recent Commonwealth election. The election outcome necessitated a change in the agreement and underscored the futility of the words in the original agreement that attempted to bind the parties legally (McDonald, 1999). The states are not prepared to cede over-riding legal powers over forests to the Commonwealth, as happened in the case of the legislation regarding corporations. A change in the Constitution is the only other solution to provide permanent legal clarity. Previous attempts to effect other changes in the Constitution give no cause for confidence that such a change could be achieved.

## *THE FUTURE FOR WOOD PRODUCTION IN NATIVE FORESTS*

The future for wood production in Australian native forests remains unclear because of the inherent problems of a federation in which the community at large has long been steeped in the argument of "one tenure-one use." This argument tends to exacerbate the polarisation of the issues (see Collins, 1999; Stewart and Jones, 2003). The conservation movement, notwithstanding some softening of attitudes to commercial utilisation of regrowth native forests recently (Neales, 2005), is therefore unlikely to forego it readily in any future political debates.

The regional forest agreement process cost at least half a billion dollars on the part of the Commonwealth, and probably close to that amount by all the states involved. Its failure to win widespread acceptance may have dampened political and bureaucratic enthusiasm for the ongoing process, as there has been little evidence of a collective will to initiate the joint 5-yearly reviews that are now overdue in many of the regions. This suggests that a major reshaping of the regional forest agreement process is desirable, and some recommendations about the future process may therefore be timely. These deal with *Broadening the debate*, *Utilising the agreement framework*, and *Relationships between the Commonwealth and States*.

### *Broadening the Debate*

The challenge for all the principal parties involved is to inform the community at large about the limitations of the "one tenure-one use" argument and the advantages that a broader ecosystem approach to sustainable forest management can provide. Those advantages apply as much to reserves as to off-reserves and to the intelligent integration of management across both public and private forests (Stewart and Jones, 2003). The issues of the tradeoffs between conservation and recreation, or between the conservation of particular species across tenures, or concerning fire management across tenures, or of the state resources available for conservation management, are likely to increase, not diminish, in the future.

In the absence of precise scientific values about tradeoffs, many issues are inherently ideological in nature and therefore will have to be resolved through the political process. That process will be more effective if the data, audits, reviews, and research findings concerning forests are more widely promulgated, especially in forms more suited to use by the media and informing the community. This part of the process needs to be directed much more effectively (Mobbs, 2003) to informing the community about the major findings and options.

### *Utilising the Framework*

A technocratic, predominantly top-down process has not solved and will not solve all the issues concerning forests. Although two of the major areas of contention–reservation of remaining old-growth forests and clearing of forests–will probably have been resolved or become minimal issues in the next 5 to 10 years, other issues such as silvicultural and plantation practices, water and chemical use, regeneration methods, recreation, and aesthetics will take their place, albeit sometimes on a more localised basis.

The lower order technocratic framework of the regional forest agreement process is therefore worth retaining in terms of the emphases on 5-yearly joint evaluation of the environmental management system through third-party audits, independent panel reviews, steering committee identification of major policy changes (but not the choice between them), and on future research needs.

The Commonwealth would be best involved by monitoring progress through the evaluation processes and seeking to raise the standards progressively, state by state. The present regional forest agreements have set an initial benchmark for each region, and evaluations of progress will provide ample material to continually improve at each review, mindful of reducing the differences between those states and territories, but not prescriptive, unless essential to trans-boundary issues.

### *Relationships Between Commonwealth and States*

The remaining issue is then relationship of the Commonwealth to the states in the overall process. Any major choice between options should be resolved by the political rather than the bureaucratic process, whether at Commonwealth or state level. In the event that the matter has commanded Commonwealth attention, then logically its decision must overrule any state determination.

In the recent Commonwealth election, the Commonwealth government, having been elected in part to implement a particular forest policy, then negotiated the implementation with the Tasmanian Government. Hopefully this heralds a de facto recognition for all states and territories that the Commonwealth can and should be able to set the minimum requirements for forest policy and negotiate the implementation with the state without the divisive and expensive threat of Constitutional challenge.

## REFERENCES


Alexander, J.S.A., D.J. Scotts and R. Loyn. 2002. Impacts of timber harvesting on mammals, reptiles and nocturnal birds in native hardwood forests of East Gippsland, Victoria: a retrospective approach. *Australian Forestry* 65(3):182-210.

Bradshaw, F.J. 2002. Forest structural goals: recommendations to the Department of Conservation and Land Management. 42pp. Retrieved July 21, 2005 from www.calm.wa.gov.au/forest facts/index.

Bradshaw, F.J and M.E. Rayner. 1997a. Age structure of the karri forest: 1. Defining and mapping structural development stages. *Australian Forestry* 60 (3):178-187.

Bradshaw, F.J and M.E. Rayner. 1997b. Age structure of the karri forest: 2. Projections of future forest structure and implications for management. *Australian Forestry* 60(3):188-195.

Brown, A.J. 2001. Beyond public native forest logging: National Forest Policy and regional forest agreements after south east Queensland. *Environmental and Planning Law Journal* 18(2):189-210.

Burrows, N., P. Christensen, S. Hopper, J. Ruprecht and J. Young. 2002. Towards ecologically sustainable forest management in Western Australia: a review of Draft Jarrah Silviculture: Guideline 1/02. Panel Report Part 2. Report for the Conservation Commission. Conservation Commission, Perth, W.A.

Centre for Social Research, 1997. Aboriginal consultation project. Vols 1 and 2. Report for Western Australia Regional Forest Agreement Committee. Centre for Social Research, Edith Cowan University, Perth, 234 pp.

Coakes, S. 1998. Valuing the social dimension: social assessment in the regional forest agreement process. *Australian Journal of Environmental Management* 5:47-54.

Collins, R. 1999. Ecopolitics and the media. In: pp. 76-87. C. Star (ed.). Green politics in grey times, Proceedings of the Ecopolitics XI Conference (1999). Department of Political Science, University of Melbourne, Melbourne.

Commonwealth of Australia, 1992, 1995. National Forest Policy Statement: a new focus for Australia's forests. Retrieved July 21, 2005 from www.daff.gov.au.

Conacher, A. and Conacher, J. 2000. Environmental planning and management in Australia. Oxford University Press, Melbourne, 460 pp.

Dargavel, J. (ed.) 1995. Fashioning Australia's forests. Oxford University Press, Melbourne.

Dargavel, J. 1998a. Politics, policy and process in the forests. *Australian Journal of Environmental Management* 5(1):25-30.

Dargavel, J. 1998b. Regional Forest Agreements and the public interest. *Australian Journal of Environmental Management* 5(3):133-4.

Davey, S. 2005. Science and sustainable forest management: an Australian perspective. Paper presented XXI IUFRO World Congress, 8-13 August, 2005, Brisbane Australia. Abstract, *International Forestry Review* 7(5):257.

Dovers. S. 2003. Are forests different as a policy challenge? In: pp. 15-30. D.B. Lindenmayer and J.F. Franklin (eds.). Towards forest sustainability. Commonwealth Scientific and Industrial Research Organisation (CSIRO) Publishing, Collingwood, Victoria.

Ferguson, I.S. 1985. Report of the Board of Inquiry into the Timber Industry in Victoria, 2 Vols. Victorian Goverment Printer, Melbourne.

Ferguson, I.S. 1996. Sustainable forest management. Oxford University Press, Melbourne. 162 pp.

Ferguson, I.S. 1998. Valuing different forest uses. In: pp. 357-365. Proceedings Outlook 98. Commodity markets and resource management, Volume 3. Bureau of Agricultural and Resource Economics, Canberra.

Ferguson, I. 2005. Australian forestry: beyond one tenure, one use. In: pp. 147-164. J. Sayer and S. Maginnis (eds.). Forests in landscapes: ecosystem approaches to sustainability. Earthscan, London.

Ferguson, I., M. Adams, J. Bradshaw, S. Davey, R. McCormack and J. Young. 2003. Calculating sustained yield for the Forest Management Plan (2004-2013): stage 3 report. Report for the Conservation Commission of Western Australia by the Independent Panel, May, 2003. Conservation Commission, Perth. 47 pp. Retrieved July 27, 2005, from www.conservation.wa.gov.au/files/docs/125.pdf.

Ferguson, I., J. Bradshaw, S. Cork, B. Egloff, R. Hill, R. McCormack and J. Raison. 1997. Assessment of ecologically sustainable forest management systems and processes: final report. Background Report Part G, Tasmania-Commonwealth Regional Forest Agreement. Public Land Use Commission, Hobart, 134 pp.

Ford, R., K. Williams, I. Bishop and T. Webb. 2005. Information effects on social acceptance of alternatives to clearfelling. Unpublished paper presented at Institute of Foresters of Australia Conference, Mt. Gambier, May 2005, 10 pp.

Gerrand, A., R.J. Keenan, P. Kanowski and R. Stanton. 2003. Australian forest plantations: an overview of industry, environmental and community issues and benefits. *Australian Forestry* 66(1):1-8.

Hickey, J. and M. Brown. 2003. Towards ecological forestry in Tasmania. In: pp. 31-46. D.B. Lindenmayer and J.F. Franklin (eds.). Towards forest sustainability. CSIRO Publishing, Collingwood, Victoria.

Hilmer, F., G. Tapperel and M. Rayner. 1993. National Competition Policy: report of the Independent Committee of Inquiry. Australian Government Publishing Service, Canberra.

Hoggett, J. 2005. The use and value of national parks: Does more mean worse? *IPA Backgrounder* May 17(2), 15pp. Retrieved July 27, 2005 from www.ipa.org.au/files/IPABackgrounder17-2.pdf.

Hollander, R. 2004. Changing places?: Commonwealth and state government performance and regional forest agreements. Refereed paper presented at the Australasian Political Studies Association Conference, University of Adelaide, 29 September–1 October 2004, 35pp. Retrieved August 6, 2005 from www.adelaide.edu.au/apsa/docs_papers/Pub%20Pol/Hollander.pdf.

Horwitz, P. and M. Calver. 1998. Credible science? Evaluating the regional forest agreement process in Western Australia. *Australian Journal of Environmental Management* 5:213-225.

JANIS, 1997. Nationally agreed criteria for the establishment of a comprehensive, adequate and representative reserve system for forests in Australia. A Report by the Joint ANZECC/MCFFA National Forest Policy Statement Implementation Subcommittee. Commonwealth of Australia, Canberra. 20 pp.

Kirkpatrick, J.B. 1998. Nature conservation and the regional forest agreement process. *Australian Journal of Environmental Management* 5:31-7.

Lee, K.M. and Abbott, I.A. 2004. Precautionary forest management; a case study from Western Australian legislation, policies, management plans, codes of practice and manuals for the period 1919-1999. *Australian Forestry* 67(2):114-121.

Lennon, J. 1998. History, cultural heritage and the regional forest agreement process. *Australian Journal of Environmental Management* 5:38-46.

Lindenmayer, D. 2003. Integrating wildlife conservation and wood production in Victorian montane ash forests. In: pp. 47-72. D.B. Lindenmayer and J.F. Franklin, (eds.). Towards forest sustainability. CSIRO Publishing, Collingwood, Victoria.

McCormack, R. 1996. A review of Forest Practices Codes in Australia. In D. Dykstra and R. Heinrich. (eds.). Forest Codes of Practice: contributing to environmentally sound forest operations. Food and Agricultural Organization (FAO) Forestry Paper 133, FAO, Rome. Chapter retrieved on June 3, 2005 from www.fao.org/documents/show_cdr.asp?url_file = /docrep/w3646e0f.htm.

McDonald, J. 1999. Regional forest (dis)agreements: the RFA process and sustainable forest management. *Bond Law Review* 11(2). 24 pp. Retrieved July 27, 2005, from www.bond.edu.au/law/blr/vol11-2/contents11-2.htm.

McKinnell, F.H., E.R. Hopkins and J.E.D. Fox. 1991. Forest management in Australia. Surrey Beatty and Sons Pty. Ltd., Chipping Norton, N.S.W. 380 pp.

Mercer, D. 2000. *A question of balance*. The Federation Press, Annandale, N.S.W. 366 pp.

Mobbs, C. 2003. Chapter 5. National forest policy and regional forest agreements. In: pp. 90-110. S. Dovers. and S.W. River. (eds.). Managing Australia's environment. The Federation Press, Annandale, N.S.W. 564 pp.

Musselwhite, G. and G. Herath, 2004. A chaos theory interpretation of community perceptions of Australian forest policy. *Forest Economics and Policy* 6(6):595-604.

National Forest Inventory, 2003. Australia's state of the forests report, 2003. Bureau of Rural Sciences, Canberra. 72 pp.

Neales, S. 2005. Green goals shifting on pragmatism and science. In: pp. 10-13. Forest and Wood Products Research and Development Corporation, Forests for tomorrow. Forest and Wood Products Research and Development Corporation, Melbourne, 52 pp.

Public Land Use Commission, 1995a. Inquiry into Tasmanian Crown Land Classifications. Background Report March 1995. Public Land Use Commission, Hobart.

Public Land Use Commission, 1995b. Inquiry into Tasmanian Crown Land Classifications. Final Recommendations Report 15 November 1995. Public Land Use Commission, Hobart.

Stewart, J. and G. Jones. 2003. Renegotiating the environment: the power of politics. The Federation Press, Annandale, N.S.W. 182 pp. [especially Chapter 3 (pp. 45-71). The green battlefield: regional forest agreements in three states.]

Strelein, G.J., L.W. Sage and P.W. Blankendaal. 2004. Rate of spread of *Phytopthora cinnamomi* in jarrah (*Eucalyptus marginata*) bioregion of south-western Australia. Unpublished paper presented at IUFRO Conference on *Phytophthera* in forests and natural ecosystems. 11-17 September, 2004, Freising, Germany, and Innsbruck, Austria.

Toyne, P. 1994. The reluctant nation: environment, law and politics in Australia. ABC Books, Sydney.

Toyne, P. 2002. Environment management–current risks and future strategies. Paper presented at ABARE Outlook 2002 Conference. ABARE, Canberra, 6 pp.

Turner, B.J., Chikumbo, O. and S.M. Davey. 2002. Optimisation modelling of sustainable forest management at the regional level; an Australian example. *Ecological Modelling* 153(2002):157-1179.

Vanclay, J.K. and B.J. Turner. 2001. Evaluation of data and methods for estimating the sustainable yield of sawlogs in Victoria: Report of the Expert Data Reference Group. Department of Sustainability and Environment, Melbourne. 97 pp.

Wilkinson, G. 1999. Codes of forest practice as regulatory tools for sustainable forest management. In: pp. 43-60. R.C. Ellis and P.J. Smethurst (eds.). Practising forestry today. Proc. 18th Biennial Conference, Institute of Foresters of Australia. Hobart, 3-8 October 1999.

Williams, K., R. Ford, I. Bishop and T. Webb. 2005. Understanding the acceptability of forest management options. Unpublished paper presented at Institute of Foresters of Australia Conference, Mt. Gambier, May 2005. 10 pp.

Zammit, C. 1999. Regional forest agreements: a model for integrated resource management. Paper presented at: Integrating the environment and the economy: trade aspects and performance measurement. Joint seminar between the European Commission and Environment Australia–Canberra, October 1999, Land Use Research Centre, University of South Queensland. 5 pp. Retrieved July 27, 2005, from www.usq.edu.au/lurc/TextDocs/ECpaper.pdf.

doi:10.1300/J091v24n01_02

# Silviculture of Scottish Forests at a Time of Change

W. L. Mason

**SUMMARY.** During the 20th century, a sustained afforestation programme was carried out in Scotland, which increased the forest cover from about 4% of the country in 1900 to 17% in 2002. The expansion was achieved primarily through the establishment of even-aged plantations of fast-growing conifer species, most of which were not native to the British Isles. Many of these forests will reach financial maturity in the next two decades, thereby providing a valuable timber resource for the Scottish economy, and giving the opportunity to diversify the forests to meet the multifunctional objectives outlined in the recent Scottish Forestry Strategy. However, diversifying these forests requires the adoption of a range of silvicultural systems and not simply perpetuating the clearfelling and replanting system that has characterized plantation management. These new approaches include greater use of continuous cover forestry and the restoration of native woodlands on sites previ-

---

W. L. Mason is Head of Forest Management Division at Forest Research, Northern Research Station, Roslin, Midlothian, Scotland, EH25 9SY UK (E-mail: Bill.mason@forestry.gsi.gov.uk).

The author is grateful to Dr. Robert Deal for the invitation to present this paper at the IUFRO World Congress in Brisbane, Australia, and for his help with the preparation of the manuscript. The author acknowledges the inputs from Dr. Gary Kerr and the helpful comments of Dr. Bob McIntosh and Simon Hodge of Forestry Commission Scotland on an early draft of this paper. Dr. Alexis Achim and two anonymous referees made helpful suggestions that have improved the final draft.

[Haworth co-indexing entry note]: "Silviculture of Scottish Forests at a Time of Change." Mason, W. L. Co-published simultaneously in *Journal of Sustainable Forestry* (Haworth Food & Agricultural Products Press, an imprint of The Haworth Press, Inc.) Vol. 24, No. 1, 2007, pp. 41-57; and: *Sustainable Forestry Management and Wood Production in a Global Economy* (ed: Robert L. Deal, Rachel White, and Gary L. Benson) Haworth Food & Agricultural Products Press, an imprint of The Haworth Press, Inc., 2007, pp. 41-57. Single or multiple copies of this article are available for a fee from The Haworth Document Delivery Service [1-800-HAWORTH, 9:00 a.m. - 5:00 p.m. (EST). E-mail address: docdelivery@haworthpress. com].

Available online at http://jsf.haworthpress.com

doi:10.1300/J091v24n01_03

ously planted with conifers. A sustained period of low timber prices is also influencing this process, as managers are more aware both of the high replanting costs involved in plantation silviculture and of the subsidies given to manage stands for nonmarket benefits. The effect of these changes upon the structure and nature of Scottish forests is unclear, and a sectoral analysis should be undertaken to explore the potential impacts. doi:10.1300/J091v24n01_03 *[Article copies available for a fee from The Haworth Document Delivery Service: 1-800-HAWORTH. E-mail address: <docdelivery@haworthpress.com> Website: <http://www.HaworthPress.com>* 

**KEYWORDS.** Silviculture, plantations, wood production, continuous cover forestry, native woodlands, Scotland

## *INTRODUCTION*

At the beginning of the twentieth century, the forest cover of Great Britain (i.e., the countries of England, Scotland, and Wales) had declined to about 4% of the land area following centuries of unsustainable management and exploitation of the natural forests. A major objective of national forestry policy throughout the last century was to increase the forest area by means of a sustained programme of afforestation, generally on land that was marginal for agriculture, and particularly in the upland zone of northern and western Britain. Conifer species were the main species used in afforestation because of their faster growth rates and their greater tolerance of difficult site conditions during plantation establishment (Zehetmayr, 1954, 1960). Nonnative conifer species such as Sitka spruce (*Picea sitchensis* (Bong.) Carr.), lodgepole pine (*Pinus contorta* Douglas ex Loud.), Corsican pine (*Pinus nigra* var. *maritima* Ait. Melville), larches (*Larix* spp.) and Douglas-fir (*Pseudotsuga menziesii* (Mirb.) Franco) were extensively planted because Scots pine (*Pinus sylvestris* L.), the only timber-producing native conifer, proved more sensitive to site conditions and relatively slow growing. As a consequence of this afforestation programme, by 2002 there were over 2.7 million ha of forests in Britain composing about 11.6% of the land area. There are substantial differences in forest cover between the three countries, with Scotland having the highest at 17% cover and England the lowest at only 8.5% (Forestry Commission, 2004a). This expansion represents one of the major changes in the rural landscape of Britain in recorded history.

Sustainable Forest Management (SFM) in Great Britain, as elsewhere in Europe, is governed by the guidelines agreed upon at the 1993 Helsinki Ministerial Conference, and by the formulation of these guidelines into "Pan-European Criteria" (PEC) at the 1998 Lisbon Ministerial Conference. The approach to implementing these principles in Scotland, and in the rest of the United Kingdom (UK), was formulated in the 1998 UK Forestry Standard (Forestry Commission, 2004b). There are slight differences of emphasis between the PEC and the UK Standard, in that the latter places lesser emphasis upon forest fires and management for soil protection, and pays greater attention to management of nonnative species. This reflects the different climatic conditions and forestry traditions within the UK. In addition, the Standard is not directly formulated upon the six PEC, but instead is based upon the interaction between forest management and four basic aspects of the forest ecosystem (physical, biological, human, and cultural) which are considered compatible with the PEC (Forestry Commission, 2004b). The adoption of the UK Standard has led in turn to the development of a voluntary certification standard for the forest industry (the United Kingdom Woodland Assurance Standard [UKWAS]), which came into being in 2000 as result of an extensive consultation between the industry and a wide range of stakeholders (Cashore et al., 2003). The UKWAS was recognized by the Forestry Stewardship Council (FSC) in 1999 so that the FSC label is increasingly used on British forest products (Forestry Commission, 2004b).

In 1997, as a consequence of changes in governance within the United Kingdom, forestry policy in Scotland and its implementation was devolved to the restored Scottish Parliament (an equivalent process took place in England and Wales). In 2000, the first Scottish Forestry Strategy was published (Forestry Commission, 2000b) with an overarching principle of contributing to national sustainability through sustainable forest management. Four major objectives were listed in support of this principle, namely those of integrating forestry with other rural land uses, of contributing to the well-being of the people of Scotland, of implementing management practices that enjoyed public support, and of respecting diversity in forest types. Five strategic priorities were defined:

- Maximize the value of the Scottish wood resource.
- Create a diverse forest resource for the future.
- Make a positive contribution to the environment.

- Create opportunities for the enjoyment of trees, woods, and forests.
- Help communities benefit from their forests.

A further aspiration was to increase the forest area to around 25% of land area by 2050. The aims of the strategy are promoted in private forests (65% of the forest area) through a system of grant subsidy and regulation, and by Forestry Commission management in the public forests. In general terms, the grant framework provides greater financial support for forest management in support of the last four objectives than for management geared to improving timber production (Forestry Commission, 2005b). The support for timber production is concentrated on aspects that would improve wood quality (e.g., use of genetically improved material, formative pruning) with the implicit assumption that revenues from harvesting will ensure continuity of the resource.

Because of the forest history described above, the forest area in Scotland has some unusual features compared to the rest of Europe and other parts of the world where SFM is being actively pursued. Plantation establishment with timber production as the primary function is the predominant forest type. The conifer plantations are highly productive in European terms with average growth rates of about 14 $m^3$ $ha^{-1}$ $year^{-1}$. The plantations are predominately composed of nonnative species so that Sitka spruce, which was first introduced in 1851, now makes up 47% of the forests, with other introduced conifers accounting for a further 25%, while the native Scots pine and mixed broadleaves compose the remainder. Ancient seminatural woodlands of these native species with high conservation and cultural values amount to less than 5% of the forest area. Despite the regular structure of many plantation forests, they are increasingly valued for recreation, and recent studies have shown them to make a substantial contribution to biodiversity objectives (Humphrey et al., 2003). Substantial progress has been made toward implementation of SFM in Scotland with over 45% of forests (including all Forestry Commission managed areas) having achieved certification under UKWAS (Forestry Commission, 2004a).

However, despite the achievement of developing a productive forest estate that provides a range of market and nonmarket benefits, uncertainties over the future management of these plantations have arisen for a number of reasons. Because the conifer stands are comparatively young with around 80% being 50 years or less of age, softwood timber production is anticipated to double to around 10 million $m^3$ by 2012 and to remain at this level for a further decade (Smith et al., 2001). From about 2007 onward, the private sector is forecast to produce 55 to 60% of this supply. The wood and fibre properties of Sitka spruce, the dominant element of this increase in production, are attractive for a range of end uses, and the forest resource is conveniently located for major markets in southern Britain and northern Europe. However, there are a number of quality issues that need to be addressed if this increased supply is to be successfully marketed (Hubert et al., 2004). As the plantations mature, future management options are influenced by the prices that are being paid for timber, the costs of conventional management after harvesting, as well as the higher subsidies being given for native woodland restoration, and other options with increased nonmarket benefits.

The purpose of this paper is to review silvicultural practices in conifer plantations in Scotland and to outline the alternative approaches that are being considered as a result of the difficult financial situation confronting many growers. The potential consequences for timber supplies are examined and ways of maintaining a sustained flow of timber products while meeting SFM requirements are discussed. The focus is on the management of Sitka spruce plantations because of its importance in the Scottish forest economy, but other species are considered where relevant.

## *CURRENT PLANTATION MANAGEMENT*

Until the late 1990s, essentially the only silvicultural system used in conifer plantations was patch clearfelling with regeneration through planting (Matthews, 1989). Rotation lengths were based upon predicted ages of maximum mean annual volume increment, with actual felling age determined by the use of discounted cash flow methods to appraise stand management strategies (Johnston et al., 1967). As a consequence, a normal rotation would be between 40 and 60 years, depending upon species, site productivity, discount rate, and windthrow risk. The latter is an important constraint on management, especially on soils with limited rooting depth in the exposed oceanic climate of parts of upland Scotland. A nonthinning regime is adopted on higher risk sites, which in some parts of the country may exceed 50% of the area (Quine et al., 1995). At time of clearfelling, yields of between 350 and 500 $m^3$ $ha^{-1}$ can be anticipated, depending upon site productivity and thinning history.. Some 45 to 50% of the volume is suitable for sawlogs that can be used for construction, fencing, and similar markets whereas the smaller roundwood is used for a variety of board products and for paper (McIntosh, 1995).

Forest design plans are used to plan the management of the felling process so that the extensive uniform areas of stands of similar ages are restructured to increase structural and species diversity. This is achieved by ensuring that the shape and sizes of potential coupes reflect the principles of good landscape design, while also adjusting the sequence of felling over time so that the time difference between adjacent coupes is five years or greater (Hibberd, 1985). The latter is an attempt to minimize the visual impact of clearfelling. When felled sites are replanted ("restocked"), care is taken to increase the amount of native tree and shrub species compared to the first rotation, as well as to protect special habitats and to take other measures to enhance biodiversity such as retaining any standing or fallen deadwood. Thus, in one forest area close to the Scottish border, the main species used in restocking were Sitka spruce (65%), a range of other conifers (13%), various broadleaved species (7%), plus 15% retained as permanent open space. Analysis of this approach in the mid-1990s showed a positive net present value of about £900 $ha^{-1}$, although this was some £500 $ha^{-1}$ less than would have been achieved by a completely commercial management strategy based upon a combination of 85% Sitka spruce and 15% open space (McIntosh, 1995).

The sequence and type of operations involved in the restocking of felled stands is outlined in Table 1, indicating that an investment of about £1500 $ha^{-1}$ is normally necessary to get stands to the "free-to-grow" stage. Additional costs will be incurred if deer fencing is necessary, although this is generally not necessary where Sitka spruce is the major species. Although fertilizer applications were standard in the establishment of first-rotation conifer plantations (Taylor, 1991), recent studies have suggested little requirement for such applications in the second rotation, provided that the species are well adapted to the site. In practice, this means avoiding the use of pure Sitka spruce on sites that are deficient in nitrogen, where either pine species or nursing mixtures between pine and spruce should be favoured (Smith and McKay, 2002).

The reason for the comparatively close tree spacing used (about 1.9 m) is that a number of studies have shown a decline in the yields of structural-grade timber in Sitka spruce once spacing widened beyond 2 m (Macdonald and Hubert, 2002). At present, similar spacing standards are recommended for other conifer species, although this may not be appropriate for species such as the larches and Douglas-fir with inherently stronger timbers than Sitka spruce (Lavers, 1983). Recent research has suggested that some silvicultural practices used in the establishment of first-rotation plantations (e.g., wide spacing, faster early growth be-

TABLE 1. A list of the major operations involved in Sitka spruce restocking and their associated costs (adapted from Forestry Commission, 2005b and I. Murgatroyd, Forest Research, pers. comm. 25 July 2005).

| **Operation** | **Year** | **Cost ha$^{-1}$ (£2005)** | **Comments and assumptions** |
|---|---|---|---|
| Ground preparation | −1 | £360 | Creating mounds (2700 planting spots per ha) from spoil trenches using an excavator with spot brash rake |
| Drainage | −1 | £70 | 60 pence per lineal metre, 120m per ha |
| Manual planting | 0 | £165 | £61 per 1000 plants (2700 plants per ha) |
| Plant cost | 0 | £356 | £132 per 1000 plants (2700 bare root plants per ha treated against damage by the weevil *Hylobius abietis*) |
| First top up spray | 1 | £100 | Treatment against *Hylobius* |
| Second top up spray | 2 | £100 | As above |
| Beat up (i.e., replacing failures) | 2 | £140 | Assume 20% beat up with a plant cost of £70 and manual planting cost of £70 |
| Manual herbicide weeding | 1-3 | £100 per year | Significant range, costs are for atrazine and glyphosate applications |

Notes: 1. No allowance is made for fencing against deer (£6-7 per lineal metre).
2. No allowance is made for remedial fertiliser applications–see text for further details.

cause of fertilizer inputs, planting of more exposed sites) have resulted in Sitka spruce trees of poorer stem form (i.e., less straight, heavier branching), leading to lesser timber performance and lower production of quality sawlogs (Hubert et al., 2004). This problem has also been influenced by the lack of thinning on more exposed sites, as other studies have shown that thinning increased the volume of sawlog production compared to nonthin regimes (Cameron, 2002). These results have highlighted an urgent need to integrate predictions of timber quality into current production forecasts that only provide information on volumes and assortment sizes. The possibility of lower sawlog quality is a major concern because the general assumption has been that increased amounts of spruce timber will be able to meet construction-grade timber specifications and gain market share from imports, given that the lower value markets for logs in Britain are nearing saturation (Hubert et al., 2004). The only options to improve quality in existing plantations will be through careful thinning and by extending normal rotation lengths to increase the amount of mature wood, but these may only be feasible on sites of lesser windthrow risk.

Improvements in timber quality and production in the second rotation can be obtained by greater use of genetically improved material of the major conifers, which should offer potential gains over unimproved material of 10 to 20% depending upon species (Lee, 2004). Indeed, for Sitka spruce, volume gains could range from 21 to 30%, owing to improved stem form and faster growth (Lee and Matthews, 2004). These predictions are supported by preliminary findings from a sawmill production study where the recovery of top-quality "green" sawlogs from three improved half-sibling families was 20 to 40 % higher compared with unimproved Sitka spruce (B.A. Gardiner, Forest Research, pers. comm., August 1, 2005). As a result of these predictions, cuttings of genetically improved Sitka spruce have been widely planted when restocking for the second rotation. In addition, the fast early growth of improved material often results in reduced establishment costs (Lee, 2004).

## *TIMBER PRICES AND PLANTATION PROFITABILITY*

Finding a market for the rising volume of timber available from Scottish forests is clearly a key factor affecting the future development of the sector. The raw material has to gain increased market share from imports, which account for nearly 85% of the United Kingdom domestic demand for timber (Thomson, 2004). The important sawn-timber section of the market is dominated by imports from Scandinavia, Russia, and the Baltic States, and Scottish sawlogs have to be competitive in terms of price and quality. In 2003, imports of coniferous industrial roundwood from these three areas were priced at about £35 $m^{-3}$ (FAO, 2005), thus setting a benchmark for Scottish grown timber. Average harvesting costs (i.e., fell and extract to roadside) are about £10 $m^{-3}$ (I. Murgatroyd, Forest Research, pers. comm., 25 July 2005), which leaves a comparatively small price margin for the grower. The marketing situation is also complicated by a sustained decline in softwood sawlog timber prices since 1996, partly owing to low-cost imports from Eastern Europe, such that the price index in March 2005 was at 67% of values a decade ago, although prices appear to have recovered slightly in the last year (Forestry Commission, 2005a). A further complication is that an increasing percentage of Scottish timber will be harvested from private forests where growers are sensitive to short-term fluctuations in timber prices. One consequence of this decline in prices has been that the returns from private-sector investment in Sitka spruce plantations have been negative for an extended period between 1996 and 2003 (Table 2; Forestry Commission, 2005b). Unfortunately, management costs have not shown a similar decline, so the profitability of forestry in remoter parts of Scotland has become marginal and the continuity of timber supplies from such locations is questionable. For example, a regional analysis suggested that the returns on plantations in North Scotland, which are further from markets, had been negative throughout the period 1992-2004, while all other areas in Scotland had shown a positive return (Forestry Commission, 2005b).

Recently, a more detailed examination of the current profitability of plantation management in Scotland was undertaken as part of an analysis of future options for the forests managed by Forestry Commission Scotland (FCS) (CJC Consulting, 2004). This distinguished between sites that were unprofitable to harvest and those where harvesting was profitable at current prices, but restocking was not. The first category was typified by sites in remote locations, with difficult access and on steep or rugged terrain, and of low productivity. The last feature was used as a surrogate for timber quality and included all sites producing less than 10 $m^3$ $ha^{-1}$ $year^{-1}$. About 15% of the FCS plantations were thought to fall into the first category, although some could be improved by the use of improved genetic material. Evaluation of sites that might be unprofitable to restock under conventional management (Table 1) was more problematic because of the uncertainty over future prices and over the range and magnitude of nonmarket benefits that might be affected by the restocking

TABLE 2. Three-year rolling annualised returns on private sector Sitka spruce plantations in Great Britain (adapted from IPD, 2005)

| Period | Percent per annum |
|---|---|
| 1992-1995 | 4.4 |
| 1993-1996 | 9.9 |
| 1994-1997 | 7.9 |
| 1995-1998 | 4.5 |
| 1996-1999 | −3.0 |
| 1997-2000 | −5.2 |
| 1998-2001 | −5.4 |
| 1999-2002 | −3.2 |
| 2000-2003 | −1.7 |
| 2001-2004 | 1.9 |

decision. A marginal site for conventional restocking ranged from 18 $m^3$ $ha^{-1}$ $year^{-1}$ at current prices and no allowance for nonmarket benefits to 12 $m^3$ $ha^{-1}$ $year^{-1}$ with prices at twice current levels and an inclusion of carbon-fixing benefits of about £300 $ha^{-1}$.

Because managers are sensitive to prices, a sustained period of low returns may make them more receptive to considering and implementing alternative management strategies (Haynes, 2006). One option is to seek to reduce establishment costs by greater use of natural regeneration or by favouring less intensive site management. Examination of Table 1 suggests that savings of £500 $ha^{-1}$ or more can be achieved if natural regeneration is used to replace planting, either in part or in whole, and the impact is sufficient to change the marginal site for restocking from 18 to 14 $m^3$ $ha^{-1}$ $year^{-1}$ without allowing for any price increase or nonmarket benefits. Natural regeneration of Sitka spruce and of a range of other conifers is increasingly reported within Scottish plantation forests (Nixon and Worrell, 1999). Another option that may be considered is to convert from a plantation to a native woodland type, which despite much lower productivity, will attract a higher rate of public subsidy. For example, restocking with conifers currently attracts 60% of standard costs whereas the equivalent figure when restoring native woodlands is 90% of standard costs (Forestry Commission, 2005b).

Figure 1 outlines the decision process that managers undertake when evaluating future options for a stand. It shows how the patch clearfelling and replanting approach based on intensive site management that has been the norm in Scottish plantation silviculture for at least 50 years (Davies, 1979; McIntosh, 1995) may no longer be preferable under certain conditions. For example, clearfelling in an area of high landscape importance would represent a major visual intrusion so that some retention of tree cover is essential. In other circumstances, areas of native woodland that had been underplanted with conifers may now be restored to meet national biodiversity objectives (Thompson et al., 2003). As a consequence, a range of outcomes are possible including transformation to continuous cover forestry (CCF), conversion to native woodland, as well as continuation of conventional plantation management. In the following section, some of the main silvicultural aspects that affect this decision process are outlined.

## *ALTERNATIVE SILVICULTURAL PRACTICES*

When managed on rotations that are financially optimal for timber production, Scottish plantation forests consist almost entirely of stands

FIGURE 1. A schematic outline of the decision process for conifer plantation stands nearing financial maturity in Scotland (adapted from CJC, 2004). Boxes with bold type indicate the main elements of the decision process.

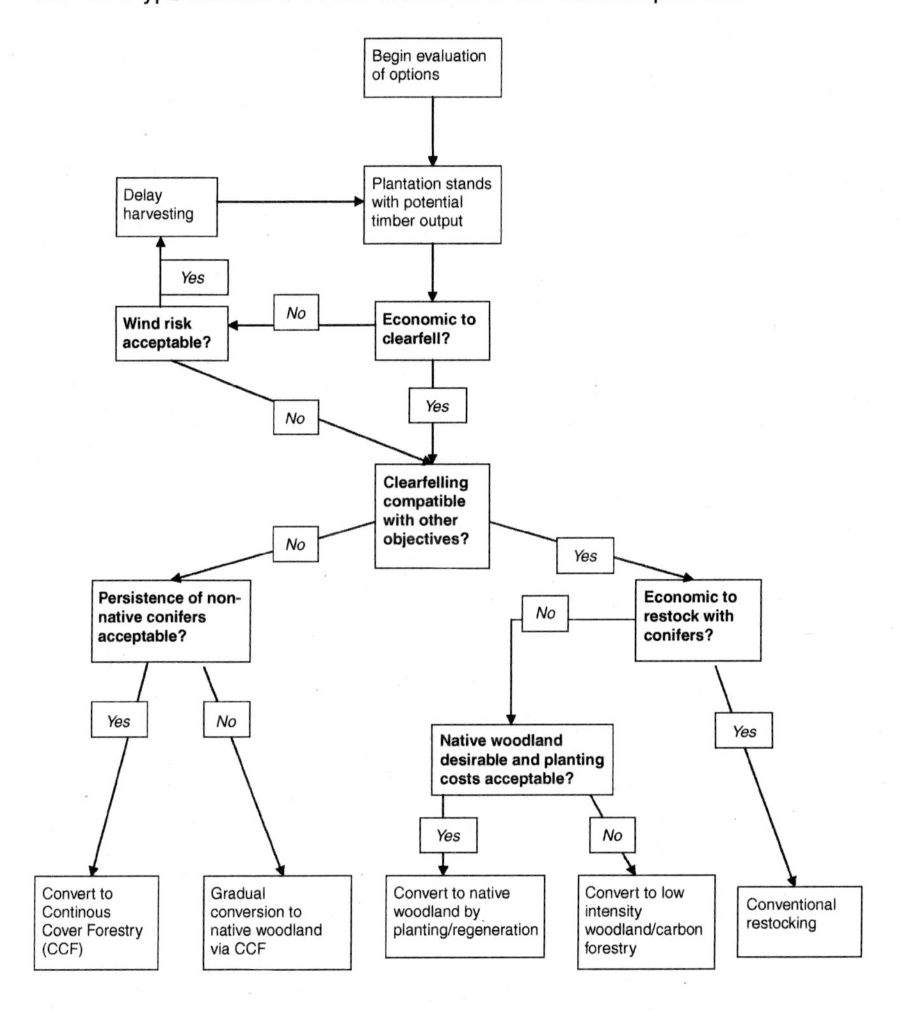


in the "stem exclusion" and "stand initiation" phases (*sensu* Oliver and Larson, 1996). As a result, they lack some structural features known to be both beneficial to biodiversity and attractive to visitors such as the presence of older, veteran trees, gaps, irregular structures, and standing and fallen deadwood (Peterken et al., 1992; Humphrey et al., 2003). Therefore, meeting the Scottish Forestry Strategy's aim of increasing the diversity of the forest resource requires a readiness to retain more stands beyond normal rotation age and, eventually, increasing the percentage of stands in the "understorey reinitiation" and "old growth" phases. It is important to recognize that discussion about alternative management strategies for plantation forests in Scotland has a long history. The possibility of diversifying structural and species composition of plantations was raised nearly 50 years ago by Anderson (1960). He instigated a number of operational trials of the process of transformation to CCF, the best known being at Glentress Forest near Edinburgh, recently described by Wilson et al. (1999). A point highlighted by these trials was the length of time required to transform regular stands to an irregular structure; since Wilson et al. (op.cit.) suggested that a further 20 years may be needed to complete this process. Perhaps for this reason, in the late 1990s, Helliwell (1997) reported that only a few thousand hectares of irregular forest existed in Great Britain. However, in recent years there has been nearly a tenfold increase in the area of forest designated for CCF management in Scotland (Mason, 2003), partially stimulated by the requirement of the UKWAS protocol that "in windfirm conifer plantations, lower impact silvicultural systems are increasingly favoured" (Forestry Commission, 2000a). For most purposes, lower impact silvicultural systems can be considered to be synonymous with CCF (Malcolm et al., 2001).

Silvicultural aspects of the greater use of CCF have been examined by a number of authors (Malcolm et al., 2001; Mason, 2002; Mason et al., 2004). In brief, the main considerations include: the need for revised thinning regimes to promote a more favourable light climate for natural regeneration of species that are either light demanding or intermediate in shade tolerance; the need to ensure that such thinning does not expose the stands to greater risks of wind damage in an oceanic climate; a lack of knowledge of the effects of CCF upon timber assortments and wood properties; and consequent difficulties of being able to model and predict the consequences of these changes in stand structure on the flows of wood from a forest or region. For example, wider use of CCF will probably result in a greater range of species being grown and a greater variety of product sizes being harvested, including increasing volumes of large (> 60-cm dbh [diameter breast height, 1.3 m]) dimension timber, which is beyond the processing capacity of many existing mills (Macdonald and Gardiner, 2005). Such changes can affect the economic sustainability of the sector. Thus in Wales, where a new woodland management regime has recently been adopted with a more ambitious commitment to CCF and to species diversification, a recent analysis suggested that the consequence could be a 20% decline in volume production and a 30% fall in revenues (Jaakko Poyry, 2004). The same report noted that improved forest planning systems are essential to monitor the effects of major changes to management practices and provide a link between practice and policy.

There will be situations where biodiversity or landscape considerations dictate that continuation with nonnative conifers as the dominant element within a particular locality is unacceptable. Such examples generally involve the felling of the conifers and their progressive replacement by native species either through natural colonisation or by planting or through a combination of the two options. Problems that can arise here include the high costs associated with restoring native woodland cover (CJC Consulting, 2004) and a lack of adequate silvicultural knowledge about tending native species, especially broadleaves (Worrell et al., 2004). In forests in remote locations, a range of solutions may develop with a mixture of native or more "open woodland" structures being promoted because of savings on establishment costs and greater provision of nonmarket benefits (CJC Consulting, 2004). Other issues that can be problematic include the scale at which native woodlands should be developed within a conifer plantation matrix and the need to develop a more holistic vision of the desired future structure of Scottish forests, which encompasses a range of forest types (Mason et al., 1999).

## *CONCLUSIONS*

The recent prolonged period of historically low timber prices has stimulated considerable changes in the management of the extensive conifer plantation resource in Scotland. There is a progressive move away from a single silvicultural approach based upon patch clearfelling and replanting toward a diversity of silvicultural regimes that will result in more varied forests better suited to meeting multipurpose objectives (Heitzman, 2003). However, these changes will inevitably have impacts on timber production over time and, as importantly, on wood properties. The desire to diversify the forest environment will lead to greater use of continuous cover forestry (CCF) which could be implemented on up to 25% of the forest area, based on assumptions about soil suitability and wind climate (Mason, 2003). Furthermore, meeting the strategic priorities of maintaining and enhancing native woodland areas means that areas of plantation that were established on native woodland sites will be restored to native tree species when the conifers reach financial maturity. For these reasons, the proportion of conifer plantation is likely to decline over time from the present 70% of Scottish forests to possibly 50% or less of the area. It seems likely that the volume of softwood timber production will also decrease. However, this may be partially or wholly offset by greater use of genetically improved conifers (Lee, 2004).

Understanding the interaction of all the above factors, not to mention market trends and climate change, requires a sectoral approach that explores the effects of different management strategies through the forest-industry wood chain. There is an urgent need to undertake an examination in Scotland of the interactions between management for wood production and for other objectives. Similar studies have been undertaken in other parts of the world (Monserud et al., 2003), and the recent findings from Wales (Jaakko Poyry, 2004) indicate that policy changes can have unanticipated impacts on sector viability. For example, it is possible that there could be a divergence in management approach and ultimately forest structure between the state and private sectors as has occurred in the Pacific Northwest of the USA with older stands found on federal lands and shorter rotation plantations in private ownership (Monserud et al., 2003). Such a study will necessitate the development of forest management planning tools that are capable of linking stand level objectives and operations with the higher forest level goals where aspirations for SFM are usually formulated (Seely et al., 2004). Such understanding will need to be formulated in a range of interactive decision support tools, of which the wind risk model, ForestGALES (Dunham et al., 2000) and the Ecological Site Classification system (Pyatt et al., 2001) are but the first examples in Scottish forestry.

Although the commitment to SFM, the desire to manage forests for multiple objectives, and the aim to substantially increase the forest area are not unique to Scotland, what is unusual and interesting is the challenge of implementing these policies in a country where the whole forest area, and not just the productive resource (c.f., Chile, France, New Zealand, Scandinavia), is dominated by plantation forestry. A probable consequence of Scottish forestry policy is a diversification toward a more complex situation where at least four strategic options (plantation silviculture, continuous cover forestry, restoration of native woodlands, and minimum management) can be discerned. This will result in more varied forests composed of a wider range of species producing timber of varied sizes, and probably of more variable properties. A further implication is that management systems will become more complex as foresters strive to provide more diverse forest structures to meet societal aspirations. These trends may also result in some zoning within the forest resource with some areas being dedicated to timber production and others to recreation or other nonmarket benefits, although this may only be evident at a comparatively small scale. The extent to which these changes can be accommodated without undermining the value of the existing Scottish wood resource is critically dependent upon our ability to monitor, understand and predict the interactions among site, species, silviculture, wood quality, and markets that underpin profitable forest management.

## REFERENCES

Anderson, M.L. 1960. Norway spruce-silver fir-beech mixed selection forest. Is it possible to reproduce this in Scotland? *Scottish Forestry* 14:87-93.

Cameron, A.D. 2002. Importance of early selective thinning in the development of long-term stand stability and improved log quality: a review. *Forestry* 75:25-35.

Cashore, B., G. Auld and D. Newson. 2003. Forest certification (eco-labelling) programs and their policy making authority: explaining divergence among North American and European case studies. *Forest Policy and Economics* 5:225-247.

CJC Consulting. 2004. Economic analysis of the contribution of the forest Estate managed by Forestry Commission Scotland. Retrieved June 21, 2005, from http://www.forestry.gov.uk/pdf/FCSforestestatefinal2.pdf/$FILE/FCSforestestatefinal2.pdf

Davies, E.J.M. 1979. The future development of even-aged plantations: management implications. In: pp. 465-480. E.D. Ford, D.C. Malcolm and J. Atterson (eds.). The ecology of even-aged forest plantations. Institute of Terrestrial Ecology, Cambridge.

Dunham, R., B.A. Gardiner, C.P. Quine and J.C. Suarez. 2000. ForestGALES, a PC based wind risk model for British forests: users' guide. Forestry Commission, Edinburgh.

Food and Agriculture Organization [FAO]. 2005. Forestry trade flow-bilateral trade matrices. Retrieved October 30, 2005, from http://faostat.fao.org/faostat/collections?version=ext&hasbulk=0&subset=forestry.

Forestry Commission. 2000a. Certification standard for the UK woodland assurance scheme. UKWAS Support Unit, Edinburgh.

Forestry Commission. 2000b. The Scottish forestry strategy. Edinburgh.

Forestry Commission. 2004a. Forestry statistics 2004. Retrieved June 21, 2005, from http://www.forestry.gov.uk/pdf/fcfs004.pdf/$FILE/fcfs004.pdf.

Forestry Commission. 2004b. The UK Forestry Standard: the government's approach to sustainable forestry. Edinburgh.

Forestry Commission. 2005a. Forestry timber prices in Great Britain. Retrieved June 21, 2005, from http://www.forestry.gov.uk/pdf/tpi0305rev.pdf/$FILE/tpi0305rev.pdf.

Forestry Commission. 2005b. The Scottish Forestry Grants Scheme: a quick guide. Retrieved June 21, 2005, from http://www.forestry.gov.uk/pdf/SFGSquickguideFINAL.pdf.

Haynes, R.W. 2006. Integrating Concerns About Wood Production and Sustainable Forest Management in the United States. *Journal of Sustainable Forestry* 24(1): 1-18.

Heitzman, E. 2003. New forestry in Scotland. *Journal of Forestry* 101:36-39.

Helliwell, D.R. 1997. Dauerwald. *Forestry* 70:375-380.

Hibberd, B.G. 1985. Restructuring of plantations in Kielder Forest District. *Forestry* 58:119-129.

Hubert, J., B.A., Gardiner, E. Macdonald and S. Mochan. 2004. Stem straightness in Sitka spruce. *Forest Research Annual Report and Accounts* 2002-2003:62-69.

Humphrey, J.W., R. Ferris and C.P. Quine (eds.). 2003. Biodiversity in Britain's planted forests. Forestry Commission, Edinburgh.

IPD UK forestry index 2005. Retrieved August 1, 2005, from http://www.ipindex.co.uk/downloads/indices/UKForestry.

Jaakko Poyry. 2004. Welsh forest industry: mapping and benchmarking the forest industry. Retrieved June 21, 2005 from http://www.forestry.gov.uk/pdf/walesjpcfullreport.pdf/$FILE/walesjpcfullreport.pdf.

Johnston, D.R., A.J. Grayson and R.T. Bradley. 1967. Forest planning. Faber and Faber Ltd., London.

Lavers, G.W. 1983. The strength properties of timber. Building Research Establishment Report 241. Building Research Establishment, Watford, England.

Lee, S.J. 2004. The products of conifer tree breeding in Britain. Forestry Commission Information Note 58. Forestry Commission, Edinburgh.

Lee, S.J. and R. Matthews. 2004. An indication of the likely volume gains from improved Sitka spruce planting stock. Forestry Commission Information Note 55. Forestry Commission, Edinburgh.

Macdonald, E. and B.A. Gardiner. 2005. A review of the effects of transformation of even-aged stands to continuous cover silvicultural systems on conifer log quality and wood properties in the UK. Report to the Scottish Forestry Trust, Edinburgh.

Macdonald, E. and J. Hubert. 2002. A review of the effects of silviculture on the timber quality of Sitka spruce. *Forestry* 75:107-138.

Malcolm, D.C., W.L. Mason and G.C. Clarke. 2001. The transformation of conifer forests in Great Britain- regeneration, gap size, and silvicultural systems. *Forest Ecology and Management* 151:7-23.

Mason, W.L. 2002. Are irregular stands more windfirm? *Forestry* 75:347-355.

Mason, W.L. 2003. Continuous cover forestry: developing close-to-nature forest management in conifer plantations in upland Britain. *Scottish Forestry* 57:141-149.

Mason, W.L., C. Edwards and S.E. Hale. 2004. Survival and early seedling growth of conifers with different shade tolerance in a Sitka spruce spacing trial and relationship to understorey light climate. *Silva Fennica* 38:357-370.

Mason, W.L., D. Hardie, P. Quelch, P.R. Ratcliffe, I. Ross, A.W. Stevenson and R. Soutar. 1999. Beyond the two solitudes: the use of native species in plantation forests. *Scottish Forestry* 53:135-145.

Matthews, J.D. 1989. Silvicultural systems. Oxford University Press, Oxford.

McIntosh, R.M. 1995. The history and multi-purpose management of Kielder Forest. *Forest Ecology and Management* 79:1-11.

Monserud, R.A., R.W. Haynes and A.C. Johnson. (eds.). 2003. Compatible forest management. Kluwer Academic Publishers, Dordrecht.

Nixon, C.J. and R. Worrell. 1999. Natural regeneration of conifers in Britain. Forestry Commission Bulletin 120. Forestry Commission, Edinburgh.

Oliver, C.D. and B.R. Larson. 1996. Forest stand dynamics. McGraw-Hill, New York.

Peterken, G.F., D. Ausherman, M. Buchanan and R.T.T. Forman. 1992. Old growth conservation within British upland conifer plantations. *Forestry* 65:127-144.

Pyatt, D.G., D. Ray and J. Fletcher. 2001. An ecological site classification for Great Britain. Forestry Commission Bulletin 124. Forestry Commission, Edinburgh.

Quine, C.P., M.P. Coutts, B.A. Gardiner and D.G. Pyatt. 1995. Forests and wind: management to minimise damage. Forestry Commission Bulletin 114. HMSO, London.

Seely, B., J. Nelson., R. Wells, B. Peter, M. Meitner, A. Anderson, H. Harshaw, S. Sheppard, F.L. Bunnell, H. Kimmins and D. Harrison. 2004. The application of a hierarchical, decision-support system to evaluate multi-objective forest management strategies: a case study in northeastern British Columbia, Canada. *Forest Ecology and Management* 199:283-305.

Smith, S., J. Gilbert, and R. Coppock. 2001. New forecast of softwood availability. *Forestry and British Timber* April:20-25.

Smith, S.A. and H.M. McKay. 2002. Nutrition of Sitka spruce on upland restock sites in northern Britain. Forestry Commission Information Note 47. Forestry Commission, Edinburgh.

Taylor, C.M.A. 1991. Forest fertilisation in Great Britain. Forestry Commission Bulletin 95. HMSO, London.

Thompson, R., J. Humphrey, R. Harmer and R. Ferris. 2003. Restoration of native woodlands on ancient woodland sites. Forestry Commission Practice Guide. Forestry Commission, Edinburgh.

Thomson, M. 2004. International markets in wood products. Forestry Commission Information Note 60. Forestry Commission, Edinburgh.

Wilson, E.R., H. Whitney McIver and D.C. Malcolm. 1999. Transformation to an irregular structure of an upland conifer forest. *Forestry Chronicle* 75:407-412.

Worrell, R., D.C. Malcolm and A. Barbour. 2004. Growing broadleaves for quality timber in Scotland. *Scottish Forestry* 58:3-7.

Zehetmayr, J.W.L. 1954. Experiments in tree planting on peat. Forestry Commission Bulletin 22. HMSO, London.

Zehetmayr, J.W.L. 1960. Afforestation of upland heaths. Forestry Commission Bulletin 32. HMSO, London.

doi:10.1300/J091v24n01_03

# Changes in Wood Product Proportions in the Douglas-Fir Region with Respect to Size, Age, and Time

Robert A. Monserud
Xiaoping Zhou

**SUMMARY.** We examined both the variation and the changing proportions of different wood products obtained from trees and logs in the Douglas-fir region of the northwestern United States. Analyses are based on a large product recovery database covering over 40 years of recovery studies; 13 studies are available for Douglas-fir (*Pseudotsuga menziesii* (Mirb.) Franco). Visual lumber grades were combined into four broad value classes. We used the multinomial logistic model to estimate the yield proportion of each value class as a function of age, diameter, and their interaction. We also examine changes in wood product proportions with respect to future projections of forest management, harvesting trends, and sustainability. We see a clear shift away from appearance grades in the 1960s to construction grades by the late 1980s. This corresponds to a concomitant shift from high quality old-growth trees to young-growth plantations (age 50 to 60) of much smaller diameter. The projected relative proportions among the four product value classes is not expected to change much over current proportions, with

---

Robert A. Monserud is Research Team Leader and Xiaoping Zhou is Forest Analyst for the U.S. Department of Agriculture, Forest Service, Pacific Northwest Research Station, P.O. Box 3890, Portland, OR 97208 USA (E-mail: rmonserud@fs.fed.us).

[Haworth co-indexing entry note]: "Changes in Wood Product Proportions in the Douglas-Fir Region with Respect to Size, Age, and Time." Monserud, Robert A., and Xiaoping Zhou. Co-published simultaneously in *Journal of Sustainable Forestry* (Haworth Food & Agricultural Products Press, an imprint of The Haworth Press, Inc.) Vol. 24, No. 1, 2007, pp. 59-83; and: *Sustainable Forestry Management and Wood Production in a Global Economy* (ed: Robert L. Deal, Rachel White, and Gary L. Benson) Haworth Food & Agricultural Products Press, an imprint of The Haworth Press, Inc., 2007, pp. 59-83. Single or multiple copies of this article are available for a fee from The Haworth Document Delivery Service [1-800-HAWORTH, 9:00 a.m. - 5:00 p.m. (EST). E-mail address: docdelivery@haworthpress.com].

Available online at http://jsf.haworthpress.com
doi:10.1300/J091v24n01_04

construction lumber a dominant 70 to 80% of the total. Thus, we expect to remain at this new equilibrium, where very little appearance grade lumber is manufactured in the Douglas-fir region and price premiums have disappeared. doi:10.1300/J091v24n01_04 *[Article copies available for a fee from The Haworth Document Delivery Service: 1-800-HAWORTH. E-mail address: <docdelivery@haworthpress.com> Website: <http://www.HaworthPress.com>.]*

**KEYWORDS.** Yield proportion, value class, construction grades, multinomial logistic, Douglas-fir, *Pseudotsuga menziesii*

## *INTRODUCTION*

We focus on the Pacific Northwest, principally western Oregon and Washington, which is also called the Douglas-fir region (McArdle et al., 1961). This mountainous region is dominated by a large temperate rainforest that extends to southeastern Alaska (Haynes et al., 2003). These coniferous forests contain the highest quality wood-producing lands on the continent (Curtis and Carey, 1996), and exhibit some of the greatest biomass accumulations and highest productivity levels of any in the world, temperate or tropical (Franklin and Dyrness, 1973; Fujimori et al., 1976; Franklin and Waring, 1981; Walter, 1985; Franklin, 1988). These forests are valued for their scenery, recreational opportunities, watershed protection, and fish and wildlife habitat as well as abundant forest products (Peterson and Monserud, 2002). Moist maritime conditions characterize the region, producing expanses of forests dominated by massive evergreen conifers, including coastal Douglas-fir (*Pseudotsuga menziesii* (Mirb.) Franco var. *menziesii*), western hemlock (*Tsuga heterophylla* (Raf.) Sarg.), western redcedar (*Thuja plicata* Donn ex D. Don), Sitka spruce (*Picea sitchensis* (Bong.) Carr.), Pacific silver fir (*Abies amabilis* Dougl. ex Forbes), and noble fir (*A. procera* Rehd.). A mediterranean climate of mild winters and relatively dry summers prevails (Walter, 1985), although summer dryness decreases markedly to the north. This climate favors needle-leaved conifers by permitting extensive photosynthesis outside the growing season and reducing transpiration losses during the summer months (Waring and Franklin, 1979). The summer climate is controlled by a large, semipermanent, high-pressure center in the Pacific Ocean, which greatly reduces the frequency and intensity of Pacific storms (Meidinger and Pojar, 1991). Infrequent catastrophic events, such as wildfires and hurricanes at intervals of several hundred years, allow for the genetic expression of enormous biomass and extremely tall and long-lived forest trees (Franklin, 1988). Such favorable conditions for forest growth result in dominance by coniferous species that are the tallest, largest, and oldest of their genera (Franklin, 1988). Unlike the other temperate rainforests of the world (see Walter, 1985), this is the only one dominated by conifers rather than hardwoods (Franklin, 1988).

Forest management in this region has been evolving for over a century. Experience in even-aged silviculture and plantation management accumulated in the past half-century is vast (Curtis et al., 1998). Methods for regenerating vigorous young stands of primary timber species following clearcut logging have been thoroughly researched and tested throughout the western Pacific Northwest (Loucks et al., 1996; Smith et al., 1997). Douglas-fir, the major timber species in the Pacific Northwest, can be grown at various densities. It rapidly responds to thinning at a variety of stand ages, with increased diameter growth as well as branch and crown development. Stocking control is important to promote vigorous growth (Barbour et al., 2003).

Silviculturists have studied the key steps in stand management with fruitful results. Nursery methods for efficiently raising healthy, superior planting stock are now common (Duryea and Dougherty, 1991), including techniques for inoculating roots with mycorrhizal fungi to promote quick establishment and sustained growth (Castellano and Molina, 1989). Average survival of seedlings has increased to 85% or better (Curtis et al., 1998). Effective methods have been developed for controlling competing shrub and nontimber vegetation, thus promoting rapid growth of established individuals (Walstad and Kuch, 1987). Various harvesting systems have been developed to reduce problems such as soil compaction (Curtis et al., 1998). Because of dependable stand establishment through the widespread sequence of clearcutting, burning, and planting, the length of commercial rotations on high-productivity lands decreased to as little as 40 to 50 years. Often, commercial thinning was eschewed in favor of earlier harvests (Curtis and Carey, 1996).

From the 1940s until the late 1960s in the United States, there was general agreement among both federal and private land managers that timber production was the primary objective in management of most forest land (Curtis et al., 1998, Peterson and Monserud, 2002). Basic assumptions were that wood production in old-growth stands was essentially static (no net growth), and that insects and disease were diminishing the amount of usable wood in these stands. It seemed desirable, therefore, to replace old-growth forests with young, rapidly growing stands (USDA FS, 1963; Curtis et al., 1998). Furthermore, in the Douglas-fir region, clearcut logging and broadcast burning were justified as mimicking the catastrophic, stand-replacing fires typical of the region before fire suppression (Halpern, 1995). Management practices during this period attempted to meet increasing wood demands by relying on the economic efficiencies of clearcutting and plantation management (Haynes et al., 2003).

In the 1990s, it became increasingly clear that the public has strongly held values and opinions regarding the appropriate use of forest land (especially public forestland) in the Pacific Northwest (Haynes et al., 2003). The decision process has become increasingly difficult on how best to manage forestlands in the region. Simply put, public opinion has often been at odds with forest management goals, especially on public land. As a consequence of this shift in societal values, the focus of management in the Pacific Northwest during the 1990s shifted from producing timber to maintaining sustainable forest ecosystems (Behan, 1990). This shift culminated in the Northwest Forest Plan (USDA and USDI, 1994a, 1994b) for western Oregon and Washington and the new Forest Practices Code of British Columbia (1994). Emphasis on timber-stand management in the Douglas-fir region declined during the 1990s, especially on public forestland. The traditional goal of efficient wood production through even-aged plantations is shifting toward multi-resource ecosystem management, with related goals favoring old-growth characteristics (Monserud, 2003), protecting endangered species, fish habitat, and promoting biodiversity (FEMAT, 1993). This trend is also occurring to the north along coastal British Columbia (Clayoquot Scientific Panel, 1995), essentially the same temperate rainforest but outside the range of coastal Douglas-fir (Meidinger and Pojar, 1991). Science-based silvicultural practices and management regimes are needed to reduce conflicts among user groups while providing concurrent production of the many values associated with forestlands on a biologically and economically sustainable basis (Curtis et al., 1998; Committee of Scientists, 1999).

Because the goals of land managers have changed, silviculturists are examining new management methods (Haynes and Monserud, 2002). Whether they are called green-tree retention, variable-density thinning, or variable retention (Franklin et al., 1997), the silvicultural treatments under examination are intermediate between the traditional extremes of even-age plantation management, uneven-age selection management, and no management (see Hummel, 2003). Often, the goal is to increase rather than decrease structural heterogeneity within a stand and a watershed (Monserud, 2002). In a strong break from the uniform plantation management of the past, these silvicultural alternatives often include mixed-species compositions and attempt to obtain uneven-aged structures (e.g., by retaining legacy trees). They explicitly consider structural and spatial diversity as values, rather than the spatially uniform stand treatments common with traditional silvicultural systems (Monserud, 2003).

The Douglas-fir region remains one of the major timber-producing regions in the world and is recognized for its productive timberlands, forest management institutions, well-organized markets, and large-scale timber processing industries (Monserud et al., 2003). Current forest conditions in the Douglas-fir region are a function both of markets for various forest products and of various regulatory actions, past and current (Haynes, 2005). The Unites States has a total of 204 million hectares of timberland in all ownerships, where timberland is defined as forestland that is capable of producing at least 1.4 $m^3\ ha^{-1}\ yr^{-1}$ of industrial wood (Smith et al., 2004). Nationally, this timberland consists of 19% national forest, 10% other public land (e.g., state land, Bureau of Land Management), 13% forest industry, and 58% nonindustrial private ownership. However, the structure of ownerships for the Pacific Northwest is quite different. The Pacific Northwest has 16.7 million hectares of timberland, of which 43% is national forest, 14% are other public land, 22% is forest industry, and 21% is nonindustrial private land. This mixture of ownerships strongly affects timber supply because of the different policies and management objectives held by the various ownerships (Haynes, 2003).

Furthermore, timber availability in Washington and Oregon has changed dramatically since the late 1980s (Zhou et al., 2005). In 1988 the region produced about 37.1 million cubic meters (15.7 billion board feet) of timber. In 2001 it produced about 17.0 million cubic meters (7.2 billion board feet) (Warren, 2003), with the decline owing mostly to harvesting reductions on National Forest System land. There are 9.44 million hectares of timberland in these west-side forests. Half of this timberland is Douglas-fir. Approximately 53% of the land in the Douglas-fir region supports stands younger than 50 years old, compared to 13% of the land with stands older than 150 years. The majority of stands less than 150 years old are on other public (e.g., state land) or private land, whereas the majority of stands greater than 150 years old are on National Forest System land (Zhou et al., 2005).

Current policy on federally administered land effectively reserves all stands over about 80 years old from harvest and in practice restricts harvest to thinning of stands younger than about 50 years old. Managers of state land in both Oregon and Washington are under pressure to harvest only younger stands and then to create structural conditions for developing old forest characteristics that may eventually be managed for habitat protection (Zhou et al., 2005). Looking ahead 50 years, Haynes (2003) estimated that Douglas-fir will remain the major timber species, accounting for 63% of total removals for 1997-2056. Relatively quickly all Douglas-fir removals from private lands will represent the 45- to 65-year age classes (Zhou et al., 2005). The national forests will provide older Douglas-fir but in relatively limited volumes.

With this context, we are interested in examining the distribution (proportion) of various wood product classes with respect to this changing age and size distribution of the Douglas-fir resource. Specifically, we are looking for the relationship between product value classes (e.g., lumber grade) and information characterizing individual trees such as size and age. We focus on the relative proportions across these product value classes. We are also looking to see if this relationship changes over time.

## MATERIALS AND METHODS

### Data

Because we are examining changes in the proportions of various wood product grades available over time, we need to use historical data collected on several different product recovery studies. The Ecologically Sustainable Production (ESP) team (and its predecessors) of the USDA Forest Service, Pacific Northwest Research Station, Portland, Oregon, has conducted product recovery studies over the past 40 years (Monserud et al., 2004), and has maintained a database of these past studies. These studies trace the processing of wood products (lumber and veneer volume and quality) from the tree in the forest to the final product on a log-by-log basis. One common objective of the studies was to obtain maximum product value recovery consistent with current industry practices. The working assumption was that if log identity could be maintained on every item through the manufacturing process (sawing or veneer slicing) then the product yield and grade proportions for sample logs with known characteristics could then be applied to a similar sample of logs or trees to predict product volumes and values when using similar processing technology. Although this database summarizes product recovery information on most of the major forest species in the western United States, we focus on Douglas-fir in western Oregon and Washington, a key species for construction lumber in the United States. We use the 13 studies available with complete records (i.e., individual tree data as well as long- and short-log information) for Douglas-fir, for a total of 1,535 trees (Table 1). Of these records, 443 lacked either diameter at breast height (dbh) or age, leaving us with 1,092 complete tree records. Height was also available on most records. Thus, the tree information includes dbh, age, height, lumber grade, and volume yield from each tree. In one sense, this is a meta-analysis across several studies, except that we have the complete raw data.

Eight of these studies were conducted in the 1960s and contain considerable numbers of large (dbh > 100 cm) old-growth trees, as does one study from 1982 (Table 1). Study objectives were to evaluate and improve the existing log grading and timber appraisal system. The remaining four studies were conducted in the 1980s and 1990s, and represent smaller and younger material more commonly found in the Douglas-fir region today (Table 1). Study objectives included estimating lumber volume and value recovery, as well as study-specific objectives such as developing a model for predicting lumber and veneer quality from young-growth tree and log characteristics. Because we are looking for continuous relationships between tree characteristics and product recovery, this collection of studies should provide us with the desired wide range of conditions.

To simplify the analysis, visual lumber grades were combined into four product value classes: (1) Appearance (Shop and Better), (2) Select Structural, (3) General Construction (No. 1 and No. 2 Lumber), and (4) Utility and Economy. Table 2 summarizes the correspondence between the various lumber grades (WWPA, 1998; Barbour et al., 2005) and these four product value classes.

The Appearance category includes Selects and Factory Lumber such as Clears and Shops (Table 2). This lumber has the highest dollar value and is typically used for finish applications. Select Structural includes the Select Structural grade and was intended to indicate the amount of lumber that might be available for higher value structural uses. Construction lumber includes No. 1 and No. 2, Standard and Construction, and No. 4 and Better Commons, plus any heavy timbers or other grades that are typically used for structural purposes. Utility and Economy also includes No. 5 Common and other grades that constitute the lowest value lumber. All the lumber produced from each study tree is then

TABLE 1. Descriptive information for Douglas-fir product recovery studies in Oregon and Washington.

| Old Study ID | New Study ID | Location | State | Year | Num. of trees | Min. dbh (cm) | Mean dbh (cm) | Max. dbh (cm) | Min. Age (yr) | Mean Age (yr) | Max. Age (yr) | Min. Height (m) | Mean Height (m) | Max. Height (m) |
|---|---|---|---|---|---|---|---|---|---|---|---|---|---|---|
| 01-72 | 18 | Chelatchie | WA | 1965 | 48 | 35.0 | 86.4 | 172.0 | 102 | 270 | 610 | 21.9 | 44.8 | 78.6 |
| 01-73 | 19 | Mapleton | OR | 1965 | 88 | 30.5 | 81.3 | 143.5 | 58 | 140 | 317 | 22.5 | 46.6 | 74.7 |
| 01-74 | 20 | Darrington | WA | 1965 | 22 | 37.1 | 96.5 | 187.7 | 46 | 220 | 530 | 36.3 | 50.0 | 63.7 |
| 01-77 | 23 | Shelton | WA | 1965 | 204 | 34.5 | 88.9 | 225.5 | 29 | 315 | 647 | 26.8 | 48.8 | 84.1 |
| 01-78 | 25 | Medford | OR | 1966 | 114 | 26.9 | 91.4 | 179.8 | 75 | 234 | 576 | 21.9 | 45.4 | 71.0 |
| 01-82 | 27 | Portland | OR | 1966 | 111 | 27.2 | 86.4 | 201.4 | 60 | 189 | 400 | 25.3 | 47.2 | 82.3 |
| 01-83 | 30 | Culp Creek | OR | 1967 | 121 | 25.9 | 99.1 | 187.2 | 87 | 234 | 394 | 21.9 | 49.4 | 67.4 |
| 01-84 | 35 | Redmond | OR | 1968 | 155 | 30.0 | 81.3 | 126.7 | 106 | 244 | 550 | 27.1 | 39.3 | 53.3 |
| 01-08 | 101 | Noti | OR | 1982 | 126 | 50.8 | 94.0 | 143.8 | 53 | 217 | 480 | 39.3 | 48.5 | 75.6 |
| 01-09 | 102 | Mapleton | OR | 1982 | 114 | 23.4 | 50.8 | 81.3 | 52 | 105 | 256 | 4.0 | 35.7 | 57.3 |
| 01-14 | 118 | Shelton | WA | 1987 | 235 | 22.9 | 43.2 | 74.0 | 25 | 45 | 88 | 7.3 | 30.5 | 49.1 |
| 01-18 | 143 | Mill City | OR/WA | 1995 | 151 | 19.3 | 35.6 | 56.6 | 20 | 44 | 90 | 14.0 | 29.0 | 41.8 |
| 01-22 | 154 | Pack Forest | WA | 1998 | 46 | 14.2 | 33.0 | 53.3 | 76 | 76 | 76 | 21.3 | 29.6 | 37.8 |

TABLE 2. Product value classes based on combined lumber grade composition (WWPA 1998; Barbour et al. 2005).

| 1. Appearance (Shop & Better) | 2. Select Structural | 3. General Construction (No. 1 & No. 2 Lumber) | 4. Utility and Economy |
|---|---|---|---|
| B + Better Select | Export Select | Number 1 | Utility |
| B Select | Structural | 1 Common | Utility Stud |
| Supreme | Select Structural | 2 & Better Common | Utility |
| C + Better Select | Select Stud | Export Standard | 5 Common |
| C Select | Export Construction | Common | Economy |
| D & Better Select | 2100 F | 1650 F | 1 × 2 Inch Side Cuts |
| Select | L-1 | Standard & Better | 4 Inch All Lengths Sf |
| Clear | Laminating Stock | Number 2 & Better | 8 Inch 16-18 Ft Sc |
| D Select | Shop Out | 1650 F Stud | Economy |
| Premium | Construction | 1450 F | Economy Stud |
| Export Clear | L-2 | Standard | Rip Stud |
| 3 Clear | L-3 | Number 2 | Skip |
| Moulding & Better | | 1200 F | Mill Run |
| Moulding | | Frame | Pallet Stock |
| Factory Select | | 2 Common | |
| 1 Shop | | Select Merchantable | |
| Pitch Select | | Construction | |
| Stained Select | | Standard | |
| 2 Shop | | 3 & Better Common | |
| 3 Shop | | 3 Common | |
| Stained Shop | | Number 3 | |
| Finger Joint | | Pet Stud | |
| | | 88 Pet Stud | |
| | | 92 Pet Stud | |
| | | 8 Ft Stud | |
| | | Stud | |
| | | 4-7" Stud | |
| | | Utility & Better | |
| | | 900 F | |
| | | 4 Common | |

grouped into these four product value classes. Totals by class are then converted to proportions (based on volume) for each study (Figure 1).

### *Analysis*

Because we are working with data from several independent studies conducted over a period of four decades, we are limited in our choice of predictor variables to those in common to all 13 studies. Diameter at breast height (dbh), height, and age are available for each tree. We then look to the relation between these variables and the proportion of each of the four product value classes. Because we are working with proportions, they must sum to one and be bounded between zero and one. The multinomial logistic model is a logical choice to estimate the yield proportion of each value class because it satisfies both of these conditions,

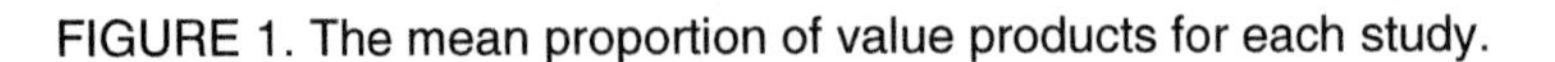
FIGURE 1. The mean proportion of value products for each study.

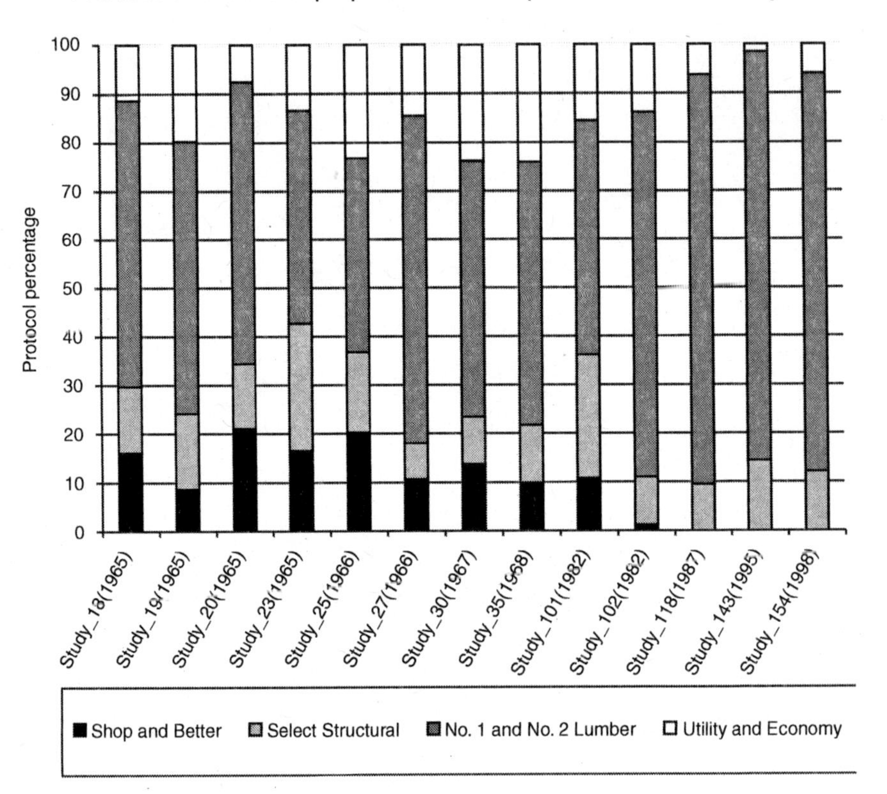


is well-behaved, and requires few limiting statistical assumptions (Teeter and Zhou, 1999). An alternate approach, the ordered probit, was used by Prestemon (1998) to predict tree grades. The basic multinomial logistic model can be expresses as (Maddala, 1983):

$$P_j = \frac{\exp(B_j'X)}{1 + \sum_{k=1}^{m-1} \exp(B_k'X)} \qquad j = 1, 2 \ldots, m-1 \qquad (1)$$

$$P_m = \frac{1}{1 + \sum_{k=1}^{m-1} \exp(B_k'X)} \qquad (2)$$

where $P_j$ is the proportion of product class $j$ across the $m = 4$ products classes, $X$ is the matrix of explanatory variables, $B_j$ is the vector of parameters associated with product class $j$. Thus, the denominator is the same sum of exponentials for all $m = 4$ classes. The numerator in each equation is class-specific, with the numerator equal to unity in the final reference class.

The model was estimated by using multinomial logistic regression (Maddala, 1983). We used the logistic procedure of the SAS Institute (1999). The parameters of the proportion functions are estimated by using the fourth group (Utility and Economy) as the arbitrary reference category.

## *RESULTS*

Maximum likelihood estimates were obtained for these multinomial logistic models. Because of the high correlation (0.82) between height and dbh, only dbh was significant when both were used in the same model. Thus, dbh and age remained as the predictor variables in $X$. In all cases dbh and age are all highly significant (beyond the $\alpha = 0.01$ level). Furthermore, their interaction (dbh·age) was also significant. In addition, the four product-specific estimates for a specific variable (dbh, age) are significantly different from each other (Table 3). The corresponding multinomial family of functions is:

$$P_1 = \frac{e^{-1.6584 + 0.00687\, dbh + 0.00213\, age + 0.00000372\, dbh \cdot age}}{D}$$

$$P_2 = \frac{e^{2.4467 - 0.0247\, dbh - 0.00205\, age + 0.000017\, dbh \cdot age}}{D}$$

$$P_3 = \frac{e^{4.2004 - 0.0257\, dbh - 0.0073\, age + 0.000044\, dbh \cdot age}}{D}$$

$$P_4 = \frac{1}{D}$$

where the denominator D is one plus the sum of the first three numerators:

$$D = 1 + e^{-1.6584+0.00689\,dbh+0.00213\,age+0.00000372\,dbh\cdot age} + e^{2.4467-0.0247\,dbh-0.00205\,age+0.000017\,dbh\cdot age} + e^{4.2004-0.0257\,dbh-0.0073\,age+0.000044\,dbh\cdot age}$$

and where $P_1$, $P_2$, $P_3$ and $P_4$ are the proportions of Shop and Better, Select Structural, Construction No. 1 and No. 2 lumber, and Utility and Economy, respectively (Table 3). The sum of these four proportions is equal to one.

The odds ratio estimates are listed in Table 4 along with the 95% confidence intervals (see Allison, 1999). With a 5 cm dbh increase, the odds or probability of getting Shop and Better lumber will increase 3.5%, while the odds of Select Structural and No. 1 and No. 2 lumber will decrease by 12%. With a 10-yr increase in age, the odds or probability of getting Shop and Better lumber will increase 2%, while the odds of Select Structural and Construction (No. 1 and No. 2) lumber will decrease by 2% and 7%, respectively. Table 4 indicates that the interaction term has little effect, on average, with an odds ratio of 0.2% or less. The linear hypothesis test for estimated coefficients ($H_0$: $\mathbf{L}\beta = c$), including

TABLE 3. Multinomial logistic parameter estimates of Douglas-fir in Pacific Northwest region.

| Parameters | Category (value class) | | |
|---|---|---|---|
| | Shop and Better | Select Structural | No. 1 & No. 2 Lumber |
| Intercept | −1.6584 | 2.4467 | 4.2004 |
| Dbh (cm) | 0.00687 | −0.0247 | −0.0257 |
| Age (yr) | 0.00213 | −0.00205 | −0.0073 |
| Dbh*Age | 0.00000372 | 0.000017 | 0.000044 |

TABLE 4. Odds ratio estimates and the 95% confidence limits.

| Effect | Group | Unit | Estimates | Lower Limit | Upper Limit |
|---|---|---|---|---|---|
| Dbh | 1 (Shop & Better) | 5 cm | 1.035 | 1.034 | 1.036 |
| Dbh | 2 (Select Structural) | 5 | 0.884 | 0.883 | 0.885 |
| Dbh | 3 (No. 1 & No. 2 Lumber) | 5 | 0.880 | 0.879 | 0.880 |
| Age | 1 (Shop & Better) | 10-year | 1.021 | 1.020 | 1.023 |
| Age | 2 (Select Structural) | 10 | 0.980 | 0.979 | 0.981 |
| Age | 3 (No. 1 & No. 2 Lumber) | 10 | 0.930 | 0.929 | 0.930 |
| Dbh*Age | 1 (Shop & Better) | 50 cm·year | 1.000 | 1.000 | 1.000 |
| Dbh*Age | 2 (Select Structural) | 50 | 1.001 | 1.001 | 1.001 |
| Dbh*Age | 3 (No. 1 & No. 2 Lumber) | 50 | 1.002 | 1.002 | 1.002 |

slope parameters for dbh, age, their interaction (dbh·age), and the intercept parameters, also show significant differences among equations.

Figure 2 shows the change in the relative proportions with respect to age, while holding dbh constant at 50 cm (Figure 2a) and 100 cm (Figure 2b). Figure 3 shows the change in the relative proportions with respect to dbh, while holding age constant at 70 years (Figure 3a) and 150 years (Figure 3b).

FIGURE 2a. Douglas-fir product proportions by age at dbh = 50 cm.

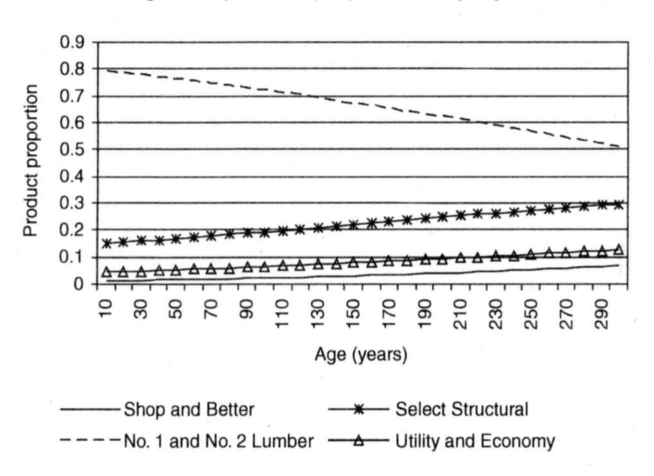


FIGURE 2b. Douglas-fir product proportions by age at dbh = 100 cm.

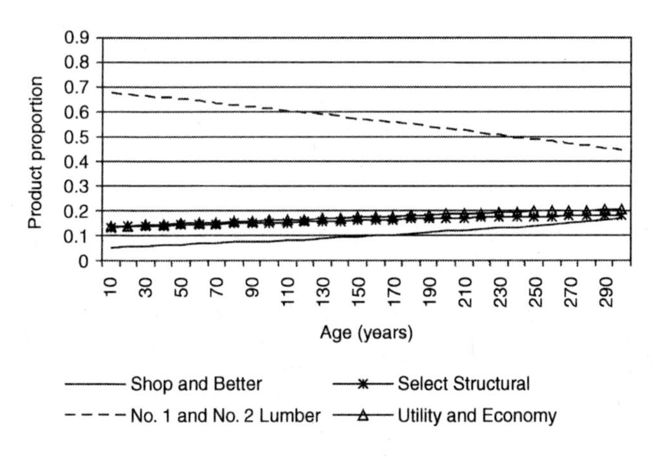

FIGURE 3a. Douglas-fir product proportions by dbh at age = 70.

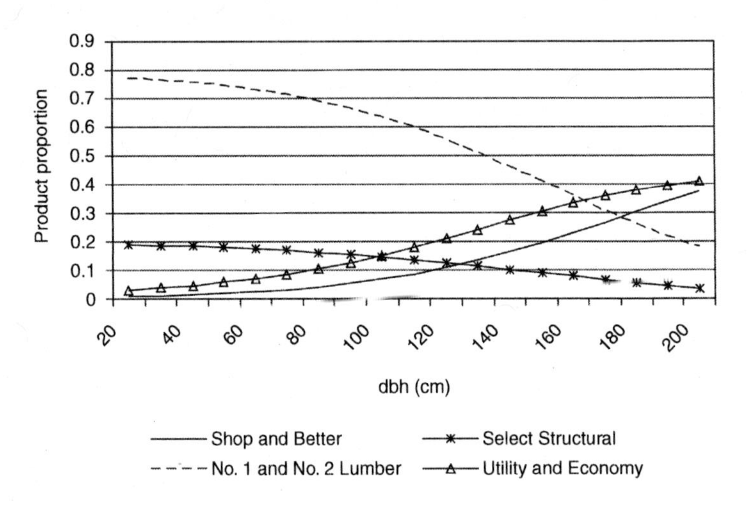


FIGURE 3b. Douglas-fir product proportions by dbh at age = 150.

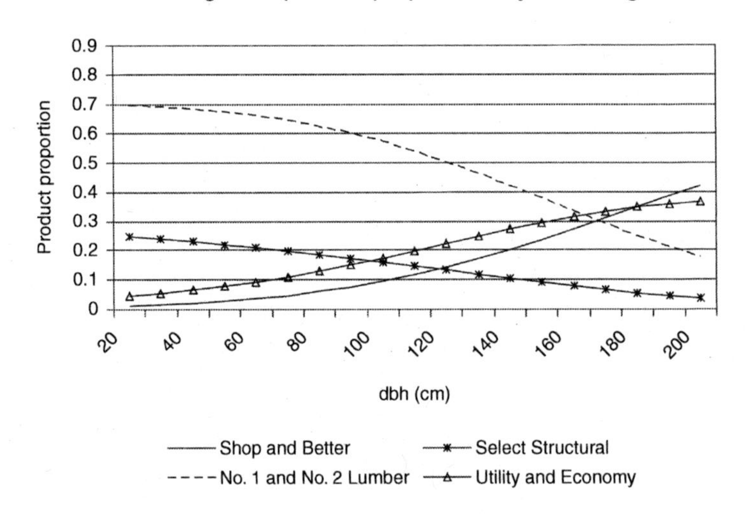


## *DISCUSSION*

The dominant feature of Figure 2 (holding dbh constant) is that Construction lumber (mostly No. 1 and No. 2 grades) dominates the three other value classes regardless of age, with the proportion of the total between 0.5 and 0.8. This simply reflects the well-known suitability and desirability of Douglas-fir as a construction material. The second feature is that the Construction proportion declines as the resource ages. With dbh fixed at 50 cm (Figure 2a), the proportion of Select Structural ranges from 0.15 to 0.3 of the total. With dbh fixed at 100 cm (Figure 2b), the proportion of the three non-Construction classes is nearly the same, with a slightly positive slope that ranges from 0.1 to 0.2 of the total. The Construction proportion ranges from 0.7 to 0.45 as age increases. A dbh of 40 to 50 cm approximates today's mean dbh at harvest, whereas a mean diameter of 100 cm reflects the past harvest sizes in studies from the 1960s (see Table 1).

Figure 3 (holding age constant) tells a more complex story. In Figure 3a, age is held at a 70-year harvest age. Although we show model behavior over the full range of diameters, only the left half of the graph is realistic; the cross of class proportions near dbh 170 cm is not reachable in 70 years. Below diameters 90 cm, we again see Construction grades dominating, with proportions between 0.8 and 0.7. In this range of diameters, Select Structural is also declining slightly (ranging from 0.20 to 0.15), while the remaining two classes increase slightly.

Figure 3b (age = 150 yrs) corresponds more to practices in the past (e.g., the 1960s), when very large old-growth trees were both plentiful and frequently harvested. This resource was characterized by narrow rings and clear boles, resulting in high wood quality that often produced a premium price over lumber from other regions (Haynes, 2005). The full range of diameters up to 200 cm is adequately represented in the numerous studies that we used from the 1960s (Table 1). Again we see the dominance of the Construction grades, but past a diameter of 170 cm its proportion of the total is passed by the highest quality products (Shop and Better) and also by the lowest quality class (Utility and Economy). These large, old trees had both high-quality clear wood and considerable defect in abundance. At this point we have three different value classes, each constituting one-third of the product mix. Note also the decline in the proportion of Select Structural as dbh increases, in both Figure 3a and 3b. Because of the decreasing age and dbh of harvested trees from the 1960s to the 1990s, the proportion of the highest value group (Shop and Better) is diminished, while more Select Structural and Construction products were produced. This is akin to reading Figure 3b from right to left.

We compared the current multinomial logistic model with a dbh·age interaction to one without the interaction. In addition to the interaction term being significant, both Akaike's information criterion (AIC) and the Bayesian information criterion (BIC) were lower (i.e., more desirable; see Allison, 1999) for the model with the interaction term than the model without it. When comparing graphs of the two models (e.g., Figure 2 or Figure 3), we found that the interaction term had the most effect in the extremes, and none near the mean. The odds ratios also bear this out (Table 4). With the interaction term included, the proportion in the Construction class was higher at young ages and small diameters (0.80 vs. 0.75) and lower at advanced ages and large diameters (0.20 vs. 0.25) than the model with no interaction. Thus, the relatively constant proportion for the Construction class is nearly 0.80 in Figures 6 and 7 instead of 0.75 without the interaction term.

## *Looking to the Future*

Current forest conditions in the Douglas-fir region are a function of both markets for various forest products and various regulatory actions, both past and current (Haynes, 2005). Because of the long time horizons common in wood production, current management is always geared to expectations of future market conditions. Haynes (2005) pointed out that the 20% price premium for Douglas-fir stumpage relative to southern pines and other regions has largely disappeared, primarily owing to the loss of the export market. The resultant focus on the domestic market shifts the emphasis to commodity grades such as Construction (No. 1 and No. 2), where no price premium exists (Haynes and Fight, 2004). Haynes (2005) found that the implication of these changes is that market incentives will favor the transition to managed stands that produce relatively uniform logs (both quality and size) with reduced market opportunities for logs that do not conform to processing standards (e.g., logs greater than 60-cm diameter). Zhou et al. (2005) expect that the mode of the age-class distribution of Douglas-fir removals will continue to decline over the next 50 years (Figure 4), to age 60 in 2020 and age 50 in 2040; currently there are two peaks, at age 60 and age 110. Based on the fifth Forest and Rangeland Renewable Resource Planning Act timber assessment (Haynes, 2003), the projection of future removals in the Douglas-fir region tends toward younger and smaller trees. The projected relative proportions among the four product value classes is not expected to change much over current proportions, with Construction lumber a dominant 70% of the total (Figure 5, which is from Zhou et al., 2005).

Implications of this expected trend to smaller and younger stands are clear. The relations graphed in the left third of both Figures 2a and 3a

FIGURE 4. Projected age class distribution of Douglas-fir removals for all ownerships (RPA 2000 base case projection) from using data from Zhou et al. (2005).

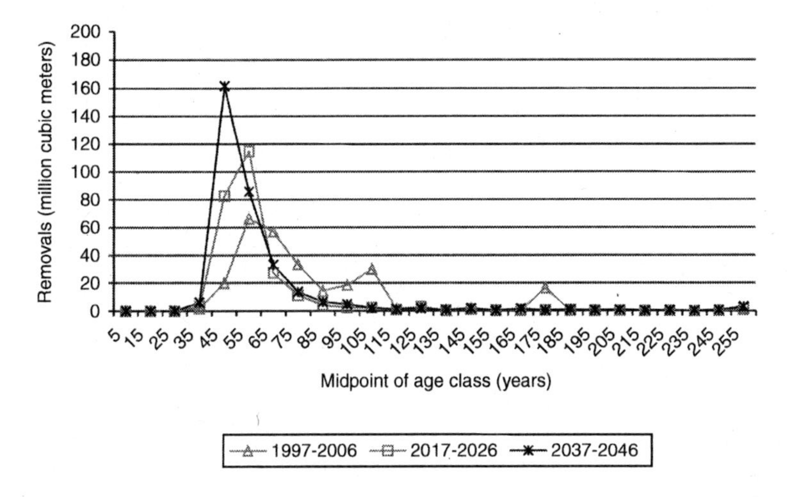


FIGURE 5. Projected percentage of each value class from Douglas-fir removals by period, all ownerships.

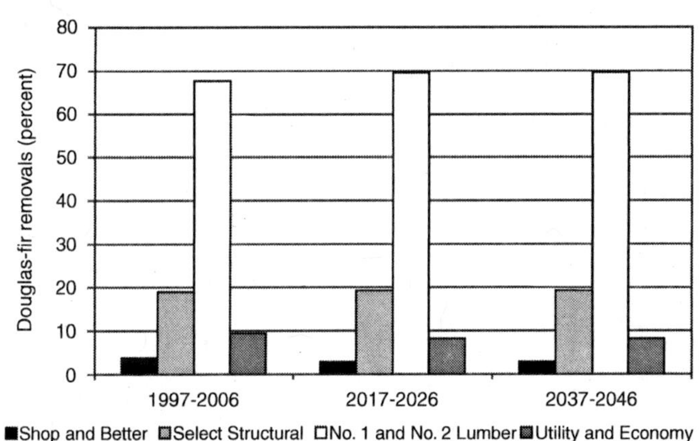


are relevant, whereas the right two-thirds of these same figures are not. Likewise, the very large diameters and advanced ages assumed in Figures 2b and 3b will not be relevant. To this end, we offer Figure 6 (product proportions by age at dbh = 40 cm) and Figure 7 (product proportions by dbh at age = 50). The much smaller range of ages and diam-

FIGURE 6. Douglas-fir product proportions by age at dbh = 40 cm.

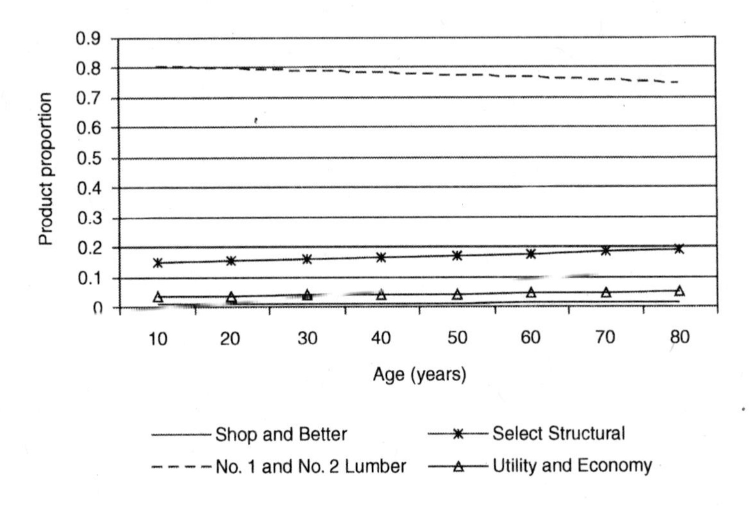


FIGURE 7. Douglas-fir product proportions by dbh at age = 50.

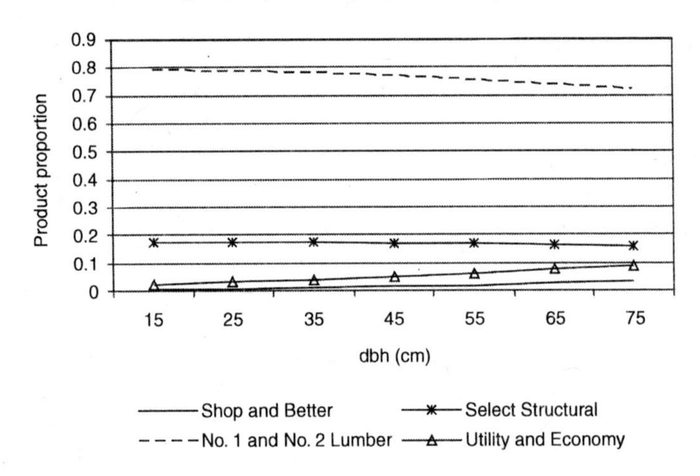


eters illustrated in Figures 6 and 7 correspond more closely with the shorter rotations expected on nonfederal land in the future. We expect that the mix of product classes will be dominated by Construction grade lumber (mostly No. 1 and No. 2 grades), with the proportion of the total between 0.7 and 0.8. The second largest class will be Select Structural, with a proportion of the total near 0.2.

Basically, the shift away from the highest quality visual grades has already occurred, with the proportion of Appearance Shop and Better grades being only 0.05 of the total. This means that as the industry continues the shift to harvesting and processing smaller, younger trees, little change in visual lumber grade yield should be expected (Zhou et al., 2005). Additional evidence of this shift to Construction grades and away from Appearance grades from the 1960s through the late 1980s is documented by Warren (1991, 2003). In recent years, the industry summaries reported by Warren (2003) have leveled off to a new equilibrium where very little appearance-grade lumber is manufactured in the Douglas-fir region, and the bulk of the production is in the general construction lumber category. This is typical of young, vigorous stands of managed timber where defects are minimal and branch size is controlled by early spacing (Barbour et al., 2005).

Wood quality implications are not so clear. As harvested trees become younger, the proportion of juvenile wood (e.g., the inner core up to age 20) will increase. Knots will also become more prominent in the resulting mix of boards, for less time will be available for trees to produce clear wood after shedding lower branches. Overall, these effects could result in a reduction in mechanical wood properties, especially regarding the suitability of this material for engineered wood products. Barbour et al. (2003) compared yields of machine stress-rated lumber from Douglas-fir in the 40-to-60-year age class and trees in the 40-and-under age class. They found a major drop in the yield of higher stress ratings of mechanically graded lumber. This result suggests that although manufacturers will be able to produce lumber with acceptable visual characteristics from a resource less than 40 years old, the lumber might not have mechanical properties that meet the requirements of the higher grades (Zhou et al., 2005). Because of the great genetic variability among Douglas-fir populations, this potential problem could possibly be ameliorated through genetic selection of planting stock (Vargas-Hernandez and Adams, 1991, 1992; Vargas-Hernandez et al., 1994; Barbour and Marshall, 2002; Rozenberg et al., 2001).

Adams and Latta (2005) examined future timber harvest potential from private lands in the Pacific Northwest. They suggested that private lands in the Pacific Northwest should be able to sustain at least recent historical harvest levels over the next 50 or more years, given unchanged policies and the anticipated levels of private management investment. They consider that these harvests could be realized with stable to rising inventories–growth would be at least as large as harvest. Accompanying such a trend, Adams and Latta (2005) expect continued concentration of industrial lands in younger age and smaller size classes and a continued decline in the average age and size of timber harvested. They suggested that shifts toward more intensive management regimes will be gradual, relatively limited, and have a relatively minor impact on harvest potential over the next 50 years. They do not anticipate a wave of intensive forest plantations with genetically modified trees managed on short rotations.

Alig (2005) examined the dynamic effect of land use changes on sustainable forestry in the Pacific Northwest. This will affect the region's progress toward sustainability at the same time that population is expected to increase. The United States is expected to add around 120 million people, an additional 40 percent, to its population in the next 50 years. The Pacific Northwest is expected to experience above-average population growth, including immigration from other regions. This will likely intensify land use pressures (Alig, 2005), especially in the forest-urban interface. In the most recent national comprehensive survey, the rate of conversion of rural land to developed land increased, with forest land the largest source (Alig, 2005). As populations in the national and regional landscapes increase, so too will their effect on sustainability options for agriculture, forestry, residential communities, biodiversity, and other land-based goods and services.

Haynes (2005) examined the role of markets and prices in providing incentive for sustainable forest management. His market analysis concluded that the majority of timberland will be lightly managed, while a small minority of acres will be heavily (or actively) managed on relatively short rotations. The net effect is that prices, although not a barrier, will not provide a strong incentive to intensify timber management by many landowners. Put another way, relatively stable price expectations are necessary for intensive timber management but not necessarily a sufficient condition for sustainable timber management (Haynes, 2005).

Regarding implications for the future (see Figures 4 and 5), the future is already here. The transition from an old-growth lumber economy with large premiums (Haynes, 2005) to dominance of Construction-grade products has been completed. The younger and smaller material harvested today (and expected in the future) is dominated by Construction-grade products. Even if future harvest ages were doubled to 100 years, relative proportions predicted by our multinomial model show little change from the current proportions illustrated in Figures 6 and 7.

It is tempting to generalize to other regions in the world. Although France (e.g., Rozenberg et al., 2001) and New Zealand have large areas of managed Douglas-fir, the species is an introduced exotic, and thus conversion from an old-growth Douglas-fir economy does not apply, although conversion from a native hardwood economy in New Zealand does. Coastal British Columbia is also experiencing a shift in forest management away from a dependence on clearcutting old-growth (Forest Practices Code of British Columbia, 1994; Clayoquot Scientific Panel, 1995), although this northern region is mostly outside the range of coastal Douglas-fir. Perhaps the southeastern United States is also a possible analogue (*sans* Douglas-fir), for this large region has experienced a broad conversion from natural southern pines to planted southern pines such as loblolly pine (*Pinus taeda* L.) (Haynes, 2003). In the end, our large historical product recovery database is unique in allowing us to examine product changes over time.

## REFERENCES

Adams, D.M. and G.S. Latta. 2005. Timber harvest potential from private lands in the Pacific Northwest: biological, investment, and policy issues. In: pp. 4-12. Deal, R.L. and S.M. White (eds.). Understanding key issues of sustainable wood production in the Pacific Northwest. Gen. Tech. Rep. PNW-GTR-626. U.S. Department of Agriculture, Forest Service, Pacific Northwest Research Station, Portland, OR.

Alig, R.J. 2005. Land use changes that impact sustainable forestry. In: pp. 28-35. Deal, R.L. and S.M. White (Eds.). Understanding key issues of sustainable wood production in the Pacific Northwest. Gen. Tech. Rep. PNW-GTR-626. U.S. Department of Agriculture, Forest Service, Pacific Northwest Research Station, Portland, OR.

Allison, P.D. 1999. Logistic regression using the SAS system: theory and application. SAS Institute, Inc, Cary, NC.

Barbour, R.J. and D.D. Marshall. 2002. Stem characteristics and wood properties: essential considerations in sustainable multipurpose forestry regimes. In: pp. 145-150. Johnson, A., R. Haynes and R.A. Monserud. (eds.). Congruent management of multiple resources: proceedings of the wood compatibility initiative workshop. Gen. Tech. Rep. PNW-GTR-563. U.S. Department of Agriculture, Forest Service, Pacific Northwest Research Station, Portland, OR.

Barbour, R.J., D.D. Marshall and E.C. Lowell. 2003. Managing for wood quality. In: pp. 299-336. Chapter 11. Monserud, R.A., R.W. Haynes and A.C. Johnson. (eds.) Compatible forest management. Kluwer Academic Publishers, Dordrecht, The Netherlands.

Barbour, R.J., R.R. Zaborske, M.H. McClellan, L. Christian and D. Golnick. 2005. Young stand management options and their implications for wood quality and other values. *Landscape and Urban Planning* 72(1-3):79-94.

Behan, R. 1990. Multiresource forest management: a paradigmatic challenge to professional forestry. *Journal of Forestry* 88(4):12-18.

Castellano, M.A. and R. Molina. 1989. Mycorrhizae. In: pp. 101-167. Vol. 5. Landis, T.D., R.W. Tinus, S.E. McDonald and J.P. Barnett (eds.). Agric. Handb. 674: The container tree nursery manual. U.S. Department of Agriculture, Forest Service, Washington, DC.

Clayoquot Scientific Panel. 1995. Scientific panel for sustainable forest practices in Clayoquot Sound, Report 5, sustainable ecosystem management in Clayoquot Sound: planning and practices. Victoria, BC.

Committee of Scientists. 1999. Sustaining the people's lands: recommendations for stewardship of the national forests and grasslands into the next century. U.S. Department of Agriculture, Washington, DC. Retrieved May 31, 2002 from http://www.fs.fed.us/forum/nepa/rule/cosreport.shtml.

Curtis, R.O. and A.B. Carey. 1996. Timber supply in the Pacific Northwest: managing for economic and ecological values in Douglas-fir forests. *Journal of Forestry* 94(9):4-7, 35-37.

Curtis, R.O., D.S. DeBell, C.A. Harrington, D.P. Lavender, J.B. St. Clair, J.C. Tappeiner and J.D. Walstad. 1998. Silviculture for multiple objectives in the Douglas-fir region. Gen. Tech. Rep. PNW-GTR-435. U.S. Department of Agriculture, Forest Service, Pacific Northwest Research Station, Portland, OR.

Duryea, M.L. and P.M. Dougherty. 1991. Forest regeneration manual. Kluwer Academic Publishers, Hingham, MA.

Forest Ecosystem Management Assessment Team (FEMAT). 1993. Forest ecosystem management: an ecological, economic, and social assessment. U.S. Department of the Interior [and others]. Washington, DC.

Forest Practices Code of British Columbia Act. 1994. Statutes of B.C., Bill 40. Queen's Printer, Victoria, BC. Retrieved April 4, 2003 from http://www.legis.gov.bc.ca/1994/3rd_read/gov40-3.htm.

Franklin, J.F. 1988. Pacific Northwest forests. In: pp. 103-130. M.G. Barbour and W.D. Billings (eds.). North American terrestrial vegetation. Cambridge University Press, New York.

Franklin, J.F. and C.T. Dyrness. 1973. Natural vegetation of Oregon and Washington. Gen. Tech. Rep. PNW-GTR-8. U.S. Department of Agriculture, Forest Service, Pacific Northwest Research Station, Portland, OR.

Franklin, J.F. and R.H. Waring. 1981. Distinctive features of the northwestern coniferous forest: development, structure, and function. In: pp. 59-86. R.H. Waring (ed.). Forests: fresh perspectives from ecosystem research. Oregon State University Press, Corvallis, OR.

Franklin, J.F., D.R. Berg, D.A. Thornburg and J.C. Tappeiner. 1997. Alternative silvicultural approaches to timber harvesting: variable retention harvest systems. In: pp. 111-140. K.A. Kohm and J.F. Franklin (eds.). Creating a forestry for the 21st century. Island Press, Washington, DC.

Fujimori, T., S. Kawanabe, H. Saito, C.C. Grier and T. Shidei. 1976. Biomass and primary production in forests of three major vegetation zones of the northwestern United States. *Journal of Japanese Forestry Society* 58:360-373.

Halpern, C.B. 1995. Response of forest communities to green-tree retention harvest: a study plan for the vegetation component of the Demonstration of Ecosystem Management Options (DEMO) study. College Forestry Resources, University of Washington, Seattle, WA.

Haynes, R.W. (tech. coord.). 2003. An analysis of the timber situation in the United States: 1952 to 2050. Gen. Tech. Rep. GTR-PNW-560. U.S. Department of Agriculture, Forest Service, Pacific Northwest Research Station, Portland, OR.

Haynes, R.W. 2005. Will markets provide sufficient incentive for sustainable forest management? In: pp. 13-19. Deal, R.L. and S.M. White (eds.). Understanding key issues of sustainable wood production in the Pacific Northwest. Gen. Tech. Rep. PNW-GTR-626. U.S. Department of Agriculture, Forest Service, Pacific Northwest Research Station, Portland, OR.

Haynes, R.W. and R.D. Fight. 2004. Reconsidering price projections for selected grades of Douglas-fir, coast hem-fir, inland hem-fir, and ponderosa pine lumber. Res. Pap. PNW-RP-561. U.S. Department of Agriculture, Forest Service, Pacific Northwest Research Station, Portland, OR.

Haynes, R.W. and R.A. Monserud. 2002. A basis for understanding compatibility among wood production and other forest values. Gen. Tech. Rep. PNW-GTR-529. U.S. Department of Agriculture, Forest Service, Pacific Northwest Research Station, Portland, OR.

Haynes, R.W., R.A. Monserud and A.C. Johnson. 2003. Compatible forest management: background and context. In: pp. 3-32. Chapter 1. Monserud, R.A., R.W. Haynes and A.C. Johnson (eds.). Compatible forest management. Kluwer Academic Publishers, Dordrecht, The Netherlands.

Hummel. S.S. 2003. Managing structural and compositional diversity with silviculture. In: pp. 85-120. Chapter 4. Monserud, R.A., R.W. Haynes and A.C. Johnson (eds.) Compatible forest management. Kluwer Academic Publishers, Dordrecht, The Netherlands.

Loucks, D.M., S.A. Knowe, L.J. Shainsky and A.A. Pancheco. 1996. Regenerating coastal forests in Oregon: an annotated bibliography of selected ecological literature. Research Contribution 14. Oregon State University, Forest Research Laboratory, Corvallis, OR.

Maddala, G.S. 1983. Limited-dependent and qualitative variables in econometrics. Cambridge University Press, New York.

McArdle, R.E., W.H. Meyer and D. Bruce. 1961. The yield of Douglas-fir in the Pacific Northwest. Tech. Bull. 201. U.S. Department of Agriculture, Washington, DC.

Meidinger, D. and J. Pojar (eds.). 1991. Ecosystems of British Columbia. Special Report Series 6. British Columbia Ministry of Forests, Victoria, BC.

Monserud, R.A. 2002. Large-scale management experiments in the moist maritime forests of the Pacific Northwest. *Landscape and Urban Planning* 59(3):159-180.

Monserud, R.A. 2003. Modeling stand growth and management. In: pp 145-176. Chapter 6. Monserud, R.A., R.W. Haynes and A.C. Johnson (eds.). Compatible forest management. Kluwer Academic Publishers, Dordrecht, The Netherlands.

Monserud, R.A., R.W. Haynes and A.C. Johnson (eds.). 2003. Compatible forest management. Kluwer Academic Publishers, Dordrecht, The Netherlands.

Monserud, R.A., E.C. Lowell, D.R. Becker, S.S. Hummel, E.M. Donoghue, R.J. Barbour, K.A. Kilborn, D.L. Nicholls, J. Roos and R.A. Cantrell. 2004. Contemporary wood utilization research needs in the Western United States. Gen. Tech. Rep. PNW-GTR-616. U.S. Department of Agriculture, Forest Service, Pacific Northwest Research Station, Portland, OR.

Peterson, C.E. and R.A. Monserud. 2002. Compatibility between wood production and other values and uses on forested lands. a problem analysis. Gen. Tech. Rep. PNW-GTR-564. U.S. Department of Agriculture, Forest Service, Pacific Northwest Research Station, Portland, OR.

Prestemon, J.P. 1998. Estimating tree grades for southern Appalachian natural forest stands. *Forest Science* 44(1):73-86.

Rozenberg, P., A. Franc, C. Bastien and C. Cahalan. 2001. Improving models of wood density by including genetic effects: a case study in Douglas-fir. *Forest Science* 58:385-394.

SAS Institute. 1999. SAS/STAT user's guide, Version 8. SAS Institute, Inc, Cary, NC.

Smith, B.W., P.D., Miles, J.S. Vissage, S.A. Pugh. 2004. Forest resources of the United States, 2002. Gen. Tech. Rep. NC-241. U.S. Department of Agriculture, Forest Service, North Central Research Station, St. Paul, MN.

Smith, J.P., R.E. Gresswell and J.P. Hayes. 1997. A research problem analysis in support of the Cooperative Forest Ecosystem Research (CFER) Program. U.S. Geological Society, Corvallis, OR.

Teeter, L. and X. Zhou. 1999. Projecting timber inventory at the product level. *Forest Science* 45:226-231.

U.S. Department of Agriculture, Forest Service. 1963. Timber trends in western Oregon and western Washington. Res. Pap. PNW-5. Pacific Northwest Forest and Range Experiment Station, Portland, OR.

U.S. Department of Agriculture, Forest Service and U.S. Department of the Interior, Bureau of Land Management [USDA and USDI]. 1994a. Final supplemental environmental impact statement on management of habitat for late-successional and old-growth forest related species within the range of the northern spotted owl. Vol. 1 and 2. Washington, DC.

U.S. Department of Agriculture, Forest Service and U.S. Department of the Interior, Bureau of Land Management. [USDA and USDI]. 1994b. Record of decision for the President's forest plan. Washington, DC.

Vargas-Hernandez J. and W.T. Adams. 1991. Genetic variation of wood density components in young coastal Douglas-fir: implications for tree breeding. *Canadian Journal of Forest Research* 21:1801-1807.

Vargas-Hernandez J., Adams W.T. 1992. Age-age correlations and early selection for wood density in young coastal Douglas-fir. *Forest Science* 38:467-478.

Vargas-Hernandez, J., W.T. Adams and R.L. Krahmer. 1994. Family variation in age trends of wood density traits in young coastal Douglas-fir. *Wood Fiber Science* 26:229-236.

Walstad, J.D., and P.J. Kuch (eds.). 1987. Forest vegetation management for conifer production. John Wiley and Sons, New York.

Walter, H. 1985. Vegetation of the Earth and ecological systems of the geobiosphere (3rd ed.). Springer-Verlag, New York.

Waring, R.F. and J.F. Franklin. 1979. Evergreen coniferous forests of the Pacific Northwest. *Science*, 204:1380-1386.

Warren, D.D. 1991. Production, prices, employment, and trade in Northwest forest industries, fourth quarter 1991. Resour. Bull. PNW-RB-187. U.S. Department of Agriculture, Forest Service, Pacific Northwest Research Station, Portland, OR.

Warren, D.D. 2003. Production, prices, employment, and trade in Northwest forest industries, all quarters 2001. Resour. Bull. PNW-RB-239. U.S. Department of Agriculture, Forest Service, Pacific Northwest Research Station, Portland, OR.

Western Wood Products Association [WWPA]. 1998. Western lumber grading rules '98. Portland, OR.

Zhou, X., R.W. Haynes, and R.J. Barbour. 2005. Projections of timber harvest in western Oregon and Washington by county, owner, forest type and age class. Gen. Tech. Rep. GTR-PNW-633. U.S. Department of Agriculture, Forest Service, Pacific Northwest Research Station, Portland, OR.

doi:10.1300/J091v24n01_04

# Measuring Sustainable Forest Management in Tierra del Fuego, Argentina

Esteban Carabelli
Hugh Bigsby
Ross Cullen
Pablo Peri

**SUMMARY.** The Tierra del Fuego government has a statutory responsibility to ensure that *Nothofagus* forests are sustainably used. To provide strategies to maximize the forest's sustainable benefits for society, this research develops indicators of sustainable management of lenga forests. Multicriteria methods are used to integrate different perspectives regarding environmental, social, and economic aspects of the forest's management. Starting with a range of internationally accepted criteria and indicator (C&I) schemes, a local set of C&I was developed to assess the forest's sustainability. The relative importance of the C&I,

Esteban Carabelli is Coordinator of a Forest Landscape Restoration Project for Fundacion Vida Silvestre Argentina, in Andresito Municipality, Provincia de Misiones, Argentina.

Hugh Bigsby is Associate Professor at Lincoln University, Canterbury, New Zealand (E-mail: bigsbyh@lincoln.ac.nz).

Ross Cullen is Professor of Resource Economics at Lincoln University, Canterbury, New Zealand.

Pablo Peri is Senior Lecturer at the Universidad Nacional de la Patagonia Austral, Argentina and a researcher with the Instituto Nacional de Tecnología Agropecuaria, Argentina.

[Haworth co-indexing entry note]: "Measuring Sustainable Forest Management in Tierra del Fuego, Argentina." Carabelli, Esteban et al. Co-published simultaneously in *Journal of Sustainable Forestry* (Haworth Food & Agricultural Products Press, an imprint of The Haworth Press, Inc.) Vol. 24, No. 1, 2007, pp. 85-108; and: *Sustainable Forestry Management and Wood Production in a Global Economy* (ed: Robert L. Deal, Rachel White, and Gary L. Benson) Haworth Food & Agricultural Products Press, an imprint of The Haworth Press, Inc., 2007, pp. 85-108. Single or multiple copies of this article are available for a fee from The Haworth Document Delivery Service [1-800-HAWORTH, 9:00 a.m. - 5:00 p.m. (EST). E-mail address: docdelivery@haworthpress.com].

Available online at http://jsf.haworthpress.com

doi:10.1300/J091v24n01_05

and the degree to which sustainability was being achieved were assessed using a survey completed at a series of workshops and a follow-up survey. The results show that current management scores an aggregate 2.51 on a range from 1.0 to 5.0, where 1.0 for any indicator means that the indicator is close to a critical threshold value (ability of the ecosystem to recover) and 5.0 is far from this threshold value. The results also show that the key areas to work on are in policy and planning, resource administration, monitoring of resource use, and public information about, and involvement in, forest use. doi:10.1300/J091v24n01_05 *[Article copies available for a fee from The Haworth Document Delivery Service: 1-800-HAWORTH. E-mail address: <docdelivery@haworthpress.com> Website: <http://www.HaworthPress.com>* 

**KEYWORDS.** Sustainability assessment, sustainable forest management, multicriteria analysis, criteria and indicators of sustainability

## *INTRODUCTION*

*Nothofagus* forests occur in only a few countries of the southern hemisphere, including Argentina, Australia, Chile, and New Zealand. Its distribution ranges from the Mediterranean-type climate region of central Chile to the subantarctic latitudes of Tierra del Fuego (Veblen et al., 1996). Tierra del Fuego is one of the five Argentinean provinces that make up the Argentinean Andean Patagonia Region and contains the largest proportion of these forests of any province in Argentina (35%) (Bava, 1999). Although the province of Tierra del Fuego includes Grande de Tierra del Fuego Island, the Argentinean Antarctic Zone and minor islands, this study includes only Grande de Tierra del Fuego Island (referred to as Tierra del Fuego). More than 700,000 hectares (ha) of Tierra del Fuego are covered by native forests, which are mainly concentrated in the centre and south parts of the island.

Three main species of the *Nothofagus* genus (Fagacea family) are most common in Tierra del Fuego. *Nothofagus pumilio* (Poepp et Endl.) Krasser, locally called lenga, is the most common species on the island, covering nearly half of the total forested area and is the most important species for wood production in Tierra del Fuego (Bava, 1999; Gobierno de Tierra del Fuego, 2003). *Nothofagus antarctica* (Forst. fil.) Oerst, locally called ñire, is located in the north of the forest distribution on the island, covering 182,000 ha (25.5% of the total forest area). Its main economic use is cattle and sheep grazing in silvopastoral systems.

*Nothofagus betuloides* (Mirb.) Blume, locally called guindo, is mainly located in the south zone of the forest distribution on the island, for example along the Beagle Channel (Dimitri et al., 1998). This means that two forest zones can be identified in Tierra del Fuego, based on ecological characteristics and current uses. The north zone is mainly used for sheep and cattle grazing, timber, recreation and firewood. The south zone is mainly used for tourism, timber, and natural resources conservation (Peri and Martínez-Pastur, 2003). Most of the lenga processed in Tierra del Fuego is extracted from state-owned forests, and silvopastoral uses are on privately-owned forest lands. On state-owned forests, harvesting is by annual cutting permits, or 10-year and 20-year concessions with companies responsible for silviculture and management under government supervision.

Based on current estimates, nearly 100,000 ha of forests have been used for timber production or have been degraded by other factors. The harvested area was calculated to be about 45,000 ha in 2000 (Collado, 2001), and it is estimated to be close to 50,000 ha now (Collado, L., pers. com., 3 December 2004). The state of the natural regeneration in these harvested sites is unknown. Another 20,000 ha of forests have been cleared in the past for the introduction of grazing animals. Introduced beavers have affected about 20,000 ha of forest close to watercourses (Martínez-Pastur, G., pers. com. 20 November 2004). Other areas have been affected by fires and the expansion of urban areas in Ushuaia and Tolhuin. Another 181,000 ha of ñire forest have not been assessed yet, and a proportion of them may have been used or degraded (Collado, 2001). There are also a number of issues in lenga forest management, especially in relation to wood production and processing, that may affect long-term sustainability. These factors include the lack of long-term policies and planning, weak control from the government, incomplete implementation of management plans and silvicultural practices, careless field operations that result in standing tree damage, and livestock and guanaco damage to regeneration (Martínez-Pastur et al., 1998, 2003, 2004; Schmidt et al., 1998; Bava, 1999; Lencinas et al., 2003; Pulido et al., 2000; Carabelli, 2002; Cellini et al., 2003; Gea Izquierdo et al., 2003).

These issues raise the question of how sustainable the management of *Nothofagus* forests in Tierra del Fuego really is, and if there are problems, how to address them. The objective of this study is to develop a method of measuring how sustainable the current management of the *Nothofagus* forests of Tierra del Fuego is, and based on this, determine what should be priorities for sustainable management. In order to address this objective, the study develops a measure of sustainability that can combine the many factors that have the potential to influence sustainable forest management and applies this to *Nothofagus* forests in Tierra del Fuego.

## *MEASURING SUSTAINABILITY*

Criteria and indicators (C&I) have been developed to better define sustainable forest management (SFM) and to assist with measuring change in forest condition and output of goods and services from forests (Lamerts van Bueren and Blom, 1997; Prabhu et al., 1996; Raison et al., 2001) and are now used in over 150 countries (FAO, 2003). Indicators are quantitative or qualitative measures or components of the different economic, social, and environmental functions, components, and characteristics that make up forest management. The C&I have been defined for different forests types, purposes, and scales. At the global and national level they may be used to guide countrywide policies, regulations, and legislation regarding productive, protective, and social roles of forests and forest ecosystems, and at the Forest Management Unit (FMU) level they are focused on more specific and local concerns.

The C&I are typically hierarchically organised with principles at the top, a number of criteria under each principle, and indicators under each criterion (Mendoza and Prabhu, 2000). A criterion is a standard by which a judgement can be made, and is an intermediate point to which the information provided by indicators can be integrated. Principles are general truths or laws that in forestry are seen as providing the primary framework for managing forests in a sustainable fashion (Mendoza et al., 1999).

The C&I have been used previously in Argentina. The Montreal Process C&I was adopted in 1995 (Montreal Process Working Group, 1998) and a working group has recently been created to deal with region-specific issues with Uruguay and Chile (Grupo de Trabajo sobre Criterios e Indicadores del Proceso de Montreal, 2002). The Forest Stewardship Council (FSC) C&I scheme is also being implemented, and about 43,000 ha of plantation forests and 81,300 ha of native forests have been certified under the FSC scheme (Fundación Vida Silvestre Argentina, 2003; Forest Stewardship Council, 2004). The Andean-Patagonia Forestry Research and Advisory Centre (CIEFAP) and the Chubut Forest Service organised two workshops to define C&I for assessing multiple-use forest management plans in the Chubut Province by using the Center for International Forestry Research's (CIFOR) generic template as the main source of C&I (Carabelli et al., 2000, 2003) The National Institute of Agricultural Research (INTA) also used the CIFOR C&I for a FMU in the Chubut Province of Patagonia (Rusch et al., 2001). This previous application of C&I in Argentina can provide the basis for measuring sustainability in Tierra del Fuego.

The C&I system provides a diverse set of indicators of sustainability. As a set of indicators, they individually provide a variety of views of what is happening at the forest level. However, without a method for aggregating individual indicators, it is difficult to obtain an overall picture of what is happening. Multicriteria analysis is one technique for incorporating a diversity of variables, both quantitative and qualitative, into a single assessment. The advantage of multicriteria analysis is that it enables a rigorous selection of the most preferred choice in a context where several criteria apply simultaneously (Mendoza et al., 1999; Mendoza and Prabhu, 2000). In a C&I system that is composed of a number of criteria, each of which contains a number of indicators organised into groups of similar indicators (Figure 1), the basic format of multicriteria analysis is described by equation 1 (Mendoza and Prabhu, 2000).

$$S = \sum_j \sum_k \sum_i (W_{jki} \times s_{jki}) \qquad (1)$$

where,

$S$ is the overall sustainability score,
$j$ is a criterion,
$k$ is a group of indicators in a criterion,
$i$ is an indicator in a group,
$W_{jki}$ is the relative weight for an indicator $i$ in group $k$ in criterion $j$, and
$s_{jki}$ is the individual sustainability score for an indicator $i$ in group $k$ in criterion $j$.

In addition to the calculation of an overall sustainability score, the elements of equation 1 can be used to focus actions on the most critical indicators, including those with the highest weights and those with the lowest and highest sustainability scores. A cumulative impact of all indicators can also be calculated to compare the performance of different management units, or the same management unit at different moments.

FIGURE 1. Criteria and indicators hierarchical structure (adapted from Lammerts van Bueren and Blom, 1997).

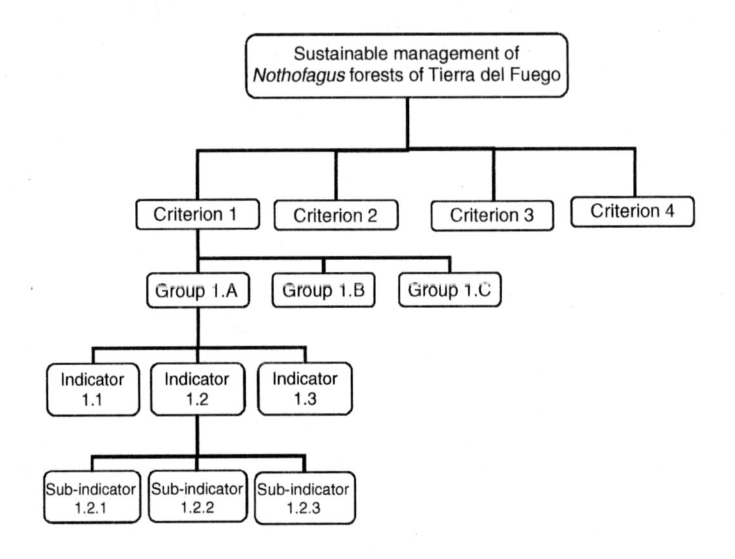


## *METHODOLOGY*

The sustainability score requires two values. First, the relative weights of indicators in a group, groups within a criterion, and the criterion within the overall sustainability score must each be established. Second, a score indicating the degree to which an indicator is contributing to sustainable forestry must be determined for each indicator. The set of criteria was developed in Stage I of the study, the relative weights in Stage II, and the scores indicating the degree of sustainability in Stage III.

### *Stage I: Identification of Criteria and Indicators*

The purpose of Stage I is to create a set of C&I applicable to Tierra del Fuego forests. The process for this was to identify useful indicators from existing C&I sets and use these as a starting point for developing local C&I. The schemes selected as a source of potential indicators were:

- Criteria and Indicators for the Conservation and Sustainable Management of Temperate and Boreal Forests (Montreal Process Working Group, 1998).

- FSC Principles and Criteria (Forest Stewardship Council, 2000).
- The CIFOR Criteria and Indicators Generic Template (CIFOR C&I Team, 1999); and,
- Criteria, indicators, and verifiers for the development of different uses in forested lands (Carabelli et al., 2000).

The indicator selection process was carried out according to the method suggested by Prabhu et al. (1999). This method involves a two-step process using a group of experts who first analyse potential indicators in an individual selection phase, and then work together in a group selection phase where individual decisions are discussed with others. For this study, a group of 10 local experts were selected, aiming for diversity of professions and interests across forestry, tourism, grazing, conservation, policy, and planning. Each expert received a database with 306 indicators, criteria, and principles from the different schemes. They were first asked to classify each C&I according to one of the following categories: (1) policy, planning, and legal framework; (2) social issues; (3) forest management; (4) physical and environmental issues; and (5) economic and finance issues. They were then asked to analyse the usefulness of each indicator by considering four questions.

- Is this indicator closely and unambiguously related to the assessment goal?
- Is it easy to detect, record, and interpret?
- Does it provide a summary or integrative measure?
- Does it have an adequate response range to changes in level of stress?

Each indicator received a score from 1 to 5, with 1 meaning no/bad/unimportant and 5 meaning yes/good/important. Once scored, the experts were then asked to decide whether each indicator should be selected for further consideration or discarded (1 if selected and 0 if discarded). In doing their selections, they had to ensure that all forest aspects were represented, while trying to limit the number of indicators to the average of 75 to 100 indicators in other systems. They were also asked to consider the availability of information, or the technical and economic resources required to generate information to assess a particular indicator. Ideally the final number of indicators should satisfy both conditions of representativeness and feasibility.

The experts were given 10 days to complete the individual phase. The reponses regarding which group an indicator belonged to, its importance score, and whether it should be retained were then collated and analysed by the researchers. The most frequent discipline for each indicator was identified and used to cluster indicators into the five groups. Indicators were then retained or dropped from further consideration on the basis of a mean importance score higher than or equal to 0.5 (50% agreeement for an indicator in a discipline).

The second part of Stage I was carried out in a workshop in November, 2003. The experts were divided into five subgroups corresponding to the categories of indicators, and according to their interests and backgrounds. Each group had to first determine a cutoff point to eliminate less preferred indicators. To do this, each subgroup was supplied with a printed version of all C&I under that category containing the average scores for importance and selection, and all the qualitative comments from each respondent. Secondly, they were asked to review each indicator against another five questions.

- Is this indicator diagnostically specific?
- Does it apply to users?
- Is the indicator precisely defined?
- Will it produce a replicable result?
- How relevant is it?

At the same time, each group was given the freedom to rewrite or combine existing indicators, or create new indicators relevant to Tierra del Fuego. The indicators selected in each subgroup were then compiled, and indicators appearing in more than one category were analysed to remove duplication. The final set of indicators would then be used in Stage II.

### *Stage II: Assessment of the Relative Importance of C&I*

The purpose of this stage is to establish a weight (w) for each indicator, group, and criterion that reflects its relative importance in the sustainable management of *Nothofagus* forests as determined by a range of stakeholders. The weighting was done by using a rating model, similar to the analytical hierarchical process (AHP) originally developed by Thomas Saaty (1980), but mathematically simpler (Taylor, 2002). The AHP and pairwise comparisons were originally thought to be the most appropriate method, but feedback from Stage I participants and other stakeholders suggested the use of a simpler rating method. The relative importance of each indicator was determined by first allocating 100 points between all the indicators in each group of indicators. For example, if there were 5 indicators in a group and each indicator was felt to have the same relative importance, each would be allocated 20 points (100/5). Scores were then divided by 100 to create a percentage weighting. This process was repeated at the group level in each criterion, and at the criterion level. Following Figure 1, the relative weight of an indicator in the complete set of C&I would be the weight of the indicator in its group multiplied by the weight of its group in the criterion and multplied by the weight of its criterion among other criteria. The points for all indicators in a group, all groups in a criterion and all criteria in the sustainability index would thus each sum to 1.0.

### *Stage III: C&I Scores Indicating the Degree of Sustainability*

In the absence of published statistics for quantitative measures that could be used to measure the degree of sustainability in this study, each indicator was given an individual sustainability score based on how close it was perceived to be to its Critical Threshold Value (CTV) of sustainability. The CTV concept is based on the assumption that ecosystems are flexible and capable of absorbing impacts up to a certain limit, but that further stress can degrade the resource (Mendoza et al., 2002). Previous studies suggested that instead of a single target data point as a reference value, a range from more conservative to more liberal values are more appropriate to handle different stakeholder opinions (Wright et al., 2002). Using a Likert scale, each indicator was given a score between 1 and 5, with 1 meaning it is close to this limit and 5 meaning it is far from the CTV. The CTV for some indicators is objective and measurable (area or volume), and data are readily available. In other cases, objective data may not be available or the assessment is inherently more subjective, and the CTV must be determined based on perceptions.

### *Stage II and III Surveys*

The process for determining weightings and individual sustainability scores was to use a written survey that was completed during two half-day workshops held on consecutive days. The first half-day workshop dealt with relative weightings and the second one with individual sustainability scores. The use of workshops was considered to offer advantages over postal surveys in terms of the complicated nature of the process and in ensuring that the process was participatory. At each workshop there was first a discussion of the meaning of sustainable management, followed by distribution of the survey form. Participants were encouraged to discuss the significance and potential impact of C&I on sustainable forest management in groups, but then to complete the survey individually. As suggested by Mendoza et al. (2002), participants were asked to not divulge their responses. Respondents were also told that they did not need to assign a value to an indicator if they did not feel confident about their level of knowledge.

Three, two-day workshops were held in the cities of Tolhuin, Ushuaia, and Rio Grande in Tierra del Fuego, in November and December, 2003. A key objective was to make the selection of participants as wide and comprehensive as possible, covering the forest industry, government, researchers, and the general public. About 3 months before the workshops, individuals in government and non-government organisations (NGOs), universities, and forest companies were contacted and asked about their interest in participating and to identify other stakeholders that may be interested in participating. One month from the workshops, local media (radio and print) and posters were used to advertise the workshops more widely. This process resulted in 31 participants in Ushuaia, 22 in Rio Grande, and 16 in Tolhuin. Of the participants, 25% were from government, 25% from the forest industry, 25% involved in research and education, 10% involved in tourism, and the remainder (15%) from the community or NGOs. When the workshops were held, there was insufficient time to complete the weights at the criteria and group levels (only individual indicators were completed), and a follow-up survey was done by using email. Surveys were able to be sent to 56 of the original participants, of which 13 (23%) provided useable surveys.

## *RESULTS*

### *Criteria and Indicators for Tierra del Fuego*

The objective of this phase was to create a local set of indicators from the combined indicators of four different schemes. By using the process discussed previously, indicators were gradually eliminated, combined with other indicators, or rewritten. The final set of indicators contained 4 criteria, 12 groups, 89 indicators, and 13 sub-indicators. The indicators of Criterion 1, focusing on the legal, institutional, and economic framework for forest conservation and sustainable management, are presented in Table 1. This criterion was organised into three groups covering legal, institutional, and economic aspects. A total of 17 indicators and 8 sub-indicators are included in this criterion, with 10 indicators bcing included in the institutional framework group.

The indicators of Criterion 2 measure the maintenance and enhancement of long-term multiple socioeconomic benefits (Table 2). Five main groups were defined in this criterion, covering employment, manufacture and consumption of forest products, recreation and tourism, education and research, and community needs.

The 18 indicators of Criterion 3 were grouped into one of two categories: biodiversity or protection (Table 3). The maintenance of forest ecosystem integrity (including productivity, biodiversity, and health) is the focal point of this criterion.

Criterion 4 has 23 indicators that deal with planning, monitoring, and assessment of the forest resource use (Table 4). The indicators of this criterion were sorted into one of two groups, long-term or short- to medium-term planning. The indicators in the former group are more focused on the regional level, whereas the latter group emphasized FMU-level indicators.

### *Weights and Individual Sustainabilty Scores for C&I*

Weights for indicators within each criterion are shown in Tables 1, 2, 3 and 4. As was discussed previously, participants were not required to provide a score where they felt they did not have enough information. As a result, each participant provided a mean of 44 weights. The mean number of responses per indicator or sub-indicator was 18, with Criterion 2 (socioeconomic factors) having the most respondents (20) and Criterion 1 having the fewest (16). For most respondents, the indicators in Criterion 2 were the easiest to understand and score.

A summary of the relative importance of the criterion and group levels are shown in Table 5. The relative importance of the four criteria are similar. At the group level, the institutional framework was assessed as substantially more important (44%) than the other two groups in Criterion 1. Community needs (24%) and education and research (22%) groups of indicators were the most important ones in Criterion 2. The two groups of Criterion 3 were evenly weighted. The long-term regional plan group of indicators was higher (60%) than the other group in Criterion 4. These results may be biased because none of the 13 respondents were from the industry sector, although the other three sectors (academics, community, and government) were represented.

TABLE 1. Criterion 1–legal, institutional, and economic framework for forest conservation and sustainable management

| C-G-I | # | Description | Weight (W) | Score (S) |
|---|---|---|---|---|
| **C** | **1** | | **0.27** | **2.44** |
| **G** | **1.A** | **Legal framework (laws, regulations, guidelines)** | **0.28** | **2.63** |
| I | 1.1 | Legal framework protects access to forest and forest resources. | 0.21 | 3.52 |
| I | 1.2 | A permanent forest estate (PFE), which includes both protection and production forests and is the basis for sustainable management, exists and is protected by law. | 0.13 | 2.75 |
| I | 1.3 | The actual legal framework protects access right to the forest resources through a transparent system for the concessions allocation. | 0.16 | 2.18 |
| I | 1.4 | Extent to which the legal framework supports the conservation and sustainable management of forests, including: | 0.30 | 3.04 |
| I | 1.4.1 | Clarifies property rights, provides for appropriate land tenure arrangements and provides means of resolving property disputes by due process. | 0.19 | 2.64 |
| I | 1.4.2 | Provides for periodic forest-related planning, assessment, and policy review that recognizes the range of forest values, coordinating with relevant sectors. | 0.24 | 2.09 |
| I | 1.4.3 | Provides opportunities for public participation in public policy and decision-making related to forests and public access to information. | 0.16 | 2.05 |
| I | 1.4.4 | Encourages best practice codes for forest management. | 0.20 | 2.42 |
| I | 1.4.5 | Provides for the management of forests to conserve special environmental, cultural, social, and/or scientific values. | 0.21 | 2.81 |
| I | 1.5 | The provisions of binding international agreements (CITES, ILO, Montreal Process and Convention on Biological Diversity), are respected. | 0.21 | 2.78 |
| **G** | **1.B** | **Institutional framework** | **0.44** | **2.23** |
| I | 1.6 | Number of government staff assigned to control forest activities is adequate. | 0.08 | 2.07 |
| I | 1.7 | The government efficiently implements the forest management plan and collects stumpage payments. | 0.21 | 1.86 |
| I | 1.8 | The government is equitable and transparent administering the forest resource. | 0.11 | 1.70 |
| I | 1.9 | Institutions responsible for forest research are adequately funded and staffed. | 0.10 | 2.25 |
| I | 1.10 | Anti-corruption provisions have been implemented. | 0.11 | 1.44 |
| I | 1.11 | Nonforestry policies do not distort forest management. | 0.06 | 2.61 |
| I | 1.12 | Extent to which the institutional framework supports the conservation and sustainable management of forests, including the capacity to: | 0.13 | 2.67 |
| I | 1.12.1 | Develop and maintain efficient physical infrastructure to facilitate the supply of forest products and services and support forest management. | 0.53 | 1.90 |
| I | 1.12.2 | Enforce laws, regulations and guidelines. | 0.47 | 1.81 |

| C-G- I | # | Description | Weight (W) | Score (S) |
|---|---|---|---|---|
| I | 1.13 | Governmental organizations promote community awareness and public participation, organize training programs, make forest-related information available and implement periodic forest-related planning and policy reviews. | 0.05 | 1.36 |
| I | 1.14 | Non-governmental organisations promote community awareness and public participation, organize training programs, make forest-related information available and implement periodic forest-related planning and policy reviews. | 0.06 | 3.58 |
| I | 1.15 | Forest conversion is avoided, and if done should be of public priority, have an environmental impact assessment study, and provide long-term benefits. | 0.08 | 3.48 |
| **G** | **1.C** | **Economic framework (policies and measures)** | **0.28** | **2.67** |
| I | 1.16 | Absence of excessive capital mobility (promoting "cut and run"). | 0.44 | 2.91 |
| I | 1.17 | Extent to which the economic framework (economic policies and measures) supports the conservation and sustainable management of forests through: | 0.56 | 2.33 |
| I | 1.17.1 | Non-discriminatory trade policies for forest products. | 1.00 | 2.76 |

Note: Shortened version of the original Spanish text
C-G-I stands for criterion, group and indicator, and # stands for number of indicator

As discussed previously, the individual sustainability score was evaluated by using a value between 1 and 5 (1 = far from sustainability or close to CTV and 5 = close to sustainability). The weights and scores assigned by respondents to each indicator and sub-indicator were averaged and the results are shown in Tables 1, 2, 3 and 4.

### *Overall Sustainability Score Calculation*

An overall sustainability score for the whole set of C&I was calculated by using the weighting results from Stage II and the individual sustainability scores from Stage III shown in Tables 1, 2, 3, and 4. The weight column in each of these tables contains the respondent's answer regarding the relative importance of indicators. The adjusted weight is calculated by multiplying the weight of a component by the weight of the next higher level of the hierarchy. This determines the relative importance assigned to any indicator in relation to the importance of the group and criterion that it belongs to. The score of criterion and group was calculated using the weighted scores of the indicators included in them.

The overall sustainability score was 2.51 out of a maximum possible sustainability score of 5.00 (all indicators and sub-indicators assessed as being close to sustainability would give a score of 5). Since the range

TABLE 2. Criterion 2–maintenance and enhancement of long-term multiple socioeconomic benefits to meet the needs of society

| C-G-I | # | Description | Weight (W) | Score (S) |
|---|---|---|---|---|
| **C** | **2** | | **0.26** | **2.38** |
| **G** | **2.A** | **Employment** | **0.18** | **2.57** |
| I | 2.1 | Number and gravity of working injuries in the main employment categories related to the forest sector diminish from year to year. | 0.12 | 2.58 |
| I | 2.2 | The communities within, or adjacent to, the forest management area are given opportunities for employment, training, and other services. | 0.21 | 3.10 |
| I | 2.3 | Forest management meets or exceed all applicable laws and/or regulations covering health and safety of employees and their families. | 0.12 | 2.02 |
| I | 2.4 | Workers and staff have adequate training to implement management. | 0.13 | 2.44 |
| I | 2.5 | Wages and other benefits conform to national and/or International Labour Organisation (ILO) standards. | 0.13 | 3.10 |
| I | 2.6 | Working injuries are fairly compensated and according to the regulations. | 0.12 | 2.45 |
| I | 2.7 | Stakeholders have a reasonable share of economic benefits from forest use. | 0.17 | 2.31 |
| **G** | **2.B** | **Forest products consumption and production** | **0.21** | **2.36** |
| I | 2.8 | Investments in forest health and management are carried out. | 0.17 | 1.30 |
| I | 2.9 | Investments in timber technology are carried out. | 0.18 | 3.18 |
| I | 2.10 | The various forest products are used in an optimal and equitable way. | 0.16 | 2.33 |
| I | 2.11 | Adequate supply and consumption of wood and wood products. | 0.13 | 2.90 |
| I | 2.12 | The value and quantity of wood products locally consumed in relation to the wood products exported is adequate. | 0.11 | 2.49 |
| I | 2.13 | Extent to which the forest products for export are processed. | 0.15 | 1.48 |
| I | 2.14 | All applicable and legally prescribed fees, royalties, taxes and other charges are paid. | 0.10 | 2.86 |
| **G** | **2.C** | **Recreation and tourism** | **0.14** | **2.79** |
| I | 2.15 | Adequate forest area and facilities available for recreation and tourism. | 0.34 | 2.72 |
| I | 2.16 | Recreation and tourism activities do not affect the resource. | 0.38 | 2.67 |
| I | 2.17 | Investments in recreation and tourism goods and services are carried out. | 0.28 | 2.99 |
| **G** | **2.D** | **Education and research** | **0.22** | **2.25** |
| I | 2.18 | Forest workers receive adequate training and supervision. | 0.21 | 1.97 |
| I | 2.19 | The people are educated about natural resource management. | 0.24 | 2.06 |
| I | 2.20 | Capacity to measure and monitor changes in the management of forests. | 0.28 | 2.38 |
| I | 2.21 | Capacity to conduct and apply research and development aimed at improving forest management and delivery of forest goods and services. | 0.27 | 2.60 |

| C-G-I | # | Description | Weight (W) | Score (S) |
|---|---|---|---|---|
| **G** | **2.E** | **Community needs** | **0.24** | **2.19** |
| I | 2.22 | Opportunities for employment and training for local, forest-dependent people. | 0.11 | 2.28 |
| I | 2.23 | Local people perceive access to forest resources is secure and fair. | 0.10 | 2.24 |
| I | 2.24 | Mechanisms for sharing benefits are seen as fair by local communities. | 0.11 | 1.91 |
| I | 2.25 | People link their future with management of forest resources. | 0.13 | 2.16 |
| I | 2.26 | Destruction of natural resources by local communities is rare and decreases. | 0.13 | 1.99 |
| I | 2.27 | Local stakeholders meet with satisfactory frequency, representation, and quality of interaction. | 0.09 | 1.48 |
| I | 2.28 | Contributions made by all stakeholders are mutually respected and valued. | 0.07 | 1.93 |
| I | 2.29 | Management staff recognise the interests and rights of other stakeholders. | 0.09 | 2.29 |
| I | 2.30 | Level of conflict is acceptable to stakeholders. | 0.08 | 2.64 |
| I | 2.31 | Plans/maps of forest use exist and are made available to the public. | 0.09 | 3.02 |

Note: Shortened version of the original Spanish text
C-G-I stands for criterion, group and indicator, and # stands for number of indicator

of the sustainability score is 1.0 to 5.0 with a midpoint of 3.0, this score is below 50% of the maximum sustainable forest management score. There is no other benchmark to compare this particular score to that would give a sense of whether this score was particularly good or bad, so it provides only a benchmark for future change in Tierra del Fuego.

### *Using the Sustainabilty Scores to Improve Sustainability*

One approach is to use the calculations in the sustainability score to identify the most critical indicator (lowest individual sustainability score) and address the issues it relates to (Mendoza et al., 2002). By using this approach, one would conclude investment in forest health and management to be the centre of efforts because this indicator had the lowest score. However, its relative importance (weight) indicated that it was not the most important of its group of indicators.

Another approach is to consider the whole set of indicators and other sources of information, as well as the score, to identify issues with the current management (Mendoza et al., 2002). This cumulative assessment requires the results of the relative importance assessment of indicators to calculate an overall sustainability score. Because an overall

TABLE 3. Criterion 3–maintenance of forest ecosystem integrity (including productivity, biodiversity, health)

| C-G-I | # | Description | Weight (W) | Score (S) |
|---|---|---|---|---|
| **C** | **3** | | **0.23** | **2.94** |
| **G** | **3.A** | **Biodiversity (ecosystems, species, habitats, genetics)** | **0.48** | **3.07** |
| I | 3.1 | Ecosystem function is maintained. | 0.13 | 2.88 |
| I | 3.2 | Extent of area by forest type and growth phase relative to total forest area. | 0.11 | 2.94 |
| I | 3.3 | Extent of areas by forest type and growth phase in protected areas. | 0.10 | 3.03 |
| I | 3.4 | Population levels of representative species from diverse habitats monitored across their range. | 0.13 | 2.53 |
| I | 3.5 | Ecological functions and values (forest regeneration; succession; genetic, species, and ecosystem diversity; natural cycles that affect the productivity of the forest ecosystem) are maintained intact, enhanced, or restored. | 0.14 | 2.93 |
| I | 3.6 | Levels of genetic diversity are maintained within critical limits. | 0.11 | 3.84 |
| I | 3.7 | Ecologically sensitive areas, especially buffer zones along watercourses, are protected. | 0.15 | 3.33 |
| I | 3.8 | Representative areas, especially sites of ecological importance, are protected and appropriately managed. | 0.14 | 3.05 |
| **G** | **3.B** | **Protection** | **0.52** | **2.83** |
| I | 3.9 | Representative samples of existing ecosystems within the landscape are protected in their natural state, appropriate to the scale and intensity of operations, and the uniqueness of the affected resources. | 0.11 | 3.39 |
| I | 3.10 | The processes that maintain biodiversity in managed forests are conserved. | 0.10 | 2.74 |
| I | 3.11 | Conservation of the processes that maintain genetic variation. | 0.08 | 3.05 |
| I | 3.12 | Change in diversity of habitat as a result of human interventions are maintained within critical limits as defined by natural variation and/or regional conservation objectives. | 0.13 | 2.63 |
| I | 3.13 | The richness/diversity of selected groups show no significant changes. | 0.09 | 2.71 |
| I | 3.14 | Area and percentage of forest affected by processes or agents beyond the natural occurrences, e.g., by fire, storm, wind, beavers, grazing. | 0.11 | 2.75 |
| I | 3.15 | Area and percentage of forest land managed primarily for protective functions, e.g., watersheds, flood protection, avalanche protection, riparian zones. | 0.10 | 3.04 |
| I | 3.16 | Safeguards exist that protect rare, threatened, and endangered species and their habitats. Conservation zones and protection areas have been established, appropriate to the scale and intensity of forest management and the uniqueness of the affected resources. Inappropriate hunting, fishing, trapping and collecting are controlled. | 0.14 | 2.44 |
| I | 3.17 | Population sizes and demographic structures of selected species do not show significant change, and demographically and ecologically critical life-cycle stages continue to be represented. | 0.07 | 2.90 |
| I | 3.18 | Erosion and other forms of soil degradation are minimised. | 0.07 | 2.68 |

Note: Shortened version of the original Spanish text
C-G-I stands for criterion, group and indicator, and # stands for number of indicator

TABLE 4. Criterion 4–planning, monitoring and assessment of the forest resource use

| C-G-I | # | Description | Weight (W) | Score (S) |
|---|---|---|---|---|
| **C** | **4** | | **0.24** | **2.38** |
| **G** | **4.A** | **Long-term regional plan** | **0.60** | **2.17** |
| I | 4.1 | There is a regional land use plan that integrates different forested land uses, and gives attention to social, economic, environmental, and cultural values. | 0.19 | 1.77 |
| I | 4.2 | Forest management planning and implementation takes into account economic variables, as well as the results of the environmental and social impact assessments. | 0.08 | 2.21 |
| I | 4.3 | Annual harvested area of forest and extracted volume of wood products is not over the values stated as sustainable. | 0.08 | 2.15 |
| I | 4.4 | Forest management, processing technology and commercial activities promote the optimum sustainable use of the forest and, generate benefits at the local level. | 0.11 | 2.47 |
| I | 4.5 | New technologies and silvicultural techniques are incorporated to improve forest goods and services, and for the assessment of the socioeconomic consequences. | 0.06 | 2.50 |
| I | 4.6 | Relevant information is available to measure or describe indicators for sustainable forest management and to update forest inventories. | 0.04 | 2.82 |
| I | 4.7 | Methodologies are developed to integrate environmental and social measures into markets, public policies, and national accounting systems. | 0.05 | 2.02 |
| I | 4.8 | Forest management areas are protected from illegal harvesting, settlement, and other unauthorized activities. | 0.11 | 2.18 |
| I | 4.9 | Forest management minimizes waste associated with harvesting and on-site processing operations and avoids damage to other forest resources. | 0.04 | 2.45 |
| I | 4.10 | Annual monitoring is conducted to assess the effectiveness of the measures employed to maintain or enhance the applicable conservation attributes. | 0.10 | 1.81 |
| I | 4.11 | The ñire (*Nothofagus antartica*) forests are well characterised and used according to a plan that guarantees the ecosystem's sustainability. | 0.05 | 1.91 |
| I | 4.12 | The system of concession allocation is transparent. | 0.08 | 1.69 |
| **G** | **4.B** | **Short- to medium-term plans** | **0.40** | **2.55** |
| I | 4.13 | The short- to medium-term management plans: | 0.17 | 2.83 |
| I | 4.13.1 | Clearly provide objectives and a description of the area. | 0.25 | 3.82 |
| I | 4.13.2 | Specify management systems, silvicultural practices, and harvesting techniques. | 0.19 | 3.48 |
| I | 4.13.3 | Specify actions to mitigate environmental impacts during harvesting, road building, or other mechanical disturbances. | 0.22 | 3.21 |
| I | 4.13.4 | Periodically are revised to incorporate the results of monitoring or new information, as well as to respond to changing environmental, social, and economic circumstances. | 0.21 | 1.86 |
| I | 4.13.5 | Are summarised and publicly available. | 0.14 | 1.75 |

TABLE 4 (continued)

| C-G-I | # | Description | Weight (W) | Score (S) |
|---|---|---|---|---|
| I | 4.14 | The management plans are fulfilled. | 0.15 | 2.23 |
| I | 4.15 | Costs, productivity, stumpage, and revenue are seen as reasonable by companies. | 0.07 | 3.22 |
| I | 4.16 | Chemicals, containers, and wastes including fuel and oil are disposed of in an environmentally appropriate manner at off-site locations. | 0.05 | 2.53 |
| I | 4.17 | The forest management has a monitoring system, and includes data for productive, economic, environmental, and social variables. | 0.10 | 2.46 |
| I | 4.18 | Forest use areas and products have been zoned in management units according to a forest inventory, and a Continuous Forest Inventory (CFI) is undertaken. | 0.05 | 2.47 |
| I | 4.19 | Infrastructure is laid out prior to harvesting and in accordance with prescriptions. | 0.09 | 3.01 |
| I | 4.20 | Rehabilitation of degraded and impacted forest is undertaken. | 0.06 | 1.32 |
| I | 4.21 | Systems for production and transformation of forest products are efficient. | 0.10 | 2.35 |
| I | 4.22 | Documentation and record of all forest management and forest activities are kept in forms that enable monitoring. | 0.04 | 2.67 |
| I | 4.23 | Forests under use are protected from animals grazing, fires, etc. | 0.11 | 1.58 |

Note: Shortened version of the original Spanish text
C-G-I stands for criterion, group and indicator, and # stands for number of indicator

sustainability score was calculated by considering the results of 102 indicators, the progress in a single indicator will only result in a very limited improvement.

Another approach would be to use the sustainability scores and weightings at group and criterion levels, and focus efforts on improving the poorest performing groups of indicators. Figure 2 shows the relative performance of each criterion and its groups by using the mean score of indicators in a criterion and group. Criteria 1, 2, and 4 had the lowest mean score, and within these criteria, the institutional framework, community needs, and long-term regional plan groups of indicators had the lowest scores. Both groups of Criterion 3, biodiversity and protection, showed the best performance. In general, the forest ecosystem integrity and conservation indicators were evaluated as performing reasonably well. Short- to medium-term management plans contents and specifications were also scored highly. The lowest scores represented mainly two categories, forest management and resource administration. Three of the lowest indicators are related to ethical concerns in the administra-

TABLE 5. Relative importance of criteria and groups

| Level in the hierarchy | Criterion and group description | Criterion-level weights (%) | Group-level weights (%) |
|---|---|---|---|
| Criterion 1 | Legal, institutional and economic framework for forest conservation and sustainable management | 27 | |
| Group A | Legal framework | | 28 |
| Group B | Institutional framework | | 44 |
| Group C | Economic framework | | 28 |
| Criterion 2 | Maintenance and enhancement of long-term multiple socioeconomic benefits to meet the needs of society | 26 | |
| Group A | Employment | | 19 |
| Group B | Forest products consumption and production | | 21 |
| Group C | Recreation and tourism | | 14 |
| Group D | Education and research | | 22 |
| Group E | Community needs | | 24 |
| Criterion 3 | Maintenance of forest ecosystem integrity (including productivity, biodiversity, health). | 23 | |
| Group A | Biodiversity | | 48 |
| Group B | Protection | | 52 |
| Criterion 4 | Planning, monitoring and assessment of the forest resource use | 24 | |
| Group A | Long-term regional plan | | 60 |
| Group B | Short- to medium-term plans | | 40 |

tion of public forests and a lack of public information on forest management or public participation in management.

In the end, the most important factors affecting sustainable management of *Nothofagus* forest were selected by selecting indicators with a mean individual sustainability score less than or equal to 2.1 or a stakeholder group score less than or equal to 1.5, and by considering the performance of groups of indicators in Figure 2. By using this approach, four main areas of priority were defined: (1) Policy and planning, (2) Resource administration and utilization, (3) Monitoring, and (4) Community (Table 6). The policy and planning issues revolve around the development of management plans and calculation of the sustainable level of timber harvest for the timber industry. Resource administration issues involve government enforcement or activity in maintaining forests through monitoring activities of the timber industry and pastoralists, and monitoring covers government efforts at maintaining adequate in-

FIGURE 2. Criterion and group scores calculated as the mean of their indicators (high scores represent a performance closer to sustainability).

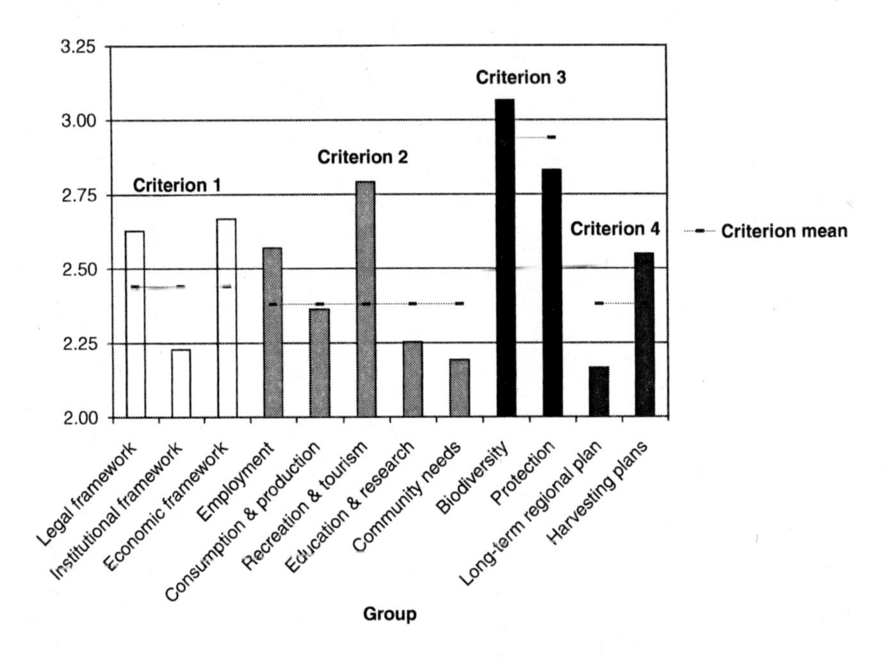


formation to assess sustainability at local and regional levels. At the community level, the key issues are involving a wider range of stakeholders in the management of forests through dissemination of information and public processes.

## *CONCLUSIONS*

Multicriteria analysis was used to incorporate a diverse range of variables related to forest management, such as forest uses and environmental qualities, into an index of sustainability. This involved the development of a list of indicators of sustainability applicable to Tierra del Fuego by using a multidisciplinary group of local experts, and a weighting of the importance of each indicator and a scoring of its degree of sustainability by forest stakeholders.

The results show that current management of *Nothofagus* forests in Tiera del Fuego scores 2.51 out of a maximum score of 5.00, indicating scope for improvement and providing a benchmark for future changes.

TABLE 6. Priorities to improve *Nothofagus* management in Tierra del Fuego

| Theme | Description |
|---|---|
| Policy and planning | • A long-term regional management plan does not exist.<br>• The annual harvested area is greater than the area judged to guarantee a perpetual supply of timber. |
| Resource administration and utilization | • Poor legal enforcement was detected.<br>• Most silvicultural treatments are not carried out.<br>• Timber harvesting and wood processing are conducted carelessly wasting the resource.<br>• Health and safety issues and training are far from ideal.<br>• Silvopastoral systems currently in place in ñire (*Nothofagus antarctica*) forests do not guarantee its sustainability.<br>• Forest rehabilitation is not conducted. |
| Monitoring | • Regional scale monitoring results are not divulged, if carried out.<br>• Forest Management Unit (FMU) monitoring is limited or not undertaken. |
| Community | • Poor information availability and limited diffusion of public information.<br>• The public has very limited, if any, chance to participate in decision-making processes regarding the management of *Nothofagus* forests.<br>• Economic benefits are not fairly distributed. |

By using the weights and scores for individual indicators and a multipronged approach to improving sustainability, we can identify particular areas of focus. The relative weights allow focus on indicators that are more important, and the scores identify indicators that need the most improvement. Although individual changes to each indicator would make only minor impacts on the sustainability score, changes to a number of indicators would have an impact collectively. The results show that the key areas to work on are in policy and planning, resource administration, monitoring of resource use, and public information about and involvement in forest use.

The results show the potential for using multicriteria analysis to address issues of forest sustainability by providing a mechanism for integrating diverse measures of sustainability, which would otherwise present a difficult management problem as a set of data. In this context, the C&I set used in this study, which mirrored the size of sets used by other schemes, is too large, even with an aggregating system like multicriteria analysis. In future work, a set of about 25 to 30 indicators is believed to be more appropriate as a management tool. Similarly, the size of the C&I sets had a flow-on effect on the workshops in terms of the ease with which surveys could be completed. While this may not have had a major impact on the participation of stakeholders in this study, it would be a consideration if planning another study where a larger number of participants was desired.

## REFERENCES

Bava, J.O. 1999. Los bosques de lenga en Argentina. In: pp. 273-296. C. Donoso and A. Lara (eds.). Silvicultura de los Bosques Nativos de Chile. Editorial Universitaria, Santiago de Chile.

Carabelli, F.A. 2002. Una contribución a la planificación del uso múltiple de las áreas boscosas de Tierra del Fuego. Publicación Técnica No. 31. CIEFAP-GTZ, Esquel, Chubut.

Carabelli, F.A., M. Jaramillo, D. Szulkin-Dolhatz and M. Gomez. 2000. Criterios, indicadores y verificadores para el desarrollo de distintos usos de tierras boscosas.: CIEFAP-GTZ-Dirección General de Bosques y Parques, Esquel, Chubut.

Carabelli, F.A., M.M. Jaramillo, D. Szulkin-Dolhatz and M. Gómez. 2003. Management tools for using and preserving natural resources: criteria and indicators for multiple use of forests in Andean Patagonia of Argentina, 2003.

Cellini, J.M., G. Martínez-Pastur, R. Vukasovic, V. Lencinas, B. Díaz and E. Wabo. 2003. Hacia un manejo forestal sustentable en los bosques de Patagonia. Paper presented at the XII Congreso Mundial.

Center for International Forestry Research (CIFOR) C&I Team. 1999. *The CIFOR Criteria and Indicators Generic Template.* CIFOR, Jakarta.

Collado, L. 2001. Los bosques de Tierra del Fuego. Análisis de su estratificación mediante imágenes satelitales para el inventario forestal de la provincia. *Multequina* 10:01-16.

Dimitri, M.J., R.F.J. Leonardis and J.S. Biloni. 1998. El nuevo libro del árbol: especies forestales de la Argentina occidental (3rd ed.). El Ateneo, Buenos Aires.

Food and Agriculture Organization (FAO). 2003. Sustainable forest management. Retrieved on August 21, 2003, from http://www.fao.org/forestry/foris/webview/forestry2/index.jsp?siteId=2000andlangId.

Forest Stewardship Council (FSC). 2000. FSC principles and criteria. Retrieved August 11, 2003, from http://www.fscoax.org/principal.htm.

Forest Stewardship Council. 2004. Data on FSC-certificates. Retrieved June 9, 2004, from http://www.fsc-info.org/default.htm.

Fundación Vida Silvestre Argentina. 2003. El manejo forestal y el FSC.

Gea Izquierdo, G., G. Martínez-Pastur, J.M. Cellini, M.V. Lencinas, I. Mundo, S. Burns [et al.]. 2003. Cuatro décadas de manejo forestal en la Provincia de Tierra del Fuego. (Forty Years of Forest Management in Tierra del Fuego). (*unpublished*).

Gobierno de Tierra del Fuego. 2003. Los recursos forestales de Tierra del Fuego y su industria.

Grupo de Trabajo sobre Criterios e Indicadores del Proceso de Montreal. 2002. Primer Reporte Argentino para el Proceso de Montreal. INTA, SAGPyA, SAyDS, APN. Buenos Aires.

Lamerts van Bueren, E.M. and E.M. Blom. 1997. Hierarchical framework for the formulation of sustainable forest management standards. Tropenbos Foundation, Wageningen, Netherlands.

Lencinas, M.V., G. Martínez-Pastur, J.M. Cellini, C. Busso and E. Gallo. 2003. Manejo forestal sustentable en Patagonia: decisiones basadas en la biodiversidad. Paper presented at the XII Congreso Forestal Mundial.

Martínez-Pastur, G., P.L. Peri, M.C. Fernández, G. Staffieri and D. Rodriguez. 1998. Desarrollo de la regeneración a lo largo del ciclo del manejo forestal de un bosque de *Nothofagus pumilio*: 2. Incidencia del ramoneo de Lama guanicoe. Paper presented at the Primer Congreso Latinoamericano IUFRO.

Martínez-Pastur, G., R. Vukasovic, M.V. Lencinas, J.M. Cellini and E. Wabö. 2003. El manejo silvícola de los bosques Patagónicos: Utopía o realidad? Paper presented at the XII Congreso Forestal Mundial.

Martínez-Pastur, G., M.V. Lencinas, R. Vukasovic, P.L. Peri, B. Diaz, B. & J.M. Cellini. 2004. Turno de corta y posibilidad de los bosques de lenga (*Nothofagus pumilio*) en Tierra del Fuego (Argentina). [Lenga (*Nothofagus pumilio*) forests rotation and harvesting possibilities in Tierra del Fuego (Argentina)]. *Bosque, 25* (1):29-42.

Mendoza, G., H. Hartano, R. Prabhu and T. Villanueva. 2002. Multicriteria and critical threshold value analyses in assessing sustainable forestry: model development and application. *Journal of Sustainable Forestry* 15(2):25-62.

Mendoza, G., P. Macoun, R. Prabhu, D. Sukadri, H. Purnomo and H. Hartano. 1999. Guidelines for applying multi-criteria analysis to the assessment of criteria and indicators. CIFOR, Jakarta.

Mendoza, G. and R. Prabhu. 2000. Multiple criteria decision making approaches to assessing forest sustainability using criteria and indicators: a case study. *Forest Ecology and Management* 131:107-126.

Montreal Process Working Group. 1998. The Montreal Process. Retrieved September 28, 2003, from http://www.mpci.org/evolution_e.html.

Peri, P. L. and G. Martínez-Pastur. 2003. Denominación de origen "Madera Fueguina." Fundación Argen-INTA, Ushuaia.

Prabhu, R., C.J.P. Colfer and R.G. Dudley. 1999. Guidelines for developing, testing and selecting criteria and indicators for sustainable forest management (Toolbox No. 1). CIFOR, Jakarta.

Prabhu, R., C.J.P. Colfer, P. Venkateswarlu, L.C. Tan, R. Soekmadi and E. Wollenberg. 1996. Testing criteria and indicators for the sustainable management of forests. CIFOR, Jakarta.

Pulido, F.J., B. Diaz and G. Martínez-Pastur. 2000. Incidencia del ramoneo del guanaco sobre la regeneración temprana en bosques de lenga de Tierra del Fuego, Argentina. *Investigación Agraria, Sistemas y Recursos Forestales* 9(2):381-394.

Raison, R.J., A.G. Brown and D.W. Flinn. 2001. Criteria and indicators for sustainable forest management. CAB International, Vienna.

Rusch, V., M. Sarasola and P. Laclau. 2001. Sustentabilidad económica y social de las forestaciones en la Región Andinopatagónica. INTA, Bariloche, Argentina.

Saaty, T.L. 1980. The analytic hierarchy process. McGraw Hill, New York.

Schmidt, H., K. Peña and P. Doods. 1998. Estabilidad y efecto del guanaco (*Lama guanicoe*) sobre la regeneración en un bosque de lenga (*Nothofagus pumilio*) en Tierra del Fuego, Chile. Paper presented at the Primer Congreso Latinoamericano IUFRO, Valdivia, Chile.

Taylor, B.W. 2002. *Introduction to Management Science* (7th ed.). Prentice-Hall International, Inc., New Jersey.

Veblen, T.T., R.S. Hill and J. Read. 1996. The ecology and biogeography of *Nothofagus* forests. Yale University Press, Connecticut.

Wright, P.A., G. Alward, J.L. Colby, T.W. Hoekstra, B. Tegler and M. Turner. 2002. Monitoring for forest management unit scale sustainability: the local unit criteria and indicators development (LUCID) test (management edition). USDA Forest Service, Fort Collins, CO.

doi:10.1300/J091v24n01_05

# Increasing and Sustaining Productivity of Tropical Eucalypt Plantations Over Multiple Rotations

K. V. Sankaran
D. S. Mendham
K. C. Chacko
R. C. Pandalai
T. S. Grove
A. M. O'Connell

---

K. V. Sankaran, K. C. Chacko and R. C. Pandalai are affiliated with the Kerala Forest Research Institute, Peechi-680 653, Kerala, India.

D. S. Mendham, T. S. Grove, and A. M. O'Connell are affiliated with Ensis Forests/CSIRO, Private Bag No. 5, Wembley, WA 6913, Australia.

Address correspondence to: K. V. Sankaran at the above address (E-mail: Sankaran_kv@yahoo.com) or D. S. Mendham at the above address (E-mail: Daniel.Mendham@ensisjv.com).

The authors wish to thank the Director of the Kerala Forest Research Institute (KFRI), Dr. J. K. Sharma for his encouragement, and other colleagues in the project, including Dr. Jose Kallarackal and Dr. S. Sankar. The authors also thank technical staff at KFRI for help in the field and laboratory, and at CSIRO for assistance with laboratory analyses.

Support for this work was provided by the Australian Centre for International Agricultural Research (ACIAR), CSIRO Forestry and Forest Products, and the KFRI.

[Haworth co-indexing entry note]: "Increasing and Sustaining Productivity of Tropical Eucalypt Plantations Over Multiple Rotations." Sankaran, K. V. et al. Co-published simultaneously in *Journal of Sustainable Forestry* (Haworth Food & Agricultural Products Press, an imprint of The Haworth Press, Inc.) Vol. 24, No. 2/3, 2007, pp. 109-121; and: *Sustainable Forestry Management and Wood Production in a Global Economy* (ed: Robert L. Deal, Rachel White, and Gary L. Benson) Haworth Food & Agricultural Products Press, an imprint of The Haworth Press, Inc., 2007, pp. 109-121. Single or multiple copies of this article are available for a fee from The Haworth Document Delivery Service [1-800-HAWORTH, 9:00 a.m. - 5:00 p.m. (EST). E-mail address: docdelivery@haworthpress.com].

Available online at http://jsf.haworthpress.com

doi:10.1300/J091v24n02_01

**SUMMARY.** Eucalypt plantations in India are an important source of fiber for paper making and fuel for local villagers. Large areas of land have supported eucalypt plantations for several rotations, and productivity has generally been declining through successive rotations. In 1997, we initiated a project to examine site management options as a way to improve the productivity of these sites. We established a large experimental infrastructure at each of four sites in Kerala, consisting of up to five separate fully randomized block experiments at each site, examining inter-rotation management options (organic matter manipulation, N and P fertilizer input, legume cover cropping, weed control). Two of the sites were typical lowland plantations with *Eucalytus tereticornis*, and the other two sites were typical upland plantations with *Eucalytus grandis*. Following treatments, we monitored plantation productivity and impacts on soil and nutrient cycling for one full rotation (6.5 years). We found it is possible to increase the volume growth of *E. grandis* by up to 48%, and *E. tereticornis* by up to 268% through a combination of optimum site practices (mostly weed control and nutrient addition), but productivity responses are dependent on site-specific factors. Key outcomes of this research are presented, with special reference to application in the broader context of tropical eucalypt plantations. doi:10.1300/J091v24n02_01 *[Article copies available for a fee from The Haworth Document Delivery Service: 1-800-HAWORTH. E-mail address: <docdelivery@haworthpress.com> Website: <http://www.HaworthPress.com>* 

**KEYWORDS.** Eucalypt plantations, organic matter, harvest residues, biomass, soil carbon, *Eucalyptus grandis*, *Eucalyptus tereticornis*

## *INTRODUCTION*

Plantation forestry has expanded in the tropics and subtropics during the past two decades. This expansion has been accelerated by increasing demand for wood and wood products. Many countries are now committed to developing plantation forestry and wood-based industries as an integral part of their regional and national economic development (Nambiar, 1996). India is no exception to this. The most recent statistics indicate that the country has over 8 million ha of land under short-rotation eucalypts (FAO, 2001). However, productivity of these plantations has been relatively low (averaging less than 10 $m^3$ $ha^{-1}$ $yr^{-1}$) for most parts of the country. In Kerala State, southwest India, the main eucalypt plantation species used are *Eucalyptus tereticornis* Sm. (grown at lower elevations) and *E. grandis* Hill ex Maiden (grown at 500 to 2000 m asl), the former occupying 60% and the latter 40% of the area under eucalypts. These plantations are generally grown on a 6-year rotation. Poor productivity has resulted in abandoning eucalypt cultivation in certain parts of the state, decreasing the total area of state-owned plantations from 40,000 ha (in the 1990s) to 24,500 ha (early 2000). However, cultivation of eucalypts by small farmers is increasing. Although the areas in Kerala are relatively small, there are much more extensive areas under eucalypt plantations in neighbouring states in southern India.

The current annual demand for eucalypt pulpwood in Kerala is about 150,000 t, but there is a proposal to expand the capacity of the operations of Hindustan Newsprint Ltd. (the main stakeholder), with demand forecast to increase to 350,000 t $yr^{-1}$. It is clear that there is already a mismatch between wood supply and demand and it is essential to enhance the productivity at low-yielding sites, maintain high rates of productivity in the longer term, and increase the area of plantations managed by small-holder farmers.

Reasons for low productivity of eucalypts in Kerala are probably use of poor planting stock, weed competition, incidence of fungal diseases, water stress, and low nutrient status of soils, exacerbated by successive rotations with burning slash and minimal nutrient inputs (Sharma et al. 1985, Ghosh et al. 1989, Kallarackal and Somen 1997). In this context, the present study aimed to evaluate management of site resources (soil organic matter, nutrients, and available water) to improve productivity and maximize the profitability of eucalypt plantations in Kerala. We examined short-term (single rotation up to 6.5 years) and likely long-term impacts of several cultural treatments on eucalypt productivity at four sites. Also studied, but not reported in this paper were additional experiments regarding thinning and trenching.

## METHODS

### Study Sites

Four main experiments were established in Kerala, southwest India. Rainfall, previous land use and soil characteristics at the four experimental sites are shown in Table 1. The climate of the region is tropical warm humid with two monsoon periods. The southwest monsoon (the main monsoon) starts in early June and extends till October. The northeast monsoon brings occasional rains from December to February. The dry season begins in March and continues through May. Mean atmo-

TABLE 1. Selected properties of the experimental sites and characteristics of surface (0 to 10 cm) soil.

| | Kayampoovam (*E. tereticornis*) | Punnala (*E. tereticornis*) | Surianelli (*E. grandis*) | Vattavada (*E. grandis*) |
|---|---|---|---|---|
| Rainfall (mm $y^{-1}$) | 2700 | 2000 | 3000 | 1800 |
| Altitude (m) | 120 | 150 | 1280 | 1800 |
| Previous land use | Moist deciduous forest | Moist deciduous forest | Grassland | Semi-evergreen forest |
| Previous rotations | 1 seedling, 1 coppice | 1 seedling, 1 coppice | 2 seedling, 2 coppice | 2 seedling, 2 coppice |
| Soil texture | Light to medium clay | Sandy loam to clay loam | Medium clay to sandy loam | Clay loam to medium clay |
| pH (1:5 water) | 5.3 | 5.1 | 4.8 | 5.3 |
| Total C (mg $g^{-1}$) | 21.5 | 43.6 | 40.9 | 52.3 |
| Total N (mg $g^{-1}$) | 1.83 | 2.89 | 2.49 | 4.50 |
| Total P (mg $g^{-1}$) | 0.60 | 0.40 | 0.55 | 0.75 |

spheric temperature is 27°C (range 20 to 42°C) and relative humidity ranges between 64% (February-March) and 93% (June-July).

We established four study sites, two planted with *Eucalyptus tereticornis* and two with *E. grandis*. The parent material of soils at these sites was saprolite or saprolitic colluvium derived from Precambrian granites and gneiss. The soils are broadly classified as ferralsols (see Sankaran et al. 2000). The two *E. tereticornis* sites (Kayampoovam and Punnala) are located in the foothills of Western Ghats adjacent to the coastal plain. *Eucalyptus* was first planted at these sites in 1977 after clearing of degraded moist deciduous forest. The first rotation of eucalypt was harvested in 1991 and the coppice crop was harvested in early 1998. The two *E. grandis* sites (Surianelli and Vattavada) are located in the high ranges of the Western Ghats. The site at Surianelli was first planted in 1968. After three rotations of the first crop, the site was replanted in 1991. At Vattavada, the first plantation was established in 1958. Here again, after three rotations of the crop, the site was replanted in 1991. Stands at all sites were harvested in May-July 1998.

### *Experimental Design*

After harvesting the existing stands, the four key experiments (organic matter manipulation, N fertilization, P fertilization, and weed control) were established at all sites, and an additional legume intercropping experiment was established at the lowland (*E. tereticornis*) sites. Establishment took place during June-September 1998. A summary of experiments and applied treatments is provided in Table 2. Also installed, but not reported on in this paper were a thinning experiment (*E. grandis*), and a trenching experiment at 3 sites (not at Kayampoovam). Each experiment was a randomized block design with 3 to 6 treatments and four replicates. The plot size is 20 m × 20 m, tree spacing 2 m × 2 m (2500 stems $ha^{-1}$) with 100 trees per plot (36 measurement trees). At Kayampoovam, 18 m × 18 m plots were used owing to the restricted available area. The *E. grandis* sites were thinned to 1667 stems $ha^{-1}$ in May-June 2000. The *E. tereticornis* sites remained unthinned for the whole rotation. In the N addition experiment, five rates of N (urea) were applied with a basal addition of other major and minor nutrients. Likewise, P (super phosphate) was also applied at five rates. Basal dressing of P in the N experiment was 63 kg $ha^{-1}$ and the basal dressing of N in the P experiment was 187 kg $ha^{-1}$ (both of these being the 2nd from the top rate in the fertilizer experiments). The legume understorey treatments in the *E. tereticornis* sites were the perennial *Mucuna bracteata* DC. and *Pueraria phaseoloides* (Roxb.) Benth. and the annual *Stylosanthes hamata* Taub. Additional P fertilizer (42 kg P $ha^{-1}$) was applied to all plots in the legume experiment. The slash remaining at each site comprised the leaf, bark, branch, and stem material with a diameter greater than 2 cm over bark. This material was redistributed as uniformly as possible across each of the experiments and burnt

TABLE 2. Experimental site and treatment matrix.

| Experiment | Sites | Treatments |
|---|---|---|
| Organic matter[a] | All 4 sites | Slash burned (**BS**), zero slash (**0S**), single slash (**1S**), double slash (**2S**), leaf slash only (**L**), and burn without starter fertilizer (**B**) |
| N fertilizer[b] | All 4 sites | 5 rates of N: 0, 18, 60, 187, 375 kg/ha |
| P fertilizer[c] | All 4 sites | 5 rates of P: 0, 6.3, 21, 63, 131 kg/ha |
| Weed control | All 4 sites | No weed control except around tree base (**NW**), 1 m strip weed control (**SW**), complete weed control (**CW**). |
| Legume intercropping | 2 *E. tereticornis* sites only | *Mucuna* (**M**), *Pueraria* (**P**), *Stylosanthes* (**S**) or control (C, equivalent to the CW treatment) |

[a]All treatments except B received starter fertilizer (100 g $tree^{-1}$ N:P:K, 17:7:14).
[b]N fertilizer rates applied in years 1 and 2, and at 50% (*E. tereticornis*) and 33% (*E. grandis*) of this rate in year 4.
[c]Total P fertilizer applied, split over 4 doses in the first 2 years.

(equivalent to the BS treatment, as per standard practice) for all plots except those treatments where we were manipulating the slash. Double slash treatments were installed by distributing the slash material from the zero slash plot onto the double slash plot within each replicate block. All experiments were maintained in complete weeded condition, except for those plots in the weed control experiment where we were exploring other weeding options.

### Tree Growth Measurements

Tree stem diameter and height were measured at 0.25, 0.5, 1, 1.5, 2, 3, 4, 5.3, and 6.5 years at each of the sites. Stem volume ($v$) was calculated as the volume of a cone according to the equation $v = 1/3\pi r^2 h$, where $r$ was the radius of the tree at ground level (predicted from the diameter measured at a known height) and $h$ was the height of the tree.

### Statistical Analyses

Analysis of variance was used to assess the significance of treatment effects on tree growth data at each measurement. Least significant differences were calculated from the ANOVA outputs by using a probability level of $p = 0.05$. Where treatments were compared across experiments, and thus not statistically correct to analyze, the standard error of the mean (SEM) of the observations is presented.

## RESULTS

### Slash Manipulation and Legume Intercropping

Impacts of biomass removal on site nutrient export at these sites have previously been presented in detail by Sankaran et al. (2005). A synthesis of the quantities of nutrients exported under different levels of biomass removal reported in that paper is shown in Table 3. Significant quantities of all of the major nutrients can be exported from the sites, especially in situations where local communities use the majority of the residue biomass, which is a common practice at some of these sites. Site nutrient export rates are higher than natural replenishment rates (Sankaran et al., 2005), resulting in a predicted net decline in site nutrient capital, although we found that harvest residue retention did not significantly

TABLE 3. Cumulative export of nutrients (kg/ha) with increasing biomass removal scenarios.

| Site | Nutrient | Stem Only | + Branches | + Bark | + Understorey | + Leaves |
|---|---|---|---|---|---|---|
| Kayampoovam | N | 72 (0.77) | 114 (1.22) | 149 (1.59) | 225 (2.40) | 358 (3.82) |
| | P | 19 (0.22) | 28 (0.31) | 38 (0.43) | 46 (0.53) | 56 (0.63) |
| | K | 83 (10.01) | 130 (15.67) | 200 (24.05) | 282 (33.91) | 334 (40.12) |
| | Ca | 83 (0.95) | 227 (2.60) | 393 (4.51) | 460 (5.27) | 627 (7.18) |
| | Mg | 11 (0.63) | 24 (1.37) | 42 (2.34) | 60 (3.39) | 89 (5.02) |
| Punnala | N | 45 (0.47) | 61 (0.64) | 82 (0.86) | 142 (1.49) | 247 (2.58) |
| | P | 3 (0.11) | 5 (0.18) | 8 (0.26) | 13 (0.44) | 20 (0.68) |
| | K | 35 (5.56) | 49 (7.84) | 73 (11.64) | 139 (22.28) | 175 (27.89) |
| | Ca | 46 (4.20) | 97 (8.79) | 176 (15.94) | 238 (21.53) | 349 (31.59) |
| | Mg | 3 (0.68) | 10 (1.96) | 22 (4.44) | 45 (9.33) | 72 (14.75) |
| Surianelli | N | 24 (0.14) | 62 (0.38) | 75 (0.46) | 197 (1.21) | 344 (2.12) |
| | P | 12 (0.23) | 16 (0.30) | 20 (0.37) | 30 (0.56) | 39 (0.75) |
| | K | 38 (7.67) | 62 (12.47) | 79 (15.80) | 164 (32.89) | 205 (41.19) |
| | Ca | 58 (5.85) | 221 (22.39) | 447 (45.27) | 523 (52.99) | 674 (68.26) |
| | Mg | 9 (3.05) | 27 (9.48) | 45 (15.81) | 80 (28.52) | 116 (41.09) |
| Vattavada | N | 77 (0.47) | 130 (0.79) | 162 (0.98) | 211 (1.28) | 338 (2.05) |
| | P | 12 (0.14) | 18 (0.21) | 23 (0.27) | 30 (0.34) | 38 (0.44) |
| | K | 105 (3.47) | 165 (5.44) | 232 (7.63) | 295 (9.72) | 358 (11.81) |
| | Ca | 167 (1.47) | 335 (2.95) | 755 (6.65) | 784 (6.90) | 882 (7.77) |
| | Mg | 19 (0.99) | 40 (2.05) | 72 (3.72) | 78 (4.03) | 102 (5.26) |

Note: Values in parentheses are percentage of site pools (including biomass and soil to 1 m depth) of total N and P, and exchangeable cations (adapted from Sankaran et al. 2005).

affect plantation productivity during the first rotation after treatment (Table 4).

Legume intercropping is another practice with potential for improving site productive capacity by increasing the quality of soil organic matter and N input through N fixation. At Punnala, legume establishment initially depressed *E. tereticornis* growth to 18 months, but by age 5 and 6, the *Pueraria* and *Stylosanthes* treatments both had significant positive impacts on standing volume (Figure 1b), with an additional 20 $m^3\ ha^{-1}$ (25% improvement) over the control treatment at rotation end. Legume intercropping had no significant impact on growth of *E. tereticornis* at Kayampoovam (Figure 1a).

TABLE 4. Slash treatment impacts on 6-year productivity ($m^3$ $ha^{-1}$) at each of the experimental sites.

| Slash Treatment | Kayampoovam | Punnala | Surianelli | Vattavada |
|---|---|---|---|---|
| Zero | 113.6 | 79.4 | 182.7 | 374.0 |
| Single | 108.8 | 78.2 | 199.5 | 328.4 |
| Double | 111.5 | 90.2 | 208.1 | 368.5 |
| Leaf | 111.4 | 91.2 | 192.6 | 372.6 |
| Burn | 106.9 | 96.4 | 170.7 | 354.6 |
| Significance | ns | ns | ns | ns |
| Mean | 110.5 | 87.1 | 190.7 | 359.6 |

FIGURE 1. Plantation volume response to legume intercropping. Where treatment effects were significant, the level of significance is shown by * P < 0.05; ** P < 0.01.

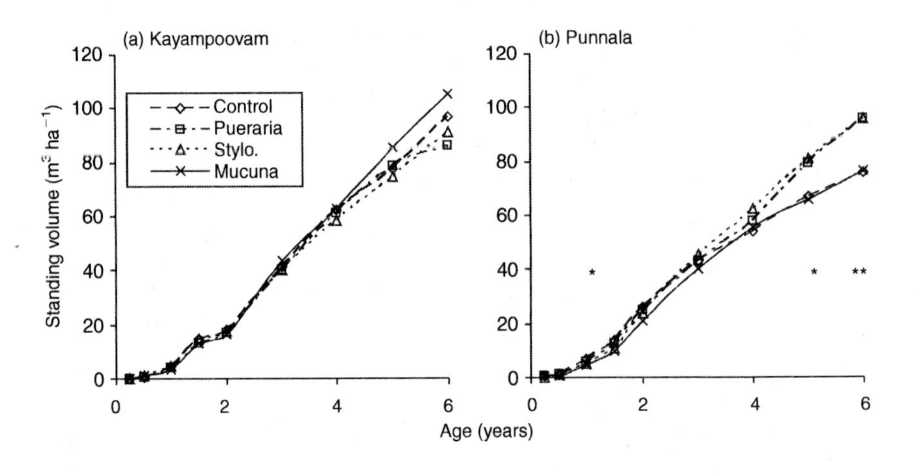


## *Weeding and Fertilizer Impacts on Productivity*

We found that nonweeded treatments always had the lowest productivity at all sites (Figure 2), and productivity increased with intensification of weeding practices. Partial weeding only marginally improved productivity over the nonweeded treatment. Fertilization with N and P had variable impacts on productivity, with only a small (nonsignificant) response at Vattavada, and a much larger response at Surianelli. The Kayampoovam site was not responsive to P fertilizer because of the high soil P status (Table 1). Diagnostic criteria are required for accurate

FIGURE 2. *Eucalyptus* plantation response to management across a range of cultural treatments. Note that we have drawn the treatments from several experiments so statistical comparison within each site is not possible. Bars represent the standard error.

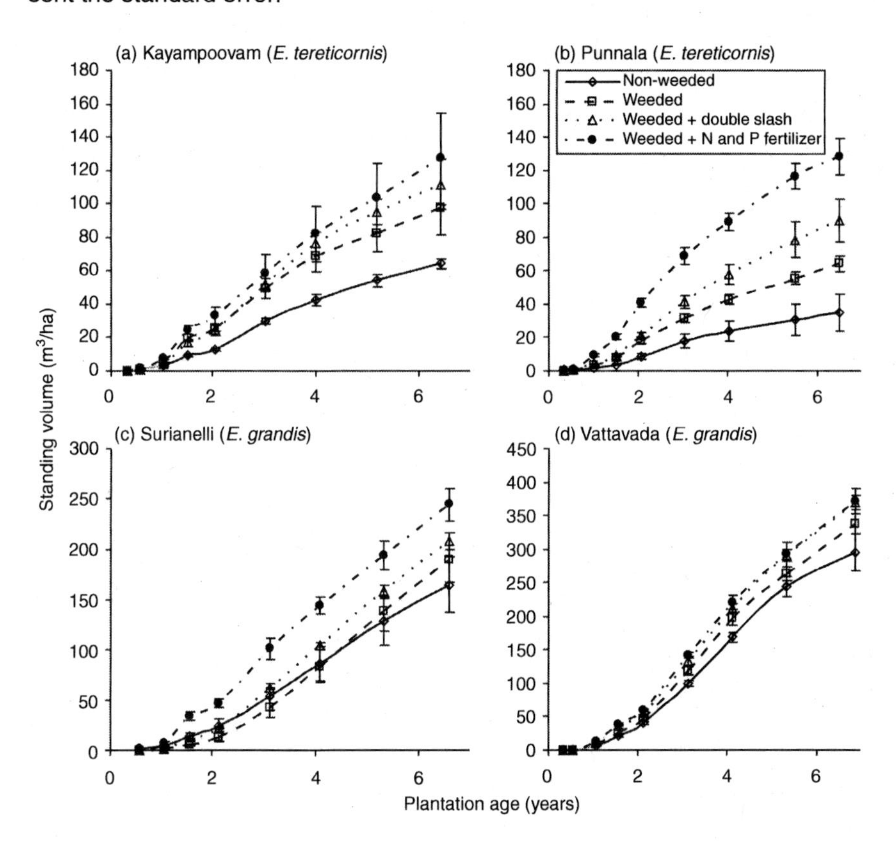


prediction of response to N and P. Across the sites, we found that the order of productivity in these treatments was nonweeded < weeded = weeded + slash retained < weeded + fertilized, with an average 73% increase in the weeded + fertilized treatment compared with the nonweeded control treatment.

## DISCUSSION

Cultural treatments can make a large difference in the productivity of eucalypt plantations in India. It is imperative that these productivity gains are realized to improve the livelihoods of rural communities that rely on these plantations. We consider that conservative site practices such as residue retention will be beneficial to eucalypt plantations, both in the short and longer term. Organic residue retention has potential to minimize depletion of the nutrient status of soils (Table 3), as well as suppress weed growth, retain moisture and reduce erosion and sedimentation (Haywood et al. 2003). Although slash retention at these sites did not affect growth in the current rotation, tree productivity enhancement owing to slash retention has been reported in several countries (e.g., *E. grandis* in Brazil and South Africa, *E. urophylla* in China, and pine species in Queensland) (Mendham et al. 2003, du Toit et al. 2004, Gonçalves et al. 2004, Simpson et al. 2004, Xu et al. 2004). The nonsignificant effect at our sites may in part be caused by the application of a starter fertilizer, and the minimal fertility effect of the small amount of slash (owing to the low productivity of the previous crop). The spatial distribution of slash also means that tree roots are not able to access nutrients released from the slash until they have developed and colonized the majority of the site. Nevertheless, given the large quantities of nutrients in slash and the potential to markedly increase productivity on these sites, it is likely that retention of harvest residues will have an increasingly significant and cumulative effect on tree growth, especially in the absence of nutritional supplementation.

We have shown that legume intercropping has good potential to deliver multiple and sustained benefits to plantation growers, including increased productivity of eucalypt plantations growing in N-enhanced soils, as well as the potential for food and fodder production in the understorey. The impact of legumes on these sites will be cumulative, with a benefit over several rotations. Legumes also help control weeds and reduce the weeding cost, especially if they are grazed with livestock. The critical management issue with legumes is the judicious choice of cover-crop species, as the ideal species needs to have an annual or creeping habit. Climbing legume species such as *Mucuna* may smother trees if not controlled regularly. Ideally, a legume cover crop will improve the soil N status, be persistent in shade, and be palatable to livestock. The lack of response to legumes at Kayampoovam may have been due to water limitation at that site where there are shallow soils and also to the relatively small response of trees to N.

Variable responses to management at the four experimental sites (Figure 2) indicate the inherent differences in the physical and chemical characteristics of soils between the sites, and perhaps interaction with climate. Of all the site management options evaluated, weeding in combination with nutrient application generally had the most significant impact on productivity. Nutrient application was more important for enhancing the productivity of eucalypts in nutrient-poor soils such as those at Surianelli and Punnala. The variability in responses across the sites indicates that nutrient application needs to be targeted to sites that are likely to be responsive. Further work is required on diagnostics for determining which sites will be responsive to applied fertilizer, and a decision-support system needs to be developed to allow farmers to apply these diagnostics.

Weed control is widely practiced in plantation forestry across the world, but it has not been possible to tap its full potential in eucalypt plantations in Kerala owing to inefficient implementation practices. Social and political issues prohibit herbicide application in several parts of India, including Kerala, so weed control depends on mechanical or manual methods. However, these methods have seldom been practiced in a thorough manner. We have demonstrated that effective weed control can result in dramatically improved productivity at the two lowland *E. tereticornis* sites. The mechanism for improved productivity through weed control is probably different at these two sites. At Kayampoovam, water is the main resource constraining plantation productivity, so weeds competed for water. At Punnala, the soils have lower inherent fertility (demonstrated by the higher C:N ratio of 15 compared to 11 at Kayampoovam, and also a strong response to N fertilizer), so weeds would have competed mainly for nutrients at that site. Both of our experiments have shown that systematic management of weeds in *E. tereticornis* results in huge benefits.

The *E. grandis* sites had a lower relative response to weed management. The cause of this may also have been different at the two sites. At Vattavada, the lack of response was due to low weed biomass associated with early canopy closure of the plantations owing to high inherent fertility and available water. Given that weeds at Surianelli, especially grasses, were a major problem, and that plantations also responded to nutrient application during the early stages, the lack of a response to weeding was surprising. At this site, it is possible that the lack of observed response may be an artifact of the experimental design, because the experimental plots were subject to wild elephant damage. Weed-retained treatments were damaged less because the weeds impeded elephant movement through those experimental plots.

## *CONCLUSION*

We have shown that a number of cultural practices can be employed to increase productivity of eucalypts in India, and that a combination of such practices will be important for improving and sustaining the productivity of these plantations. Organic-matter-conservative treatments including harvest residue retention and legume cover cropping produced small or no responses when implemented in the initial rotation cycle, but their cumulative effect will probably be important for maintaining and improving the fertility of soils and the productive potential of the sites in the long term. Practices of weeding and balanced nutrient application provide a means to optimize production and maximize the profitability of the current plantation cycle, provided that managers have the tools to determine the appropriate level of management inputs. Such tools would need to discriminate between responsive and non-responsive sites for both nutrient and weed management.

## REFERENCES

du Toit, B., S.B. Dovey, G.M. Fuller and R.A. Job. 2004. Effects of harvesting and site management on nutrient pools and stand growth in a South African eucalypt plantation. In: E.K.S. Nambiar, J. Ranger, A. Tiarks and T. Toma (eds.). Site management and productivity in tropical plantation forests: proceedings of workshops in Congo, July 2001, and China February 2003. CIFOR, Bogor, Indonesia.

Food and Agriculture Organization [FAO]. 2001. Global Forest Resources Assessment 2000. FAO Forestry Paper 140. Food and Agriculture Organization of the United Nations, Rome.

Ghosh, S.P., B.M. Kumar, S. Kabeerathumma and G.M. Nair. 1989. Productivity, soil fertility and soil erosion under cassava-based agroforestry systems. *Agroforestry Systems* 8: 67-82.

Gonçalves, J.L., J.L. Stape, J.P. Laclau, J.L. Gava and P.J. Smethurst. 2004. Silvicultural effects on the productivity and wood quality of eucalypt plantations. *Forest Ecology and Management* 193: 45-61.

Haywood, J.D., J.C. Goelz, M.A. Sword Sayer and A.E. Tiarks. 2003. Influence of fertilization, weed control, and pine litter on loblolly pine growth and understory development through 12 growing seasons. *Canadian Journal of Forest Research* 33:1974-1982.

Kallarackal, J. and C.K. Somen. 1997. Water use by *Eucalyptus tereticornis* stands of differing density in southern India. *Tree Physiology* 17: 195-203.

Mendham, D.S., A.M. O'Connell, T.S. Grove and S.J. Rance. 2003. Residue management effects on soil carbon and nutrient contents and growth of second rotation eucalypts. *Forest Ecology and Management* 181:357-372.

Nambiar, E.K.S. 1996. Sustained productivity of forests is a continuing challenge to soil science. *Soil Science Society of America Journal* 60:1629-1642.

Sankaran, K.V., K.C. Chacko, R.C. Pandalai, J. Kallarackal, C.K. Somen, J.K. Sharma, M. Balagopalan, M. Balasundaran, S. Kumaraswamy, S. Sankar, R.J. Gilkes, T.S. Grove, D. Mendham and A.M. O'Connell. 2000. Effects of site management on *Eucalyptus* plantations in the monsoonal tropics-Kerala, India. In: pp. 51-60. E.K.S. Nambiar, A. Tiarks, C. Cossalter and J. Ranger (eds.), Site management and productivity in tropical plantation forests: a progress report. Center for International Forestry Research, Bogor, Indonesia.

Sankaran, K.V., T.S. Grove, S. Kumaraswamy, V.S. Manju, D.S. Mendham and A.M. O'Connell. 2005. Export of biomass and nutrients following harvest of eucalypt plantations in Kerala, India. *Journal of Sustainable Forestry* 20:15-36.

Sharma, J.K., C. Mohanan and E.J.M. Florence. 1985. Disease survey in plantations of forest tree species grown in Kerala. KFRI Research Report No. 36. Kerala Forest Research Institute, Peechi, Kerala, India.

Simpson, J.A., T.E. Smith, P.T. Keay, D.O. Osborne, Z.H. Xu and M.I. Podberscek. 2004. Impacts of inter-rotation site management on tree growth and soil properties in the first 6.4 years of a hybrid pine planatation in subtropical Australia. In: pp. 139-149. E.K.S. Nambiar, J. Ranger, A. Tiarks and T. Toma (eds.). Site management and productivity in tropical plantation forests: proceedings of workshops in Congo, July 2001, and China February 2003. CIFOR, Bogor, Indonesia.

Xu, D.P., Z.J. Yang and N.N. Zhang. 2004. Effects of site management on tree growth and soil properties of a second-rotation plantation of *Eucalyptus urophylla* in Guandong Province, China. In: pp. 45-60. E.K.S. Nambiar, J. Ranger, A. Tiarks and T. Toma (eds.). Site management and productivity in tropical plantation forests: proceedings of workshops in Congo, July 2001, and China February 2003. CIFOR, Bogor, Indonesia.

doi:10.1300/J091v24n02_01

# Structural and Biometric Characterization of *Nothofagus betuloides* Production Forests in the Magellan Region, Chile

Gustavo Cruz Madariaga
Raúl Caprile Navarro
Álvaro Promis Baeza
Gustavo Cabello Verrugio

**SUMMARY.** This study, in the Chilean Magellan Region, undertook the structural and biometric characterization of *Nothofagus betuloides* (Mirb.) Blume (coihue de Magallanes) production forests to provide in-

---

Gustavo Cruz Madariaga is Faculty and Researcher, Departamento de Silvicultura, Facultad de Ciencias Forestales, Universidad de Chile, Casilla 9206, Santiago, Chile (E-mail: gcruz@uchile.cl).

Raúl Caprile Navarro is Research Assistant, Departmento de Silvicultura, Facultad de Ciencias Forestales, Universidad de Chile, Casilla 9206, Santiago, Chile (E-mail: rcaprile@uchile.cl).

Álvaro Promis Baeza is Faculty and Researcher, Facultad de Ciencias Forestales, Universidad de Chile (E-mail: alvaro.promis@waldbau.uni-freiburg.de).

Gustavo Cabello Verrugio is Resarch Assistant, Departmento de Silvicultura, Facultad de Ciencias Forestales, Universidad de Chile, Casilla 9206, Santiago, Chile (E-mail: gcabello@uchile.cl).

The authors thank the National Commission for Scientific and Technological Research (CONICYT) for its financial support through the Chilean FONDEF Project D02I1080 "Incorporation of *Nothofagus betuloides* forests to forest management for the purposes of diversification and production increase in Region XII (Chile)."

[Haworth co-indexing entry note]: "Structural and Biometric Characterization of *Nothofagus betuloides* Production Forests in the Magellan Region, Chile." Cruz Madariaga, Gustavo et al. Co-published simultaneously in *Journal of Sustainable Forestry* (Haworth Food & Agricultural Products Press, an imprint of The Haworth Press, Inc.) Vol. 24, No. 2/3, 2007, pp. 123-140; and: *Sustainable Forestry Management and Wood Production in a Global Economy* (ed: Robert L. Deal, Rachel White, and Gary L. Benson) Haworth Food & Agricultural Products Press, an imprint of The Haworth Press, Inc., 2007, pp. 123-140. Single or multiple copies of this article are available for a fee from The Haworth Document Delivery Service [1-800-HAWORTH, 9:00 a.m. - 5:00 p.m. (EST). E-mail address: docdelivery@haworthpress.com].

Available online at http://jsf.haworthpress.com

doi:10.1300/J091v24n02_02

formation for sustainable forest management. Six locations were surveyed, five on the continent and one on Tierra del Fuego Island. A total of 6,102 ha were photointerpreted and then checked and described in the field according to their location, environmental characteristic, and vegetation type. The degree of disturbance, stage of development, status of regeneration, and composition and cover of the understory were also described. From the surveyed locations, a total of 3,807 ha were production forests of which a 76% area had *Nothofagus pumilio* (lenga) and *N. betuloides* mixed forests. The remaining areas (24%) were pure *N. betuloides* forests. Nondisturbed, old-growth forests covered 47% of the surveyed area. The remaining area (53%) had some disturbance. In the nondisturbed forests, mean volume stocks were 438 $m^3$ $ha^{-1}$, and in disturbed forests, mean volume stocks were 316 $m^3$ $ha^{-1}$. doi:10.1300/J091v24n02_02 

**KEYWORDS.** Coihue de Magallanes, Magellan Region, *Nothofagus betuloides*, production forest

## *INTRODUCTION*

In terms of area covered, *Nothofagus betuloides* (Mirb.) Blume (coihue de Magallanes) is one of the most important species in the Chilean Magellan forests. Natural forests of *N. betuloides* cover about 1,037,960 ha in the Chilean Magellan Region. It is the second most important forest species after *Nothofagus pumilio* (Poepp. *et* Endl.) Krasser (lenga) (CONAF-CONAMA, 1999a).

There are about 280,000 ha of pure or mixed forests covered with *N. betuloides* available for commercial management for timber production. However, due to the difficult accessibility, the lack of knowledge about management, and problems in the wood drying, this species is of marginal use in the Chilean Magellan Region (CONAF-CONAMA, 1999b; Cruz et al., 2004). In addition, forests covered with *N. betuloides* have not been studied for purposes of commercial production. They have been managed only on an experimental scale (Silva, 1997; Nuñez and Salas, 2000; Martínez-Pastur et al., 2002), and rarely at an operational scale by companies, as a result of the lack of information about the potential for timber production, the area covered, forest structure, and silviculture.

Moreover, any increase in industrial consumption as a result of either an increase in industrial scale or the development of new projects under a sustainable management scheme, would necessarily imply an increase in the area of managed forests and an efficient management of timber resources. Therefore, knowledge about vegetation, structure, stand dynamics, and productivity are necessary prerequisites for establishing the silvicultural treatments, management and monitoring practices, and for providing the basis for prediction of possible future changes or for biological conservation.

The objective of this study was to perform a structural and biometric characterization of *Nothofagus betuloides* production forests, as this is a requirement for sustainable management of the species in the Magellan Region of the Chilean Patagonia.

## *PRIOR KNOWLEDGE ABOUT NOTHOFAGUS BETULOIDES FORESTS*

### *Forest Species Record*

*N. betuloides* is an evergreen and shade-intolerant species that belongs to the Fagaceae family. It is endemic in subantarctic forests and reaches heights over 35 m and a diameter at breast height (DBH) up to 1.5 m on optimal sites (Veblen et al. 1977; Moore, 1983; Rodríguez et al., 1983; Dollenz, 1995; Donoso, 1996; Gerding and Thiers, 2002; Martínez-Pastur et al., 2002).

### *Geographical Distribution*

In Chile, *N. betuloides* grows from Valdivia (40º 30′ S) to the archipelago of Cape Horn (55° 30′ S), being the most frequent tree species in the southernmost end of Patagonia (Donoso, 1981; Moore, 1983; Rodríguez et al., 1983). It occupies areas from sea level at the southern distribution to 1,200 m, at the northern limit (Kalela, 1941b; Young, 1972; Veblen et al., 1977; Del Fierro, 1998).

The survey and evaluation of native vegetation resources of Chile (CONAF-CONAMA, 1999b) found that the total area of native forests in Chile is approximately 13,430,600 ha. Of this, 13.4% corresponds to the Coihue de Magallanes forest type, being the third most important forest type in Chile.

Even though in the Chilean Magellan Region, the coihue de Magallanes forest type covers about 1,037,960 ha and it is the second most important forest type after the lenga forest type (CONAF-CONAMA, 1999b), the area covered by pure or mixed *N. betuloides* forests suitable for commercial management for timber production is estimated only at 280,000 ha (Cruz et al., 2004). This difference is explained by the forest within the National System of Protected Wild Areas of the State (SNASPE) (680,615 ha), which cannot be commercially harvested, and by the forest with both environmental and legal restrictions.

### *Ecological Aspects*

Climatically, *N. betuloides* is distributed mainly in cold marine, Patagonian humid marine, and polar marine climates. Continental climate determines the eastern boundary for *N. betuloides*. At the northern limit of its distribution, it occurs only at higher elevations, where the climate is oceanic. Absolute temperatures may oscillate between −29°C and 20°C, with an average of 6°C. Annual precipitation ranges from 416 mm $year^{-1}$ close to Punta Arenas to 2657 mm $year^{-1}$ in the Guaytecas Islands (CONAF, 1978; Del Fierro, 1998).

The soils under *N. betuloides* forests are heavily influenced by volcanic activity, topography, and climate. The predominant soil type is forest brown podzol (Díaz et al., 1960). In the mixed evergreen *N. betuloides* and *N. pumilio* forests, the local distribution of both species depends on soil water retention capacity and drainage effectiveness. Therefore, *N. pumilio* grows in areas of well-drained soils where the dominant soil formation process is podzolization, whereas *N. betuloides* grows in areas under moderately to highly waterlogged conditions in which the dominant process is organic soil formation (Díaz et al., 1960; Pisano, 1977; Gerding and Thiers, 2002).

### *Vegetation*

In the north of the distribution in Chile, out of the Magellan Region, *N. betuloides* grows associated with *N. pumilio*, forming a mixed evergreen-deciduous *Nothofagus* forest at elevations between 950 and 1,200 m (Veblen et al., 1977; Del Fierro, 1998). Also, out of its Magellan distribution, this species may commonly be associated with the following tree species: *Fitzroya cupressoides* (Mol.) Johnston, *Pilgerodendron uviferum* (D. Don) Florín, *Drimys winteri* Forst., *Weinmannia trichosperma* Cav., *Podocarpus nubigena* Lindl., *Maytenus magellanica* (Lam.)

Hook.f. and *Embothrium coccineum* J.R. *et* G. Forster, among others (Donoso, 1981, Del Fierro, 1998).

In the Magellan Region, two *Nothofagus* forest types may be recognized: evergreen and deciduous forests. At the lower altitudes of the evergreen forests, *N. betuloides* is usually found mixed with *D. winteri*, while at the higher altitudes or where the forests are exposed to strong winds, *N. betuloides* is the main tree species (Kalela, 1941b; Young, 1972; Pisano, 1977; CONAF, 1978; Moore, 1983; Dollenz, 1995; Veblen et al., 1996). In the deciduous forest type, *N. betuloides* is mainly found in forests mixed with *N. pumilio,* where the latter is the dominant species. This association is called transitional forest (Young, 1972; Pisano, 1977; Veblen et al. 1977; CONAF, 1978; Donoso, 1981; Dollenz, 1982; Moore, 1983; Pisano, 1983; Dollenz, 1985-86; Roig et al., 1985; Pisano, 1997). Stunted *N. betuloides* individuals are also found mixed with *N. antarctica* in transitional forests on peat bogs (Arroyo et al., 1996).

The ground flora of the understory of *N. betuloides* forests may also be associated with vegetative species that need great humidity, such as several important fern species including *Hymenophyllum pectinatum* Cav., *Blechnum magellanicum* (Desv.) Mett., and *Gleichenia quadripartita* (Poir.) T. Moore (Young, 1972), and in sectors with lower humidity, it is possible to find *Berberis buxifolia* Lam., *Maytenus disticha* (Hook.f.) Urban, *Fuchsia magellanica* Lam, *Gaultheria mucronata* (Lf) Hook. *et* Arn., *Ribes magellanicum* Poiret, *Chiliotrichum diffusum* (G. Forster), *Macrachaenium gracile* Hook.f., *Adenocaulon chilense* Poepp. ex Less., *Gunnera magellanica* Lam., *Ranunculus peduncularis* Sm., and *Viola magellanica* G. Forster (Pisano, 1977; Pisano, 1983; Roig, et al., 1985).

### Stand Volume and Tree Growth

The gross volume stock of pure production *N. betuloides* forests fluctuates between 300 $m^3 ha^{-1}$ and 600 $m^3 ha^{-1}$, whereas in mixed production *N. betuloides* and *N. pumilio* forests, gross volume stock is reported from 260 $m^3 ha^{-1}$ to 790 $m^3 ha^{-1}$ (Schmidt and Urzúa, 1982; Schmidt, 1990; Sievert, 1995; Silva, 1997; Trillium, 1997).

Growth studies are limited and generally refer to mixed forests, where the dominant species is *N. pumilio*. Although Kalela (1941a) established that at juvenile stages the diameter of *N. betuloides* is larger than that of *N. pumilio*, this relationship is reversed later. Different authors have concluded that, in primary or secondary forests at initial phases, *N. betuloides* trees show diameter increments of 1 mm year$^{-1}$ to 3 mm year$^{-1}$ (Young, 1972; Silva, 1997; Martínez-Pastur et al., 2002).

## *MATERIALS AND METHODS*

### *Study Areas*

This study was conducted in six *N. betuloides* forest areas throughout the Chilean Magellan Region, five on continental territory and one on Tierra del Fuego Island (Figure 1).

The climate of this region is dominated by a precipitation and humidity gradient that decreases from west to east. Precipitation originates mainly from the influence of westerly winds and ranges from 600 mm year$^{-1}$ to 1000 mm year$^{-1}$ (INIA, 1989; Rosenblüth et al., 1997; Carrasco et al., 1998).

In the study area, the natural vegetation is dominated by *N. pumilio* and *N. betuloides* forests, which pertain to the cold temperate *Notho-*

FIGURE 1. Study area and location of the study sites.

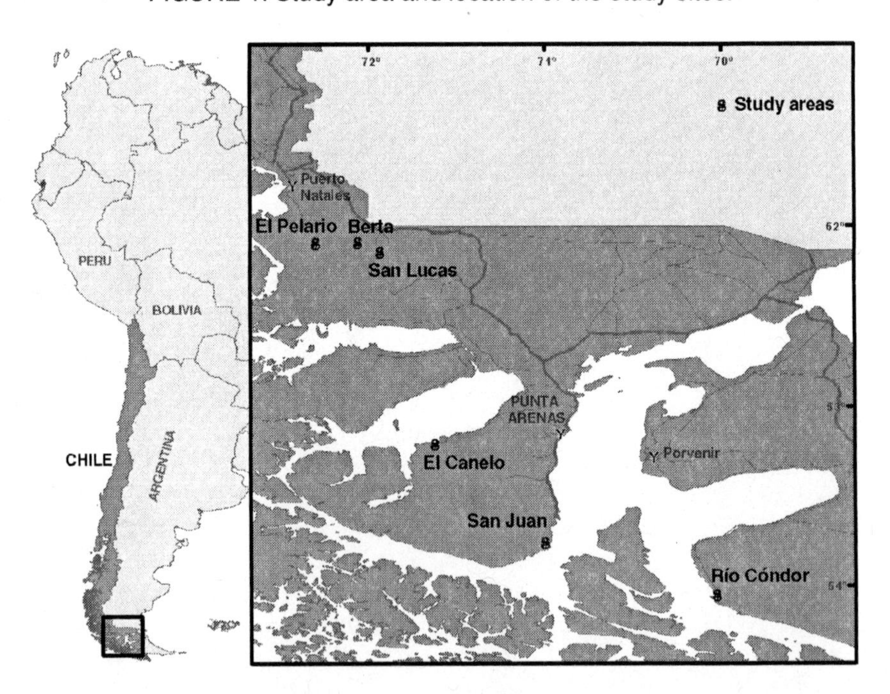

*fagus* forests (Veblen et al., 1995). They form: (1) evergreen forests, dominated principally by *N. betuloides*, which grows associated with *D. winteri*; (2) deciduous forests, dominated by *N. pumilio,* which can be associated with *N. antarctica*; and (3) mixed forests dominated by *N. betuloides* and *N. pumilio* in different proportions.

*Sphagnum* peat bogs, shrubs, and prairies are present to a lesser extent, the latter originating from the destruction of *Nothofagus* forests. *N. antarctica* grows as a shrub on some peat bog margins and indicates the transition from forests to peat bogs.

### *Selection of the Study Areas*

The selection of the study areas considered information from the survey and evaluation of the native vegetation resources of Chile (CONAF-CONAMA, 1999b). In addition, local forest management plans, ecological and vegetation studies, and roads networks were used to choose forest areas where sustainable management could be possible. Owing to the lack of pre-existing information about the total surface area for each stand structure in which *N. betuloides* was present, the selected study areas cannot be extrapolated to the whole Magellan Region. However, the selected study areas do represent adequately the entire different stand structures where *N. betuloides* exists in that region.

### *Stand Description*

Selected areas were photointerpreted and georeferenced on a 1:10,000 scale. For this purpose, aerial photographs (SAF [Chilean Aerial Photogrametric Service] panchromatic, year 1998; and color, year 1994, no metric) were used through a geographic information system, along with the support of pre-existing cartography.

The vegetation units determined from the photointerpretation process were both checked and described in the field based on environmental (i.e., slope, aspect, altitude, and soil drainage) and vegetation parameters. More detailed information, on dasometric variables, was obtained for the units dominated by forest vegetation types (stands). These were classified based on the dominant species as pure *N. betuloides* forests, *N. betuloides-N. pumilio* forests (when *N. betuloides* is more than 50% of the trees), *N. pumilio-N. betuloides* forests (when *N. pumilio* is more than 50% of the trees), and pure *N. pumilio* forests.

The forests were classified as disturbed or nondisturbed forests, based on whether or not they had undergone one of the following: harvesting (as selective logging and seedlings cut under a shelterwood system), thinning, burning, and permanent livestock grazing. All these types of disturbances, at any degree of intensity or timing of achievement, were grouped as one.

Forest structure was described in the field according to the most representative development stage. Therefore, disturbed forests were classified as initial regeneration (less than 0.5 m high), seedling (0.5 to 2 m high), sapling (2 to 5 m high), small pole (5 to 12 m high), large pole (more than 12 m high and DBH smaller than 30 cm), and sawtimber (more than 12 m high and DBH larger than 30 cm). In addition, the proportion of residual canopy, corresponding to the primary forest, was determined in disturbed forests.

Nondisturbed forests were classified as juvenile forests (which includes both the regeneration and the optimal-growth structural phases), and mature forests (which includes both the mature and the decaying structural phases).

Site quality was determined in the field by using the mean height of 3 to 5 dominant trees in a stand. This method is based on Husch et al. (2003), who considered that dominant and codominant tree heights represent well the maximum stand height, and therefore, their average is a good site quality indicator.

Each stand was classified either as production forest, restricted production forest, or protection forest. Production forests are stands having neither environmental restrictions nor legal restrictions to be potentially harvested. Production forests have a slope up to 60% and are at least 30 m from water bodies or watercourses. Restricted production forests are stands having some environmental (i.e., slow drainage, rocks, strong wind, etc.) or technical restriction (i.e., accessibility, restriction to machinery use, low timber quality, etc.) that inhibits use for timber production. Protection forests are stands having severe harvest restrictions such as having more than 60% slope, being closer than 30 m from a water body or watercourse, or being located on shallow soils or soils with several drainage problems.

The percentages of soil covered and plant heights of tree regeneration were determined and used for the description of forest regeneration. Forest understory was described according to its horizontal structure and the dominant species. Bryophytes, fungi, and lichens were not included in the survey.

For each stand, biometric information as tree density (Nha), basal area (G), and quadratic mean diameter (Dg) was determined by using the Bitterlich method of variable-radius plot sampling (Prodan et al., 1997), considering only those trees with a DBH grater than 5 cm. Stand volume (V) was computed by using a local volume function (Promis et al., 2005).

In order to determine statistical differences in the biometric values, the regeneration coverage, and the understory coverage among the different forest structures, Kruskal-Wallis and Mann-Whitney (Spiegel, 1991), nonparametric tests were performed at a 5% significance level. Dominant species in the understory were analyzed by using analysis of variance (ANOVA) of their frequencies per stand structure.

## *RESULTS*

A total of 648 vegetation units, equivalent to an area of 6,102 ha, were described. Out of this area, 84% (5,123 ha) corresponded to forest stands and 16% corresponded to nonforest land (i.e., prairies, peat bog areas).

From the total forest area, 94% (4,791 ha) was production forest, and the remaining 6% was protection forest (Table 1). Out of the total area of production forest, 79% (3,807 ha) was *N. betuloides* forests, with the remainder 21% being pure production *N. pumilio* forest. In the same way, 25% of production *N. betuloides* forest showed some type of use

TABLE 1. Summary statistics for *N. betuloides* forests by study location

| Study location | Forest area (ha) | Production forest (ha) | *N. betuloides* production forest: Without restrictions (ha) | With restrictions (ha) | Total (ha) | Site quality (m) | Elevation (m) |
|---|---|---|---|---|---|---|---|
| Berta | 672 | 656 | 312 | 247 | 559 | 13-25 | 260-520 |
| El Canelo | 874 | 844 | 262 | 84 | 346 | 12-23 | 130-290 |
| Río Cóndor | 1407 | 1294 | 1052 | 216 | 1268 | 12-29 | 0-150 |
| El Pelario | 822 | 767 | 369 | 169 | 538 | 13-25 | 200-700 |
| San Juan | 583 | 506 | 424 | 82 | 506 | 13-28 | 0-300 |
| San Lucas | 765 | 724 | 445 | 145 | 590 | 13-27 | 230-600 |
| **Total** | **5123(1)** | **4791(2)** | **2864** | **943** | **3807** | | |

(1) Includes protection forests; (2) Includes *N. pumilio* production forests.

restriction, such as steep slope, proximity to streams or water bodies, and poorly-drained soils. The production forest stands were located at altitudes not higher than 700 m. The site quality of the production *N. betuloides* forest stands shows high variability in each study area, ranging from 12 to 29 m at almost all study locations.

In the study areas, 44% of production *N. betuloides* forests are located in areas with site qualities ranging from 20 to 24 m. Most of the past logging was conducted on sites with these site qualities. Only 25% of the production forests under study are located at site qualities over 24 m (Figure 2).

Table 2 indicates that most of the *N. betuloides* forests (76%) were mixed with *N. pumilio*, whereas only 24% of the production *N. betuloides* forests were pure. Some degree of disturbance such as selective logging, shelterwood system, forest fire, and browsing was found in 53% of the total area under study. Selective logging was the most common disturbance (66% of disturbed forests). Only 21% of the disturbed forests were recently managed by using a shelterwood system. Over the total area of nondisturbed production *N. betuloides* forests, mature forests were the most representative structures. Most of the disturbed forests were either in initial regeneration or seedling stand structure, both summing to 29% of the total *N. betuloides* forests.

The regeneration coverage of both *Nothofagus* species was lower than 28% for the different stand structures (Table 3). Regeneration coverage for nondisturbed forests was slightly higher than that for disturbed forests. However, this difference is not statistically significant. No sig-

FIGURE 2. *Nothofagus betuloides* production forest distribution by site quality.

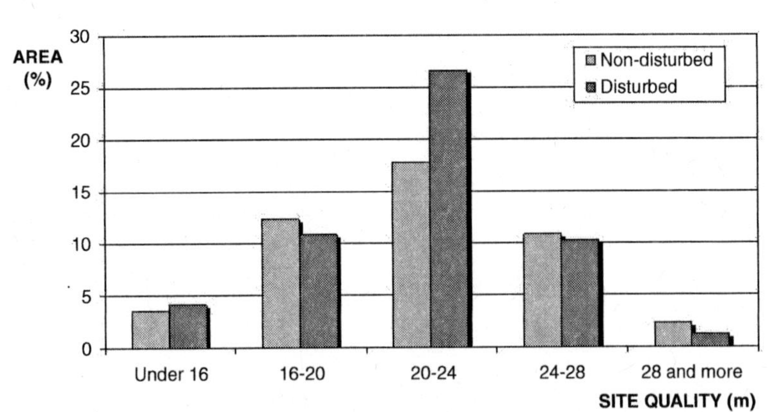

TABLE 2. Area of mixed and pure production *Nothofagus betuloides* forests by different stand structures

| Stand structure | Mixed forest | | Pure forest | | Subtotal | | Total | |
|---|---|---|---|---|---|---|---|---|
| | ha | % | ha | % | ha | % | ha | % |
| **Nondisturbed forest** | | | | | | | **1789** | **47** |
| Juvenile | 652 | 17 | 53 | 1 | 705 | 19 | | |
| Mature | 581 | 15 | 503 | 13 | 1084 | 29 | | |
| **Disturbed forests** | | | | | | | **2018** | **53** |
| Initial regeneration | 611 | 16 | - | - | 611 | 16 | | |
| Seedling | 488 | 13 | - | - | 488 | 13 | | |
| Sapling | 88 | 2 | 9 | < 1 | 97 | 3 | | |
| Pole small | 299 | 8 | 140 | 4 | 439 | 12 | | |
| Pole large | 159 | 4 | 216 | 6 | 375 | 10 | | |
| Sawtimber | 8 | < 1 | - | - | 8 | < 8 | | |
| **Total** | **2886** | **76** | **921** | **24** | **3807** | **100** | **3807** | **100** |

TABLE 3. Average coverage of different regeneration height layers by stand structures[1]

| Stand structure | Average cover (%) | | | | | | |
|---|---|---|---|---|---|---|---|
| | Regeneration height layer (cm) | | | | | | |
| | < 20 | 20-50 | 50-100 | 100-200 | > 200 | Subtotal | Total |
| **Nondisturbed forest** | | | | | | | **24.6** |
| Juvenile | 11.2 | 12.7 | 15.4 | 10.0 | 5.0 | 22.8 | |
| Mature | 10.3 | 10.1 | 11.4 | 8.3 | 11.0 | 26.0 | |
| **Disturbed forests** | | | | | | | **22.4** |
| Initial regeneration | 12.2 | 13.1 | 8.3 | 11.3 | 5.0 | 25.0 | |
| Seedling | 5.0 | 5.0 | 4.5 | 4.0 | 5.0 | 14.0 | |
| Sapling | - | - | 5.0 | 5.0 | - | 10.0 | |
| Pole small | 12.5 | 12.5 | 8.3 | 7.5 | - | 23.3 | |
| Pole large | 5.0 | 5.0 | 5.0 | 5.0 | - | 10.0 | |
| Sawtimber | 5.0 | 7.5 | 7.5 | 7.5 | - | 27.5 | |
| **Total** | **10.6** | **11.3** | **11.0** | **8.4** | **8.3** | **24.0** | |

(1) There were no statistically significant differences among the stand structures ($p < 0.05$). There were no statistically significant differences among the regeneration height layers ($p < 0.05$).

nificant coverage differences were observed among different regeneration layers and different stand structures. It was observed, however, that the largest proportion of regeneration had heights lower than 100 cm in both nondisturbed and disturbed forests.

Mixed *N. betuloides* and *N. pumilio* forests had higher average understory coverage than pure *N. betuloides* forests (Table 4). There were no significant differences in understory coverage among nondisturbed forest structures. In contrast, understory coverage tends to be larger in disturbed forests where the tree canopy allows a higher level of sunlight penetration. A total of 33 dominant species were recognized within the understory of the forests under study. The most frequent species were *Berberis ilicifolia*, *Maytenus disticha*, *Pernettya mucronata*, *Blechnum penna-marina*, *Osmorhiza chilensis*, and *Valeriana lapathifolia*, in variable proportions according to the different stand structures.

Tree density decreases with stand age in the nondisturbed forests. Although juvenile stands have higher tree density and lower quadratic mean diameter, no significant differences were observed for these variables between juvenile and mature stands (Table 5). Tree density in disturbed forests also decreases with stand development stage. At the initial regeneration and seedling development stages, the abundant regeneration established does not reach 5 cm in DBH. Therefore, as only

TABLE 4. Understory coverage for mixed and pure *Nothofagus betuloides* forests by different stand structures

| Stand structure | Mixed forest (%) | Pure forest (%) | Subtotal[(1)] (%) | Total (%) |
|---|---|---|---|---|
| **Nondisturbed forest** | | | | **50** |
| Juvenile | 51 | 38 | $49^{ac}$ | |
| Mature | 47 | 57 | $51^{ac}$ | |
| **Disturbed forests** | | | | **42** |
| Initial regeneration | 55 | - | $55^{c}$ | |
| Seedling | 51 | - | $51^{ac}$ | |
| Sapling | 55 | 5 | $44^{abc}$ | |
| Pole small | 34 | 24 | $32^{ab}$ | |
| Pole large | 26 | 14 | $19^{b}$ | |
| Sawtimber | 48 | - | $48^{abc}$ | |
| **Total** | **48** | **35** | **46** | |

[(1)] Understory coverage values followed by the same letter (a, b or c) were not significantly different ($p < 0.05$).

TABLE 5. Summary statistics for the *Nothofagus betuloides* forest by different stand structures[1]

| Stand structure | Stand density (trees ha$^{-1}$) | Quadratic mean diameter (cm) | Basal area (m$^2$ ha$^{-1}$) | Volume (m$^3$ ha$^{-1}$) |
|---|---|---|---|---|
| **Nondisturbed forest** | | | | |
| Juvenile | 1,802$^{ab}$ | 26 | 53$^{a}$ | 437$^{a}$ |
| Mature | 1,278$^{a}$ | 29 | 50$^{ab}$ | 439$^{a}$ |
| **Disturbed forests** | | | | |
| Initial regeneration | 593$^{c}$ | 36 | 37$^{c}$ | 318$^{b}$ |
| Seedling | 486$^{c}$ | 41 | 38$^{c}$ | 342$^{ab}$ |
| Sapling | 4,961$^{abc}$ | 25 | 37$^{bc}$ | 268$^{b}$ |
| Pole small | 3,651$^{ab}$ | 21 | 38$^{c}$ | 286$^{b}$ |
| Pole large | 2,749$^{b}$ | 19 | 45$^{bc}$ | 333$^{b}$ |
| Sawtimber | 945$^{abc}$ | 23 | 42$^{abc}$ | 404$^{ab}$ |

(1) Biometric values followed by the same letter (a, b, or c) were not significantly different ($p < 0.05$), for the correspondent parameter.

the largest trees remain, tree density at these development stages was significantly lower in disturbed forests. At the sapling development stage, density is higher because many juvenile trees exceed 5 cm in DBH, and there remain only a few large individuals, these being the few large residual trees (quadratic mean diameter from 36 to 41 cm) from the primary forest. Nondisturbed forests showed the greatest basal area, with mean values around the 50 m$^2$ ha$^{-1}$ for both juvenile and mature forests. In the disturbed forest, the most juvenile forest stands (initial regeneration, seedling, sapling, and pole small) had mean values smaller than the more developed second-growth forests (sawtimber and pole large).

In general the nondisturbed forests had larger volume stocks than the disturbed forests. This occurs more because of past logging than possible differences in site quality.

## *CONCLUSIONS AND DISCUSSION*

The areas under study show a high proportion of forests of which the majority corresponds to production *N. betuloides* forests in different situations (pure or mixed) and conditions (disturbed and undisturbed). In the studied areas, production *N. betuloides* forests are mostly mixed with *N. pumilio*. Of these, a portion has use restrictions owing especially to the occurrence of poorly drained soils.

In the study areas, the majority of production *N. betuloides* forests (53%) have undergone some degree of disturbance, such as harvesting, thinning, burning, and permanent livestock grazing; the rest are mostly nondisturbed forests with mature structures. Most of the disturbed forests are in the early stages of development, with initial regeneration and seedling structures. These are managed forests, recently subjected to intervention through the shelterwood system. The remaining disturbed production *N. betuloides* forests, in a more advanced development stage, correspond to forests that were subjected to selective logging during the past century.

The majority of the production *N. betuloides* forests of the study areas, in particular those that were disturbed, grow at site qualities of 20 m to 24 m, and only a quarter of them develop at better site qualities.

Regeneration coverage of the study areas shows high variability, with no clear trend among the different structures in the forests under study. This is due to the different levels of sunlight supporting the settling and development of regeneration, which is a consequence of the heterogeneity of the forest canopy.

Understory coverage is heterogeneous and tends to be denser in nondisturbed forests, particularly where the structure of the upper canopy enables the passage of more sunlight to the forest floor. Pure *N. betuloides* forests present an average understory coverage lower than that of mixed forests. The most frequent species of the understory were *Berberis ilicifolia*, *Maytenus disticha*, *Pernettya mucronata*, *Blechnum penna-marina*, *Osmorhiza chilensis*, and *Valeriana lapathifolia*, in variable proportions according to the different stand structures.

Nondisturbed forests show a larger gross volume stock than the disturbed ones. In contrast, the stand structures with a smaller volume stock are sapling and small pole.

The lack of detailed studies or data about both the stand structures and volume stocks of the *N. betuloides* production forests makes it difficult to compare the results of this study. In many cases, the scarce information available does not indicate if forests are disturbed or nondisturbed. In addition, there are differences in the characteristics of the volume functions used to estimate stocks. In spite of these restrictions, the values of basal area and volume stocks in general are similar to those reported by the literature or by forest inventories for the Magellan Region (Schmidt, 1990; Silva, 1997; Trillium, 1997).

Briefly, it has been verified that, just as with *N. pumilio* forests in the Magellan Region (Schmidt et al., 2003), disturbed *N. betuloides* forests have a very heterogeneous structure, presenting a high diversity of coverage and thus different degrees of secondary forest or regeneration development. This is generally produced by the ordinary exploitation practice applied in these forests, by selectively logging the few existing sawable trees in the stand, with poor-quality, badly-shaped, and unhealthy trees being left standing. Therefore, and similar to what happens in *N. pumilio* forests (Schmidt and Urzúa, 1982; Martínez-Pastur et al., 2002), the production potential of disturbed forests at the Magellan Region decreases even more than in the nondisturbed forests, with the future consequence that these will have worse quality than original forests.

According to Donoso (1981), *N. betuloides* forests may be managed through either selection or shelterwood silvicultural systems. However, in the scarcely managed stands found at the studied areas, the only management observed in *N. betuloides* was the shelterwood system.

In conclusion, this study has enabled structural and biometric characterization of *N. betuloides* production forests in the Chilean Magellan Region. This provides the necessary support to perform all vegetational, structural, dynamics, productivity, and biodiversity studies to establish and improve future silvicultural treatments. This must be done considering a rational and sustainable use of the resource, to recover degraded zones, or simply to carry out biological conservation projects.

## REFERENCES

Arroyo, M., C. Donoso, R. Murúa, E. Pisano, R. Schlatter and I. Serey. 1996. Toward an ecologically sustainable forestry project: concepts, analysis and recommendations. Departamento de Investigación y Desarrollo, Universidad de Chile, Santiago, Chile.

Carrasco, J., G. Casassa and A. Rivera. 1998. Climatología actual del Campo de Hielo Sur y posibles cambios por el incremento del efecto invernadero. *Anales del Instituto de la Patagonia*, Serie Ciencias Naturales (Chile) 26:119-128.

CONAF. 1978. Antecedentes forestales XII Región Magallanes y Antártida Chilena. CONAF, Ministerio de Agricultura. Unpublished report.

CONAF-CONAMA. 1999a. Catastro y evaluación de recursos vegetacionales nativos de Chile. Informe Regional Duodécima Región. Corporación Nacional Forestal. Santiago, Chile.

CONAF-CONAMA (1999b) Catastro y evaluación de recursos vegetacionales nativos de Chile. Informe nacional con variables ambientales. Santiago, Chile.

Cruz, G., R. Caprile, F. Hidalgo and G. Cabello. 2004. Clasificación y cuantificación de la superficie regional de los bosques de Coihue de Magallanes. Unpublished report. Proyecto FONDEF D02I1080. Universidad de Chile. Chile.

Del Fierro, P. 1998. Experiencia silvicultural del bosque nativo de Chile. Recopilación de antecedentes para 57 especies arbóreas y evaluación de prácticas silviculturales. GTZ-CONAF. Proyecto Manejo Sustentable del Bosque Nativo. Publicaciones Castillo S.A. Santiago, Chile.

Díaz, C., C. Avilés and R. Roberts. 1960. Los grandes grupos de suelos de la provincia de Magallanes. Estudio preliminar. In: pp. 227-308. Agricultura Técnica. Ministerio de Agricultura, Chile.

Dollenz, O. 1982. Estudios fitosociológicos en las reservas forestales Alacalufes e Isla Riesco. *Anales del Instituto de la Patagonia*. Punta Arenas. Chile. 13: 161-170.

Dollenz, O. 1985-86. Relevantamientos fitosociológicos en la Península Muñoz Gamero, Magallanes. *Anales del Instituto de la Patagonia*. Punta Arenas. Chile. 16: 55-62.

Dollenz, O. 1995. Los árboles y bosques de Magallanes. Ediciones Universidad de Magallanes. Punta Arenas, Chile.

Donoso, C. 1981. Tipos forestales de los bosques nativos chilenos. Documento de Trabajo No. 38. Investigación y Desarrollo Forestal (CONAF-PNUD-FAO), FAO, Chile.

Donoso, C. 1996. Ecology of *Nothofagus* forests in central Chile.. In: pp 271-292. T. Veblen, R. Hill and J. Read (Eds.). The ecology and biogeography of *Nothofagus* forests. Yale, New Haven, USA.

Gerding, V. and O. Thiers. 2002. Caracterización de suelos bajo bosques de *Nothofagus betuloides* (Mirb.) Blume, en Tierra del Fuego, Chile. *Revista Chilena de Historia Natural* 75:819-833.

Husch, B., T. Beers and J. Kerschaw. 2003. Forest mensuration. 4th edition. John Wiley & Sons, Inc., New Jersey, USA.

INIA. 1989. Mapa Agroclimático de Chile. Instituto de Investigaciones Agropecuarias, Santiago, Chile.

Kalela, E. 1941a. Über die Holzarten und die durch die klimatischen Verhältnisse verursachten Holzartenwechsel in den Wäldern Ostpatagoniens. *Annales Academiae Scientarium Fennicae*. Series A. IV Biologica 2. Helsinki, Finlandia.

Kalela, E. 1941b. Über die Entwicklung der herrschenden Bäume in den Bestanden verscheidener Waldtypen Ostpatagoniens. *Annales Academiae Scientarium Fennicae*. Series A. IV Biologica 2. Helsinki, Finlandia.

Martínez-Pastur, G., J. Cellini, M. Lencina, R. Vukasovic, P. Peri and S. Donoso. 2002. Response of *Nothofagus betuloides* (Mirb.) Oerst. to different thinning intensities in Tierra del Fuego, Argentina. *Interciencia* 27(12):679-685.

Moore, D. 1983. Flora of Tierra del Fuego. Anthony Nelson, Shropshire, England; Missouri Botanical Garden, St. Louis, MO, USA.

Nuñez, P. and C. Salas. 2000. Estudio dendrométrico Proyecto forestal Río Cóndor, Tierra del Fuego, XII Región, Chile. Unpublished report.

Pisano, E. 1977. Fitogeografía de Tierra del Fuego–Patagonia Chilena. I Comunidades vegetales entre las latitudes 52° y 56° S. *Anales del Instituto de la Patagonia* 8:121-250.

Pisano, E. 1983.Comunidades vegetales en el sector norte de la península Muñoz Gamero (Ultima Esperanza, Magallanes). *Anales del Instituto de la Patagonia.* Punta Arenas. Chile. 14:83-101.

Pisano, E. 1997. Los bosques de Patagonia austral y Tierra del Fuego chilenas. *Anales del Instituto de la Patagonia.* Punta Arenas. Chile. 25:9-19.

Prodan, M., R. Peters, F. Cox, and P. Real. 1997. Mensura forestal. Instituto Interamericano de Cooperación para la Agricultura, San José, Costa Rica.

Promis, A., G. Cruz, H. Schmidt, R. Caprile, and J. Caldentey. 2005. Tree general volume equations for *Nothofagus betuloides* in Chilean Patagonia forests. In: XII IUFRO World Congress. Poster Session D4. Brisbane, Australia.

Rodríguez, R., O. Matthei and M. Quezada. 1983. Flora arbórea de Chile. Universidad de Concepción, Concepción, Chile.

Roig, F., J. Anchorena, O. Dollenz, A. Faggi and E. Mendez. 1985. Las comunidades vegetales de la transecta botánica de la Patagonia austral. Primera Parte. La vegetación del área continental. In: pp. 350-456. O. Boelcke, D. Moore, F. Roig (eds.). Transecta botánica de la Patagonia austral. Consejo Nacional de Investigaciones Científicas y Técnicas (Argentina), Instituto de la Patagonia (Chile) y Royal Society (Gran Bretaña), Buenos Aires, Argentina.

Rosenblüth B, H. Fuenzalida and P. Aceituno 1997. Recent temperature variations in southern South America. *International Journal of Climatology* 17:67-85.

Schmidt, H. 1990. Antecedentes silvícolas para los bosques de Lenga y Coihue, Sector Río Cóndor, Tierra del Fuego. Informe Técnico. Santiago, Chile.

Schmidt, H., G. Cruz, A. Promis and M. Álvarez. 2003. Transformación de los bosques de lenga vírgenes e intervenidos a bosques manejados. Guía para los bosques demostrativos Universidad de Chile, Facultad de Ciencias Forestales. Publicaciones Misceláneas Forestales N° 4, Santiago, Chile.

Schmidt, H. and A. Urzúa. 1982. Transformación y manejo de los bosques de Lenga en Magallanes. Universidad de Chile. Facultad de Ciencias agrarias y forestales. *Ciencias Agrícolas* N° 11. Santiago, Chile.

Sievert, H. 1995. Estudio de crecimiento para un bosque multietáneo de Lenga (*Nothofagus pumilio*) en Aysén, XI Región. Memoria de Título. Escuela de Ciencias Forestales, Universidad de Chile.

Silva, J. 1997. Crecimiento y acumulación de biomasa en renovales de coihue de Magallanes (*Nothofagus betuloides*) en el sector del río San Juan, XII Región. Memoria de Título. Escuela de Ciencias Forestales, Universidad de Chile, Chile.

Spiegel, M. 1991. Estadística. McGraw Hill, México. 556 pages.

Trillium. 1997. Estudio de impacto ambiental. Proyecto Forestal Río Cóndor. Unpublished report. Forestal Trillium Ltda, Punta Arenas, Chile.

Veblen, T., H. Ashton, F. Schlegel and A. Veblen. 1977. Distribution and dominant of species in the understory of a mixed evergreen-deciduous *Nothofagus* forest in South-Central Chile. *Journal of Ecology* 65:815-830.

Veblen, T., C. Donoso, T. Kitzberger and A. Rebertus. 1996. Ecology of southern Chilean and Argentinean *Nothofagus* forests. In: pp. 293-353. T. Veblen, R. Hill and J. Read (eds.). The ecology and biogeography of *Nothofagus* forests. Yale, New Haven, CT, USA.

Veblen, T., T. Kitsberger, B. Burns and A. Rebertus. 1995. Perturbaciones y dinámica de regeneración en bosques andinos del sur de Chile y Argentina. In: pp. 169-198. M. Arroyo, J. Armesto and C. Villagrán (eds.). Ecología de los bosques nativos de Chile. Editorial Universitaria, Santiago, Chile.

Young, S. 1972. Subantarctic rain forest of Magallanic Chile: Distribution, composition, and age and growth rate studies of common forest trees. In: pp. 307- 322. G. Llano (ed.). Antarctic Research Series (20). American Geophysical Union, Washington, USA.

doi:10.1300/J091v24n02_02

# A Working Definition of Sustainable Forestry and Means of Achieving It at Different Spatial Scales

Chadwick Dearing Oliver
Robert L. Deal

**SUMMARY.** Sustainable forestry can be a useful concept when it includes both spatial equity and the sustainable development concept of intergenerational equity. Using this definition and criteria such as those

Chadwick Dearing Oliver is Pinchot Professor of Forestry and Environmental Studies, and Director, Global Institute of Sustainable Forestry, School of Forestry and Environmental Studies, Yale University, New Haven, CT 06511 USA (E-mail: chadwick.oliver@yale.edu).

Robert L. Deal is Research Silviulturist, USDA Forest Service, PNW Research Station, Portland, OR 97205 USA (E-mail: rdeal@fs.fed.us).

[Haworth co-indexing entry note]: "A Working Definition of Sustainable Forestry and Means of Achieving It at Different Spatial Scales." Oliver, Chadwick Dearing, and Robert L. Deal. Co-published simultaneously in *Journal of Sustainable Forestry* (Haworth Food & Agricultural Products Press, an imprint of The Haworth Press, Inc.) Vol. 24, No. 2/3, 2007, pp. 141-163; and: *Sustainable Forestry Management and Wood Production in a Global Economy* (ed: Robert L. Deal, Rachel White, and Gary L. Benson) Haworth Food & Agricultural Products Press, an imprint of The Haworth Press, Inc., 2007, pp. 141-163. Single or multiple copies of this article are available for a fee from The Haworth Document Delivery Service [1-800-HAWORTH, 9:00 a.m. - 5:00 p.m. (EST). E-mail address: docdelivery@haworthpress.com].

Available online at http://jsf.haworthpress.com

doi:10.1300/J091v24n02_03

developed by the Montreal Process, each country can examine itself in a matrix to determine if it is overly protecting or exploiting its ecosystems according to the different criteria. Then, each country can examine the ecosystems within its country to determine if all are being equitably treated. For example, the United States appears sustainable when viewed as a whole; however, there are gross differences in ecosystem conditions within the country. Self-examination by each country of its contributions and voluntary actions to rectify its excesses and deficiencies will probably be more effective than a global policing system. Local, ground-specific actions can be taken to correct the imbalances within each ecosystem within a country. This paper illustrates that sustainable forestry can be considered and analyzed at different scales, and that even in countries where broad data sets indicate acceptable progress toward sustainable forestry, local data show the heterogeneous nature of this progress. doi:10.1300/J091v24n02_03 *[Article copies available for a fee from The Haworth Document Delivery Service: 1-800-HAWORTH. E-mail address: <docdelivery@haworthpress.com> Website: <http://www.HaworthPress.com>* 

**KEYWORDS.** Sustainable forestry, sustainable forest management, ecosystem management, spatial scales

## *DEFINING SUSTAINABLE FORESTRY*

Sustainable forestry is a logical consequence of global environmental awareness and the long-term concerns foresters have had with sustaining wood resources. Sustainable forestry needs to include global, national, regional, and local perspectives. Forests are not distributed equally in the world, but the resources and the concern over their sustainability need to be shared because travel and trade connect the world and because air and water pollution as well as other forest values affect all parts of the Earth.

Global environmental concerns are rapidly shifting from a phase of making people aware of the problem to a phase of technically addressing the specific actions needed to remedy these problems. Whereas the "awareness" phase required advocates, the technical phase will require many scientifically trained individuals to create and maintain a management capacity at the global, regional, and countrywide spatial scales, as well as within-country regions to the local level.

Forests were first harvested to convert the wealth of forests to other goods and services that allowed people to live more comfortably–and ultimately to become aware of longer term concerns (Perlin 1991). An early concern was that the forests were being depleted, and forestry was created to reverse this trend. People also became concerned about the depletion of soil, water quality, habitats, a clean atmosphere, and other environmental and social values (Marsh 1864). The increasing population led to the concern that people would create an environment that would not sustain even themselves. As this concern for sustainability extended beyond forests to other environmental values, foresters have expanded their concerns to the sustainability of the environment as a whole. This expansion is reflected in the changing of many forestry schools to include other environmental studies. The expansion was motivated both by the realization that the experience, knowledge, and tools developed to sustain forests could assist in sustaining other environmental goods and services and by the realization that many other facets of the environment need to be sustained in order to sustain forests.

This paper first describes "sustainable forestry" in a way that enables specific actions to be taken to achieve it. It then describes how data can be analyzed at the global scale to determine what actions need to be taken to achieve sustainable forestry. Using the United States as an example, this paper shows how data within a country can be examined to determine and rectify imbalances in treatment of forests in different regions within a country. It then describes how people within each region can be empowered to take appropriate actions to lead to sustainable forestry.

Early, fragmented concerns about broader environmental issues in the 1960s and 1970s became unified as intergenerational equity under the concept of sustainable development:

> Sustainable development is development that meets the needs of the present without compromising the ability of future generations to meet their own needs.
>
> –Brundtland, 1987

The concept of sustainability has been extended to forests. Each ecosystem is unique and deserves to be sustained, so the concept of spatial equity has been included with that of intergenerational equity in a working definition of sustainable forestry:

> People living in one place and time should provide their "fair share" of values–neither unfairly exploiting nor depriving themselves of certain values to the detriment or benefit of people in another place or time.
>
> –Oliver, 2003

Other definitions of sustainable forestry are more specific and focus attention on issues directly related to forestry. For example, the Society of American Foresters (Helms, 1998) defined sustainable forestry as:

> The capacity of forests, ranging from stands to ecoregions, to maintain their health, productivity, diversity, and overall integrity, in the long run, in the context of human activity and use.
>
> –Helms, 1998

The spatial equity, or "fair share," definition of sustainable forestry (Oliver, 2003) described above is used as a working definition in this paper because it enables explicit analyses and actions that can improve the sustainability of forests. Two ways of discussing "sustainable forestry" have been proposed. One is the Venn Diagram approach (Figure 1), also known as the "triple bottom line" approach. This approach considers sustainable forestry as the overlapping of the acceptable ecological, economic, and social activities. Sustainable forestry ensures that all three spheres of values are sustainable, so management is conceptually done at the overlap of the three activities.

A second approach, the "matrix" approach, ensures that one ecosystem is neither hoarding nor overly exploiting the things valued in forests to the detriment of itself or other ecosystems. This approach compares the different countries and ecosystems in the world (e.g., Tables 1-6) by using data collected by international organizations and individual countries.

Fortunately, agreement is emerging on the things people value from forest ecosystems. Following the United Nations Conference on Environment and Development (UNCED) in Rio de Janeiro in 1992, government scientists met in a series of "processes" and developed several "criteria" of sustainable forestry, which they are now expanding into "indicators" as measurable subdivisions of the criteria (Burley, 2001). The criteria prove to be reassuringly similar among processes, and they provide a robust set of "values" that society seeks to sustain. The Mon-

FIGURE 1. Venn diagram approach that views sustainable forestry as the overlap of the ecologically, socially, and economically viable activities. By contrast, the matrix approach compares different ecosystems to determine if each is providing its fair share (e.g., Tables 1 through 6).

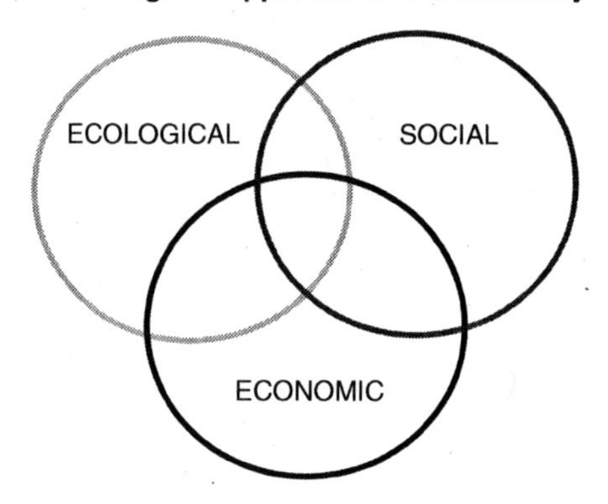


treal Process (Montreal Process Working Group, 1998) criteria are listed here.

1. Conservation of biological diversity.
2. Maintenance of productive capacity of forest ecosystems.
3. Maintenance of forest ecosystem health and vitality.
4. Maintenance of soil and water resources.
5. Maintenance of forest contribution to global carbon cycles.
6. Maintenance and enhancement of long-term socioeconomic benefits to meet the needs of society.
7. Legal, institutional, and economic framework for forest conservation and sustainable management.

Other values may emerge, and some of these values may be combined or subdivided; however, if a system is in place to sustain these values, it is highly likely that other values will also be provided, or can easily be provided as they emerge.

Providing these values equitably requires a comparison among ecosystems, as well as techniques for managing within ecosystems. The first issue is whether it is even possible to sustain all ecosystems because of the increasing human population and the large consumption of wood. Contrary to early concerns, the population may stabilize within

TABLE 1. Characteristics of selected countries (countries do not, and were not selected to show extremes; they were selected to illustrate ranges)

| | Forest area (Thousand ha) | Total forest volume (Million $m^3$) | Population density (People/$km^2$) | GNP per capita (1997 US$) |
|---|---|---|---|---|
| Australia | 154,539 | 8,506 | 2 | 19,689 |
| Belize | 1,348 | 272 | 10 | 2,547 |
| Canada | 244,571 | 29,364 | 3 | 19,267 |
| China | 163,480 | 8,437 | 137 | 668 |
| Comoros | 8 | 0 | 363 | 413 |
| Dem. Rep. of the Congo | 135,207 | 17,932 | 22 | 114 |
| Ghana | 6,335 | 311 | 86 | 384 |
| Guatemala | 2,850 | 1,012 | 102 | 1,481 |
| Honduras | 5,383 | 311 | 56 | 723 |
| Jamaica | 325 | 27 | 236 | 1,525 |
| Mauritius | 16 | 1 | 569 | 3,796 |
| Micronesia | 15 | - | 168 | 1,886 |
| Myanmar | 34,419 | 1,137 | 69 | --- |
| Niger | 1,328 | 4 | 8 | 202 |
| Thailand | 14,762 | 252 | 119 | 2,821 |
| Ukraine | 9,584 | 1,719 | 87 | 1,452 |
| United States | 225,993 | 30,838 | 30 | 28,310 |
| Uruguay | 1,292 | - | 19 | 6,076 |
| Venezuela | 49,506 | 6,629 | 27 | 3,499 |

Data from FAO, 2001

TABLE 2. Characteristics of forests of countries shown in Table 1

| | Annual forest cover change 1990-2000 (%) | Forest in Protected Areas (%) | Country-endemic endangered forest species/ million ha of forest | Ratio: standing volume/ harvested volume | Fuelwood as percentage of harvest |
|---|---|---|---|---|---|
| Australia | −0.2 | 13 | 1.0 | 267 | 24 |
| Belize | −2.7 | 37 | 3.0 | 1,450 | 67 |
| Canada | 0.0 | 5 | 0.0 | 151 | 2 |
| China | 1.1 | 3 | 0.7 | 30 | 67 |
| Comoros | 0.0 | 0 | 1,375.0 | 0 | 0 |
| Dem. Rep. of the Congo | −0.4 | 9 | 0.2 | 248 | 95 |
| Ghana | −1.9 | 9 | 0.8 | 14 | 95 |
| Guatemala | −1.9 | 35 | 5.6 | 63 | 97 |
| Honduras | −1.1 | 5 | 7.4 | 32 | 90 |
| Jamaica | −1.5 | 11 | 547.7 | 31 | 67 |
| Mauritius | 0.0 | 0 | 2,750.0 | 59 | 53 |
| Micronesia | −6.7 | 0 | 666.7 | NA | NA |
| Myanmar | −1.5 | 5 | 0.2 | 28 | 86 |
| Niger | −4.7 | 77 | 0.0 | 0 | 95 |
| Thailand | −0.8 | 23 | 0.7 | 9 | 72 |
| Ukraine | 0.3 | 6 | 0.0 | 130 | 62 |
| United States | 0.2 | 40 | 1.0 | 69 | 10 |
| Uruguay | 3.9 | 5 | 0.0 | NA | 69 |
| Venezuela | −0.4 | 66 | 1.3 | 1,333 | 75 |

Data from FAO, 2001

TABLE 3. Forest land and proportion of reserved and other nontimber forested land by ecological areas of the United States

| Region | Total forest Area (million ha) | Timberland (million ha) | Reserved forest | | Reserved and other nontimberland | |
|---|---|---|---|---|---|---|
| | | | Area (million ha) | Percentage of total | Area (million ha) | Percentage of total |
| United States: | 302 | 204 | 21 | 7 | 98 | 33 |
| Northeast | 35 | 32 | 2 | 6 | 3 | 8 |
| North Central | 34 | 33 | 1 | 3 | 2 | 5 |
| Southeast | 36 | 34 | 1 | 3 | 2 | 4 |
| South Central | 51 | 47 | 0 | 1 | 4 | 7 |
| Intermountain | 56 | 27 | 7 | 13 | 29 | 52 |
| Alaska | 52 | 5 | 4 | 8 | 47 | 90 |
| Pacific Northwest | 21 | 17 | 2 | 12 | 4 | 20 |
| Pacific Southwest | 16 | 8 | 2 | 15 | 9 | 54 |

Note: "Reserved" and "Reserved and other nontimberland" could be considered "protected areas."
Data from Smith et al., 2001.

TABLE 4. Conifer and nonconifer timber net growth (growth less mortality), harvest (removals), ratio of harvest (removals) to net growth, and ratios of standing volume to harvest volume for ecological areas of the United States

| Region | Conifers | | | | Non-conifers | | | |
|---|---|---|---|---|---|---|---|---|
| | Net growth (million $m^3$) | Removals (million $m^3$) | Removals/ net growth ratio | Standing-to-harvest volume ratio | Net growth (million $m^3$) | Removals (million $m^3$) | Removals/ net growth | Standing-to-harvest volume ratio |
| United States: | 379,261 | 284,995 | 75 | 51 | 379 | 285 | 75 | 67 |
| Northeast | 18,295 | 11,715 | 64 | 79 | 18 | 12 | 64 | 111 |
| North Central | 14,813 | 7,210 | 49 | 78 | 15 | 7 | 49 | 72 |
| Southeast | 78,687 | 83,462 | 106 | 18 | 79 | 83 | 106 | 53 |
| South Central | 88,068 | 99,982 | 114 | 15 | 88 | 100 | 114 | 42 |
| Intermountain | 54,149 | 13,619 | 25 | 259 | 54 | 14 | 25 | 658 |
| Alaska | 3,876 | 5,021 | 130 | 179 | 4 | 5 | 130 | 641 |
| Pacific Northwest | 87,234 | 45,915 | 53 | 87 | 87 | 46 | 53 | 140 |
| Pacific Southwest | 32,711 | 17,500 | 54 | 81 | 33 | 18 | 54 | 947 |

Note: The standing-to-harvest volume ratio is a rough surrogate for average rotation age.
Data from Smith et al., 2001.

TABLE 5. Age distributions for forests, by regions of the United States

| Region | Age classes (years) | | | | | | | | |
|---|---|---|---|---|---|---|---|---|---|
| | 0 to 19 | 20 to 39 | 40 to 59 | 60 to 79 | 80 to 99 | 100 to 149 | 150 to 199 | 200 and older | Uneven aged |
| | | | | | percent | | | | |
| United States: | 18 | 18 | 20 | 16 | 10 | 8 | 3 | 2 | 5 |
| Northeast | 4 | 12 | 19 | 22 | 14 | 6 | 0 | 0 | 23 |
| North Central | 14 | 18 | 25 | 21 | 11 | 6 | 0 | 0 | 4 |
| Southeast | 33 | 20 | 22 | 16 | 6 | 1 | 2 | 0 | 0 |
| South Central | 25 | 32 | 26 | 10 | 3 | 0 | 1 | 0 | 4 |
| Great Plains | 10 | 21 | 22 | 20 | 17 | 6 | 2 | 0 | 0 |
| Intermountain | 10 | 4 | 9 | 20 | 22 | 23 | 9 | 3 | 0 |
| Alaska | 4 | 11 | 17 | 9 | 9 | 10 | 8 | 31 | 2 |
| Pacific Northwest | 17 | 13 | 15 | 13 | 12 | 17 | 7 | 6 | 0 |
| Pacific Southwest | 13 | 6 | 9 | 11 | 13 | 28 | 12 | 8 | 0 |

Note: When compared with Figure 5, this table indicates that some regions may have low amounts of some forest structures that are important for animal and plant species that need these structures.
Data from Smith et al., 2001.

TABLE 6. Area of ecological provinces (Bailey, 1983) in the eastern United States remaining in forests

| Province | Forest area (million ha) | Area in forests (%) |
|---|---|---|
| Everglades | 3.6 | 19 |
| Lower Mississippi Riverine Forest | 2.8 | 28 |
| Eastern Broadleaf Forest (Continental) | 22.1 | 31 |
| Eastern Broadleaf Forest (Oceanic) | 12.6 | 56 |
| Southeastern Coastal Plain Mixed Forest | 32.2 | 59 |
| Southeastern Mixed Forest (Piedmont) | 28.8 | 63 |
| Laurentian Mixed Forest | 25.3 | 71 |
| Central Appalachian Broadleaf Forest–Coniferous Forest–Meadow | 12.0 | 72 |
| Ozark Broadleaf Forest–Meadow | 1.3 | 76 |
| Ouachita Mixed Forest–Meadow | 2.3 | 78 |
| Adirondack-New England Mixed Forest–Coniferous Forest–Alpine Meadow | 9.3 | 88 |

Data from Smith et al., 2001.

this century and the world is currently growing more wood than it is consuming, although its harvest is currently not equitably distributed (Oliver, 1999). Consequently, it will probably not be necessary to sacrifice some ecosystems to protect others–if the commodity and noncommodity values are fairly shared among ecosystems.

Global information collected during recent decades has given a strong understanding of the amount and distribution of the world's forest resources and conditions. The United Nations Food and Agriculture Organization (FAO) has classified the terrestrial world into ecological areas (Figure 2) (FAO 2001). A more detailed classification is being developed by the National Geographic Society and World Wildlife Fund (2005); however, the data are not yet readily available for this classification. The ecological zones are important because they represent broad divisions of areas that each need to be sustained. Some insights from this information are, among others:

- The ecoregion sizes, forest areas, and conditions within each country are very unequally distributed. For example, the ecoregion sizes vary from over 650 million ha to less than 10 million ha, and the forest area within each ecoregion varies from over 500 million ha to less than 10 million ha.
- Biodiversity is not equally distributed, with greater biodiversity generally at lower latitudes.
- Forest growth rates differ by ecoregion, with natural forests in moist, moderately cool areas (e.g., parts of Europe and the Pacific Northwest United States) growing as much as 8 $m^3$ $ha^{-1}$ $year^{-1}$ while those in colder, hotter, or drier regions grow much slower (FAO, 2001).
- Tree standing volume is also inequitably distributed (Figure 3), with nearly all conifers in the northern hemisphere, where they are native. Nonconifer forests are found in many places, but Africa contains the most nonconifer timber volume of any continent (FAO, 2001); the tropical forests there also contain very large trees (Huston, 1994).
- Carbon sequestered in forests is in the standing trees and in the soil. At colder latitudes, the proportion of carbon in the soil increases.
- Areas located at about 30 degrees latitude and isolated from the east coasts of continents are more likely to have seasonal water shortages (MacArthur, 1972), leading to fewer forests but with rel-

atively greater importance of forests for watersheds, and a tendency for these forests to burn.
- People are also concentrated in various places; however, most people live in forest ecosystems–areas that are forests or would be forests if people had not cleared them for agriculture or buildings (Tallis, 1991).
- Fifty-three percent of the wood consumed worldwide is being used for firewood; however, this consumption is not uniform (Oliver, 2001).
- In some regions–primarily the less developed countries–most people have rural, subsistence agriculture and grazing lifestyles; in others, people are rapidly migrating to cities; while in others, most people already live primarily in cities (Oliver, 2001).
- The economic well-being of people also differs greatly, with the per capita gross domestic product high in parts of Europe, North America, and a few other places and lower in many other regions.
- Consumption levels also differ, with the United States and western Europe consuming far more than the world average of resources (Oliver, 2003). If all countries consumed at these levels, it is unclear if the resources could be sustained.
- Other resources and values such as food and energy are also not equally distributed and have a large effect on human well-being; however, such considerations are beyond the scope of this paper.

Global data collected by the FAO and other organizations also allow more specific analyses of the biodiversity, commodities, and social conditions for each country (FAO, 2001). These data from FAO and other sources are being developed into a spreadsheet and will be available on the Yale Global Institute of Sustainable Forestry (http://research.yale.edu/gisf/) when completed. Like all data being updated periodically, their quality will improve with time; however, the current data allow an initial analysis that can be refined or disputed.

Tables 1 and 2 show some forest-related characteristics for selected countries. The countries selected are not intentionally singled out as the extremes, but are used to provide examples of information. Table 1 gives country-specific examples of some of the information described above. It shows a range in sizes and volumes of forests in different countries. Table 1 also shows that the population densities and per capita gross national product differ greatly among countries. Table 2 shows information about forest changes and proportions for the same

FIGURE 2. Global ecological zones, developed by the FAO. These can be used as ecosystem delineations at the global scale, especially if subdivided by continent. (Map from U.S.G.S. and U.N. FAO map, UN FAO 2001.)

**Global Ecoological Zones**

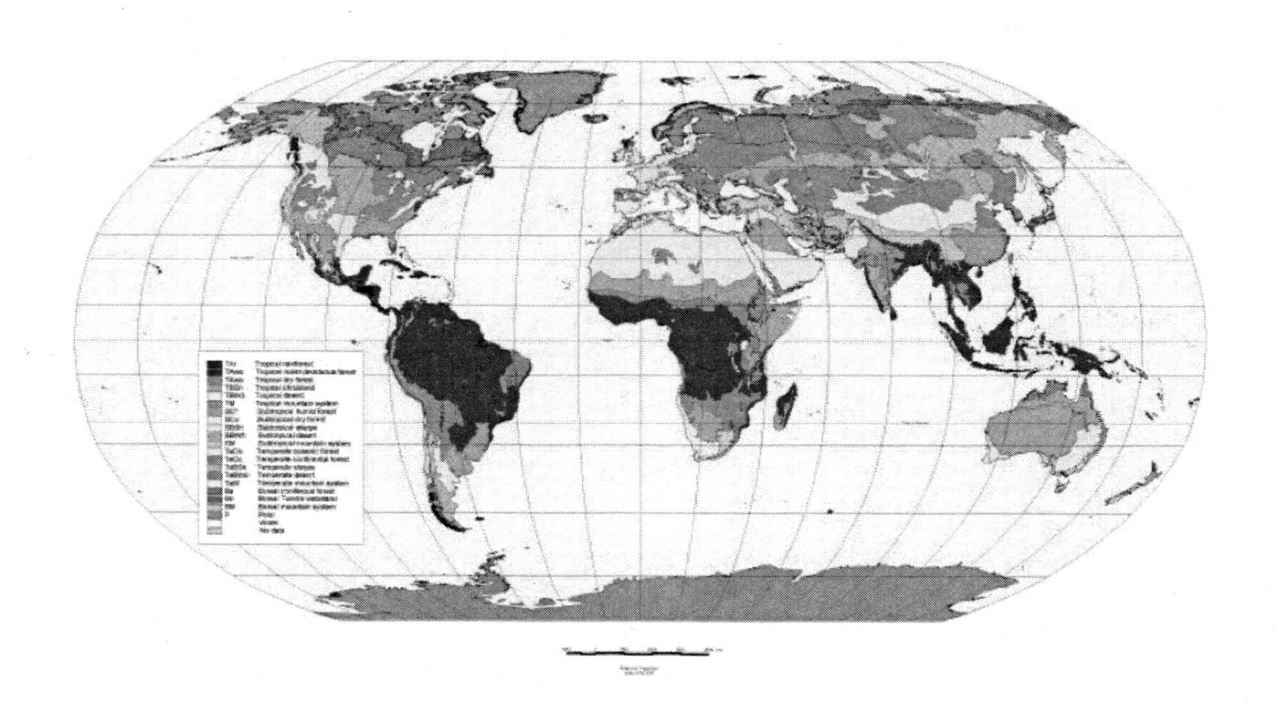

FIGURE 3. Locations of the world's accessible conifer and nonconifer timber. (Accessible is defined by FAO, 2001; map modified from U.S.G.S. and U.N. FAO map, UN FAO 2001.)

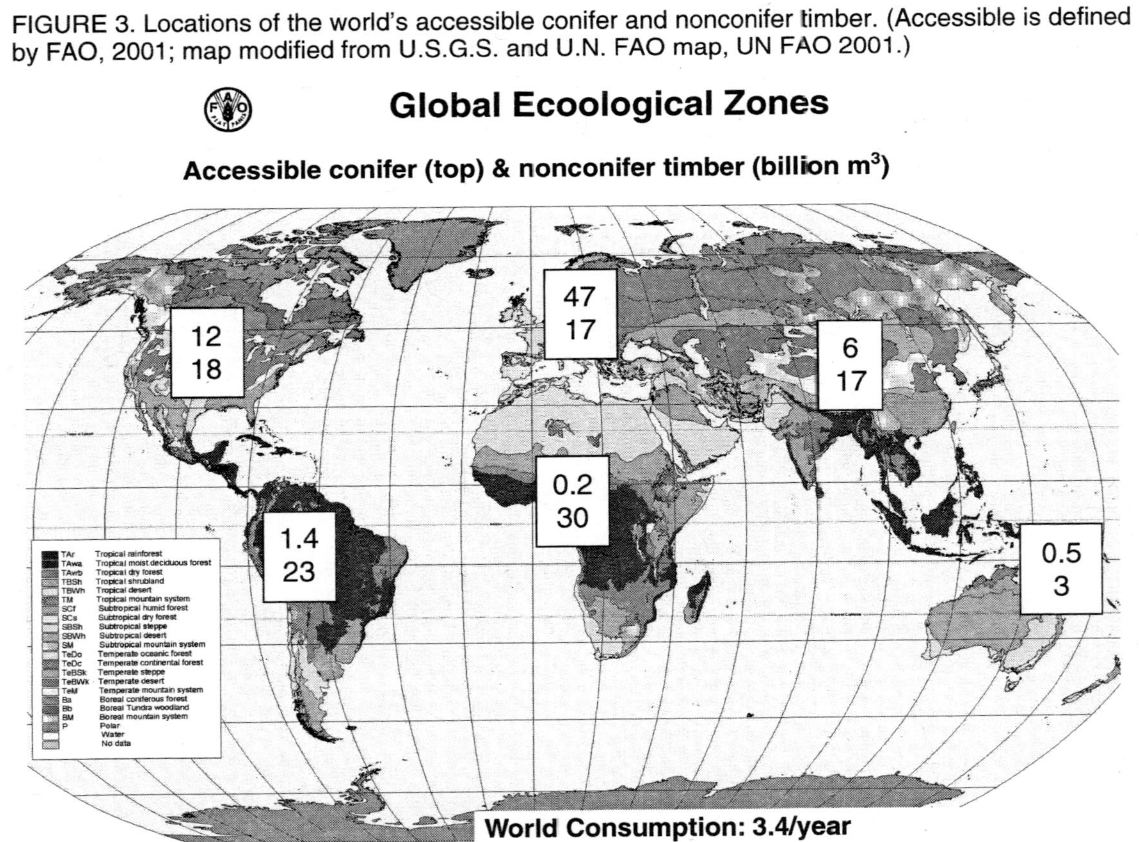

countries shown in Table 1. Several trends are suggested, although they need to be interpreted with caution, as will be discussed later:

- Some places with the greatest concentrations of country-endemic, endangered forest species tend to be islands.
- The ratio of standing forest to annual harvest is used as a rough surrogate for rotation age; growth is not included in this ratio, and so higher ratios (more relative growth before harvesting) for regions of rapid growth would result in rotation age being undervalued. Sustainable-harvest rotation ages would differ with species and other factors; however, Table 2 shows that some countries are probably excessively harvesting–with ratios of less than 50, while other countries are harvesting very little with ratios well over 200.
- Some countries are declining in forest area, while other countries are increasing. There does not seem to be a relation of harvesting rate (the standing volume/annual harvest ratio described above) to forest decline; however, there seems to be a relation of forest decline to fuelwood use. This relation is consistent with suggestions elsewhere (Oliver, 2001) that fuelwood use, rural subsistence lifestyles, and deforestation are related.
- There is a great difference in the amount of forest in protected areas, as defined by the FAO. For example, approximately 12.4 percent of the world's forests are in protected areas, on average (FAO, 2001); however, some countries have far less than this and others have far more.

Such analyses as in Tables 1 and 2 can be considered as a starting point for engaging in sustainable forestry within each country, not as a final analysis. Errors and omissions in the data may emerge as they become more continuously and consistently recorded. In addition, there may be subtleties in a nation's concerns that are masked by the numbers. For example, Table 2 suggests that Canada has relatively little forest in protected areas; however, Canada can be shown to have much larger protected areas (FAO, 2001) if a different definition of protected areas is given.

Further caution is needed in inferring solutions from these numbers because an inappropriate action that tries to improve a country's forest sustainability may exacerbate the problem, and a far more appropriate solution may be counterintuitive. For example, Ghana appears to be overharvesting its timber (Table 2). Consequently, banning wood export may appear to be an appropriate solution; however, most of Ghana's harvested wood is used for subsistence fuelwood. Another important concern is the widespread use of illegal logging in some countries. The data reported on volume harvested (Table 2) may significantly underestimate the harvesting in countries such as Honduras and Nicaragua where as much as 80% of the logging is illegal. A more appropriate solution may be to develop a value-added wood products export industry that will increase the people's standard of living and allow them to purchase hydroelectric power, rather than deforest their country for fuelwood.

There are other challenges with global analyses as well. One concern is that the delineation of ecosystems and the subdividing or grouping of species versus varieties or genera could be used to enable overprotection or exploitation of certain ecosystems. It is also difficult to develop a narrow measure of "fair share" for each value. The conditions of each ecoregion vary widely with climate change, species migrations and introductions, economic development and other social changes, and human migration.

The uneven distributions of commodity and noncommodity values, people, economies, and consumption mean that sustainable forestry will not be best achieved by simply providing and protecting an equal proportion of each ecosystem's resources. In addition, sustainable forestry is too complicated to be achieved through international centralized planning and dictating of activities. Not only would such centralized planning lead to bureaucratic inefficiencies, but there are so many subtle variations within each ecoregion and country that the uniformity of a top-down procedure would probably miss these and create more harm than good.

A more effective solution than international centralized planning is a diffuse approach that allows each country to examine its contributions and to rectify its excesses. Even in an environment of institutional downsizing, governments have an important role to play including formulating and applying policy that encourages responsible stewardship and developing a leadership role in efforts to curb illegal logging. Before such activities begin, it may not even be necessary to have a specific target of "fair share" for each value, because much of the effort for a long time will be to change the condition of those countries whose forests are in extreme conditions–too much or too little wood harvesting, rapid loss of biodiversity including endangered species, extremes of rural poverty, and others. A very important role for international institutions using this diffuse approach is to collect and make available the data that allows assessments. International public opinion and pressures will help avoid rogue behaviors.

## *WITHIN-COUNTRY ANALYSES*

It is also useful to examine conditions at increasingly smaller scales. Average data for a country can indicate its overall condition, but may hide imbalances within a country. For example, the United States appears to be managing its forests very sustainably on average (Table 2). According to the FAO data, it contains 40% of its forest area in protected areas. It has a standing volume/harvest ratio that is not so high that the forest is not allowed to grow and not so low that it is not providing a "fair share" of wood (Table 2). And, it has a relatively low number of country-endemic, endangered forest species.

Within the United States, however, the situation appears less sustainable. Depending on how protected areas were calculated, the United States overall estimated that it had either 7 or 33 percent of its forests under nontimberland classification in 2003 (Table 3), whereas the FAO reported 40% for the United States (Table 2) (FAO, 2001). According to any of these figures, the protected areas are not well distributed because most protected areas are in the western United States but over half of the forest area is in the eastern United States (Table 3).

In addition, the harvest in the mid-1990s was disproportionately high in the southeastern United States (Table 4)–an area of currently relatively little standing timber volume (Figure 4). (Calculations of growth, standing volume, and harvest for the United States shown in this paper exclude "reserves and other nontimberland," Tables 3 and 4.) An interesting anomaly in Table 4 is the apparently high percentage of net growth that is harvested in Alaska, even though little volume is harvested there. Much of Alaska is unmanaged, and there is so much natural mortality that net growth is very small. Consequently, although there appears to be a high harvest rate compared to growth, the standing volume/harvest ratio of 179 (Table 4) indicates that harvesting is not excessive. Further exacerbating the unequitable harvesting and distribution of protected areas within the United States is the fact that the south and southeastern United States are the areas of highest harvest and least protected area; however, more species–and endangered species–are found there than in almost any other region of the United States (Currie and Paquin 1987, U.S. Fish and Wildlife Service, 2005).

FIGURE 4. Standing volume of timber for regions of the United States (adapted from Smith et al., 2001). Numbers correspond to the following regions: 198 = Alaska; 265 = Pacific Northwest; 223 = Pacific Southwest; 142 = Intermountain West; 95 = North Central; 114 = Northeast; 109 = Southeast; 88 = South Central.

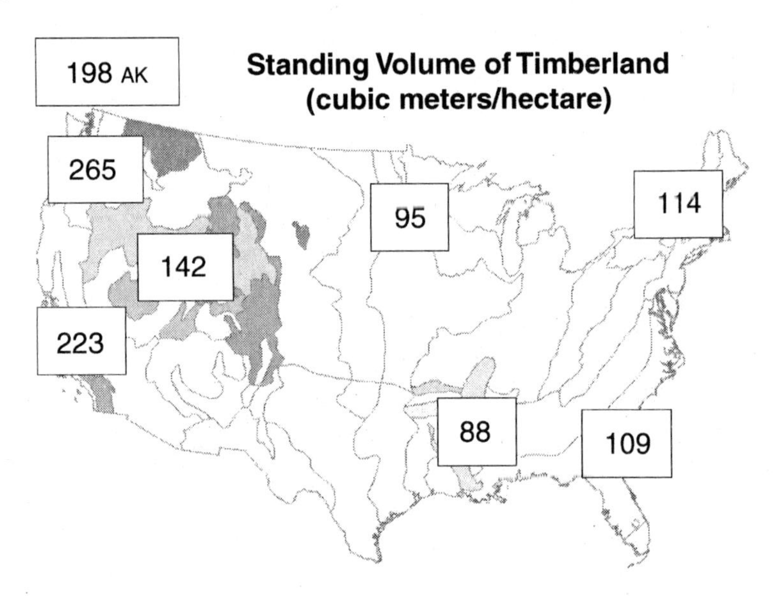


The disproportionate harvest within the United States is also reflected in the age distributions of forests in the regions. Areas in different age classes in forests can give an indication of the amount of area in each stand structure (Figure 5). Because different species depend on different stand structures, a somewhat balanced proportion of each structure is needed to maintain robust habitats for all species. Table 5 shows the age distributions of forests in different regions of the United States. As can be seen, the Southeastern and South Central areas appear to have little old forests (age classes 80 years and above, and uneven-aged forests) but extensive open structure (age class 0 to 19 years) because of the extensive harvest. The Intermountain West and Northeast have very little open structure. The Intermountain West's lack of open structure is because little harvest has been done there recently, whereas the Northeast has been primarily selectively harvested–or possibly high graded–and many uneven-aged forests and little open structure were created.

A country can face difficult challenges in its attempts to make its for-

FIGURE 5. (Top) Stand structures characteristic of most natural forests. Different species depend on each structure. (Bottom) Example of age-class distributions of area, with different age classes roughly corresponding to different structures. (Structures from Oliver and Larson, 1996; age classes from Smith et al., 2001.)

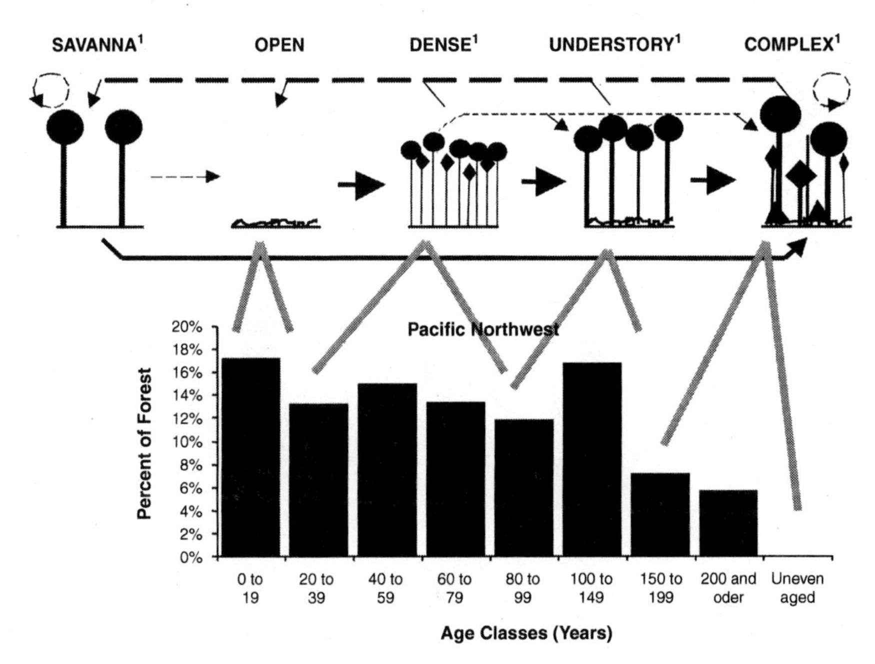


ests more sustainable. For example, to provide its fair share of wood to the world as well as habitats, protection, and other values for all of its forest ecosystems, the United States would need to readjust its forest activities in many regions. It would need to shift more of its timber harvest to its western forests, change its harvesting practices to create more openings in the northeastern United States, and establish more protected areas in the eastern United States.

## *MANAGING WITHIN AN ECOSYSTEM*

Specific activities need to occur within specific ecosystems to improve sustainable forestry. Forest ecosystems, or parts of ecosystems that cross national borders, can be managed sustainably within each country with the same spatial and intergenerational definition of sustainable forestry. The first issue is to determine the area of forest in each ecosystem. Ecosystems that are naturally small, diverse, or otherwise unique will need careful attention to ensure that diverse forest values are provided. Forests in other ecosystems may be small because of extensive clearing for other human uses; these will require efforts to restore more forests and to ensure multiple forest values are provided within the limited forest area. For example, Table 6 shows the percentage and total remaining areas of forest in each ecological province in the eastern United States (as defined by Bailey, 1983). The Lower Mississippi Riverine Forest province contains the smallest proportion and total amount of remaining forest area, with the partial exception of the Everglades. Consequently, more attention will need to be directed to restoring these forests than attending to other forests. Other ecosystems may contain larger forest areas and will require other considerations.

A robust, generalized approach to managing forests within each ecosystem has been developed by Seymour and Hunter (1999), known as the "triad" approach. This approach zones the forest into three types of management: protected areas, integrated management, and intensive plantations.

### *Protected Areas*

Protected areas are areas where the forest is either not managed or is managed to maintain an ecosystem-specific, often prehistorical condition. The original justification for protected areas was the ecological theory that assumed a balance of nature such that forests remained in a climax, or steady state condition that resembled old-growth forests when not interrupted by modern people. Although this steady state ecological theory is no longer considered valid (Botkin, 1979; Stevens, 1990; Oliver and Larson, 1996), protected areas still serve a function of maintaining some areas relatively free of human activity in case people later learn that their most conscientious activities to sustain forests have inadvertently had a damaging effect on some species or ecosystem processes.

### *Integrated Management Areas*

Integrated management is based on the current, dynamic ecological paradigm in which the forest has always been constantly changing through growth and disturbances. These disturbances and growth create a variety of stand structures (Figure 5) (Oliver and Larson, 1996). Some species depend on each of these structures for survival, so all structures are needed to maintain the natural biodiversity. Integrated management uses silvicultural operations to avoid, mimic, or recover from disturbances to maintain the natural distribution of structures.

### *Intensive Tree Plantation*

Intensive plantations for various commodities can also be established–especially in those large, stable forest ecosystems that contain enough protected areas and integrated management areas. It is questionable whether it is currently financially worthwhile to invest in these plantations in many places in the world (Oliver and Mesznik, 2005); however, there may be temporary needs for intensive plantations for noneconomic reasons. At some future time, integrated management may not be able to provide all of the world's consumption of wood, and intensive plantations may prove to be economically viable as a supplement.

## *REGIONAL IMPLEMENTATION– THE PACIFIC NORTHWEST, USA, EXAMPLE*

Even after general targets for a region within a country are established, achieving the targets requires a general vision for the region and many individual actions working in concert toward this vision. The actions are case-specific and require an assessment of the region's needs and goals and a plan to achieve the goals. The Pacific Northwest region of the United States is presented as an example of the some of the major issues related to regional implementation of sustainable forest management. The Pacific Northwest is potentially one of the world's major timber-producing regions (Haynes, 2003). The Pacific Northwest will need to continue providing its present levels of timber, or even increase its production, to prevent other regions and countries from overharvesting to compensate for its lack of harvest–that is, for the Pacific Northwest to provide its "fair share" of timber. Issues relating to the ecological, social, and economic frameworks of sustainable forestry, however, have and will play a major role in future wood production of the region. There have been both public interest in assuring that forests are being sustainably managed and a desire by landowners and forest managers to demonstrate their commitment to responsible stewardship.

To address concerns about sustainable forestry in the region, the Pacific Northwest Research Station of the United States Forest Service sponsored a 3-year Sustainable Wood Production Initiative (SWPI) (Deal and White, 2005). Researchers and cooperators involved with the initiative conducted a series of client meetings and invited a wide array of forest landowners and managers representing forest industry, small private forests, state forestry, and others interested in growing and producing wood. The most pressing need mentioned by almost all forest landowners and managers was the need to identify and understand barriers to sustainable forestry. These focus groups identified five major topics as priorities. These topics and the associated objectives of the SWPI are:

- Identify and understand the major economic, ecological, and social issues relating to wood production in the Pacific Northwest in the broad context of sustainable forestry.
- Identify barriers to sustainable forestry and assess the impacts of market incentives and environmental regulations on sustainable forest management.
- Develop a regional assessment of resource trends and market conditions including the long-term economic viability of forestry in the region.
- Identify and assess niche market opportunities for small woodland owners in the Pacific Northwest.
- Identify emerging technologies for wood products and summarize and synthesize new and existing information on wood technology.

Key issues relating to sustainable forestry included economic, ecological, and social components. The collective findings suggest that in the future, the region's wood supply will primarily come from private land, and the barriers and opportunities related to sustainable wood production will have more to do with future markets, harvest potential, land use changes, and sustainable forestry options than with traditional sustained yield outputs (Deal and White, 2005). Private lands in the Pacific Northwest should be able to sustain recent historical harvest levels over the next 50 years, but declines in sawmilling capacity and uncertain market conditions may affect wood production in the region. Public perceptions of forestry, land use changes, and alternative forestry options were also important considerations for sustainable forestry.

Once an infrastructure is in place, management within an ecosystem can be done by dividing the forests into political or geographic units, such as watersheds. And, these can be further subdivided by ownerships or other convenient means. Eventually, coordination of stand-specific prescriptions will be needed at broad scales–levels of 50 to 10,000 ha–with the exact scale depending on the ownership structure, political boundaries, and sensitivity of the ecosystem. Planning for integrated management may involve several owners, will require accounting for many values simultaneously, and will require the planning and management to be transparent. The public will expect the manager to explain the expected outcomes of the management and follow through as planned, just as an architect develops and follows a building design.

For planning and management, the challenge is develop a plan that is both flexible and specific. It needs to be flexible enough to allow adjustment to local variations created by delaying operations because of weather, responses to market fluctuations, and variations in the sampled inventory. On the other hand, the plan needs to be specific enough to ensure the forest will provide the expected values at the designated times. A general practice is to develop deterministic outcomes by using computer models (Oliver et al., 2004; Twery, 2005) but allow periodic (e.g., 5-year) windows for implementing the operations. The outcome will vary depending on which of the 5 years each operation is performed. Consequently, the variation is adjusted by the periodic monitoring and quality improvement. Eventually, forest plans will become instruments like architectural or engineering plans, with various levels of approval needed to adjust for unforeseen circumstances.

Policies will be needed to encourage the balance of management activities among landowners that enable a forest ecosystem, in total, to be sustainable–and sustainable relative to the other ecosystems within the nation and world. For example, incentives such as payment for environmental services (Daily and Ellison, 2002; Jenkins et al., 2004), carbon sequestration, or watershed protection can be implemented to encourage different management behaviors. The policies and actions will also need to address social and economic considerations of sustainability that not only are equally important but also are necessary to achieve sustainable forests.

## *CONCLUSIONS*

Data exist to begin determining how well each country is practicing sustainable forestry by comparing the country's performance with other countries. The comparison can be based on the values people want from forests based on the Montreal Process or similar criteria (Burley, 2001). Data exist in many countries to assess within-country sustainability of practices in different regions or ecosystems. Once a country or region's performance is assessed, situation-specific actions can be undertaken to make the country or region more sustainable.

## REFERENCES

Bailey, R.G. 1983. Delineation of ecosystem regions. *Environmental Management* 7:365-373.

Botkin, D.B. 1979. A grandfather clock down the staircase: stability and disturbance in natural ecosystems. In: pp. 1-10. Forests: fresh perspectives from ecosystem analysis. Proceedings of the 40th annual biology colloquium. Oregon State University Press, Corvallis, OR.

Brundtland, G.H. 1987. Our common future. World Commission on Environment and Development, United Nations. Oxford University Press, New York.

Burley, J. 2001. International initiatives for the sustainable management of forests. In: pp. 95-102. S.R.J. Sheppard and H.W. Harshaw (eds.). Forests and landscapes: linking ecology, sustainability, and aesthetics. CABI Publishing in Association with The International Union of Forest Research Organizations, New York.

Currie, D.J. and V. Paquin. 1987. Large-scale biogeographical patterns of species richness of trees. *Nature* 329:326-327.

Daily, G.C. and K. Ellison. 2002. The new economy of nature: the quest to make conservation profitable. Island Press, Washington, DC.

Deal, R.L. and S.M. White. 2005. Understanding key issues of sustainable wood production in the Pacific Northwest. Gen. Tech. Rep. PNW-GTR-626. USDA Forest Service, Pacific Northwest Research Station, Portland, OR.

Food and Agriculture Organization (FAO). 2001. Global forest resources assessment 2000. FAO Forestry Paper 140. Food and Agriculture Organization of the United Nations, Rome.

Haynes, R.W. (Tech. coord.). 2003. An analysis of the timber situation in the United States: 1952 to 2050. Gen. Tech. Rep. PNW-GTR-560. USDA Forest Service, Pacific Northwest Research Station, Portland, OR.

Helms, J.A. (Ed.). 1998. The dictionary of forestry. Society of American Foresters, Bethesda, MD.

Huston, M.A. 1994. Biodiversity: the coexistence of species on changing landscapes. Cambridge University Press, Cambridge, UK.

Jenkins, M., S.S. Scherr and M. Inbar. 2004. Markets for biodiversity services: potential roles and challenges. *Environment* 46(6):32-42.

MacArthur, R.H. 1972. Climates on a rotating Earth. In: pp. 5-19. Geographical ecology: patterns in the distribution of species. Harper & Row, New York.

Marsh, G.P. 1864. Man and nature. Reprint, edited by D. Lowenthal, 1967. Belknap Press, Cambridge, MA.

Montreal Process Working Group. 1998. The Montreal process. Retrieved October 21, 2005 from http://www.mpci.org.

National Geographic Society; World Wildlife Fund. 2005. Terrestrial ecoregions of the world. Retrieved May 10, 2005 from http://www.worldwildlife.org/science/ecoregions/terrestrial.cfm.

Oliver, C.D. 1999. The future of the forest management industry: highly mechanized plantations and reserves or a knowledge-intensive integrated approach. *Forestry Chronicle* 75(2):229-245.

Oliver, C.D. 2001. Policies and practices: options for pursuing forest sustainability. *Forestry Chronicle* 77(1):49-60.

Oliver, C.D. 2003. Sustainable forestry: what is it? How do we achieve it? *Journal of Forestry* 101(5):8-14.

Oliver, C.D. and B.C. Larson. 1996. Forest stand dynamics, updated edition. John Wiley & Sons, New York.

Oliver, C.D., J.B. McCarter, C.S. Nelson, K. Ceder, J. Hall, M. Johnson, J. Comnick, S. Humann, M. McKinley, L. Mason, P.S. Park, M. Andreu, L. Ingaramo, C. Manriquez and J. Cross. 2004. The landscape management system (LMS) and associated methods and tools for ensuring sustainable management and planning of forest landscapes. Paper presented at the International Conference on Modeling Forest Production. University of Natural Resources and Applied Life Sciences, Vienna, Austria.

Oliver, C.D. and R. Mesznik. 2005. Investing in forestry: opportunities and pitfalls of intensive plantations and other alternatives. *Journal of Sustainable Forestry* 21(4): 97-111.

Perlin, J. 1991. Forest journey: the role of wood in the development of civilizations. Harvard University Press, Cambridge, MA.

Seymour, R.S. and M.L. Hunter, Jr. 1999. Principles of ecological forestry. In: pp. 22-61. M.L. Hunter, Jr. (ed.). Maintaining biodiversity in forest ecosystems. Cambridge University Press, New York.

Smith, W.B., J.S. Vissage, D.R. Darr and R.M. Sheffield. 2001. Forest resources of the United States. Gen. Tech. Rep. NC-GTR-219. USDA Forest Service, North Central Research Station, St. Paul, MN.

Stevens, W.K. 1990, July 31. New eye on nature: the real constant is eternal turmoil. New York Times, B5-B6.

Tallis, J.H. 1991. Plant community history: long-term changes in plant distribution and diversity. Chapman and Hall, New York.

Twery, M.J. 2005. NED: a set of tools for managing nonindustrial private forests in the East. Retrieved May 10, 2005 from http://www.fs.fed.us/ne/burlington/ned/pub1.htm.

U.S. Fish and Wildlife Service. 2005. Species information: threatened and endangered animals and plants. Retrieved September 5, 2005 from http://www.fws.gov/endangered/wildlife.html.

doi:10.1300/J091v24n02_03

# Environmental Change and the Sustainability of European Forests

Peter Freer-Smith

**SUMMARY.** It is some years now since forest decline was a major public concern in Europe and was one of the principal environmental issues around which international research programs were focused. A number of internationally coordinated activities were initiated in the 1980s and 1990s and have continued until now; these contributed significantly to our current understanding of forestry and to the way in which forestry policies have developed. In short, the concept of sustainable development has had an increasing influence and is now of immeasurable value in forest policy, with sustainable forest management well established as its guiding principle. This sequence of events is examined here. The extent to which understanding has advanced is remarkable; much has changed. Arguably this period gave the first indication of the extent to which forests and ecosystems globally are threatened by environmental change. On the basis of the last 20 years, it is tempting to conclude that we now have an effective institutional framework and have made excellent progress. However, some of the recorded ecosystem responses seem

---

Peter Freer-Smith, Forest Research, Alice Holt Lodge, Farnham, Surrey, GU10 4LH, UK (E-mail: peter.freer-smith@forestry.gsi.gov.uk).

The author would like to thank FAO's Global Forest Resource Assessment and the UNECE Data Coordination Centre of ICP Forests, BFH Hamburg for permission to reproduce Figures 2, 3, and 4, and the Forestry Commission for support to attend the IUFRO XXII World Congress.

[Haworth co-indexing entry note]: "Environmental Change and the Sustainability of European Forests." Freer-Smith, Peter. Co-published simultaneously in *Journal of Sustainable Forestry* (Haworth Food & Agricultural Products Press, an imprint of The Haworth Press, Inc.) Vol. 24, No. 2/3, 2007, pp. 165-187; and: *Sustainable Forestry Management and Wood Production in a Global Economy* (ed: Robert L. Deal, Rachel White, and Gary L. Benson) Haworth Food & Agricultural Products Press, an imprint of The Haworth Press, Inc., 2007, pp. 165-187. Single or multiple copies of this article are available for a fee from The Haworth Document Delivery Service [1-800-HAWORTH, 9:00 a.m. - 5:00 p.m. (EST). E-mail address: docdelivery@haworthpress.com].

Available online at http://jsf.haworthpress.com

doi:10.1300/J091v24n02_04

anomalous; there are surprises in system responses, e.g., the linkage between sulphur and nitrogen depositions and forest growth. Even more importantly some specific pollutant problems remain and will intensify, and climate change has become an environmental issue of overwhelming importance. doi:10.1300/J091v24n02_04 *[Article copies available for a fee from The Haworth Document Delivery Service: 1-800-HAWORTH. E-mail address: <docdelivery@haworthpress.com> Website: <http://www.HaworthPress.com> © 2007 by The Haworth Press, Inc. All rights reserved.]*

**KEYWORDS.** Forest declines, pollution, climate change, biodiversity evaluation, EU monitoring schemes, European forestry conventions, forest management

## *INTRODUCTION*

In order for woodlands and forests to support sustainable development in rural and periurban areas, it is necessary for their managers to have sustainability as an objective and to have an understanding of how to achieve sustainable management. Some years ago society might have been content to see sustainable forestry as the removal of roundwood and residues from forests at rates that did not exceed the overall increment of growing stock sustained by silviculture, forest improvement, and afforestation. It remains true that the ability to grow successive rotations without loss of productivity not only contributes to the economics of forestry but, perhaps more importantly, is a long-term indicator of forest condition and sustainability (see for example Evans, 1994). Indeed there has been considerable interest in the recording of total forest increment as a parameter against which to view total removal, with a general conclusion for European forests that average annual increments have in recent years tended to exceed removal (Spiecker et al.,1996)–a balance that has the potential to influence forest condition very significantly. However, this production-orientated view of sustainability no longer applies, as we now have a considerable body of biogeochemical data that show that sustainable forest management is far more complex than the simple achievement of sustained roundwood supplies.

Figure 1 illustrates a contemporary view of sustainable forest management (Forestry Commission, 2003). It is recognized that in addition to traditional timber products, forests provide social, environmental, and economic benefits. Not all the economic, environmental, and social benefits are specified in Figure 1; protection of freshwater supply catch-

FIGURE 1. The concept of sustainable forest management (Forestry Commission, 2003)

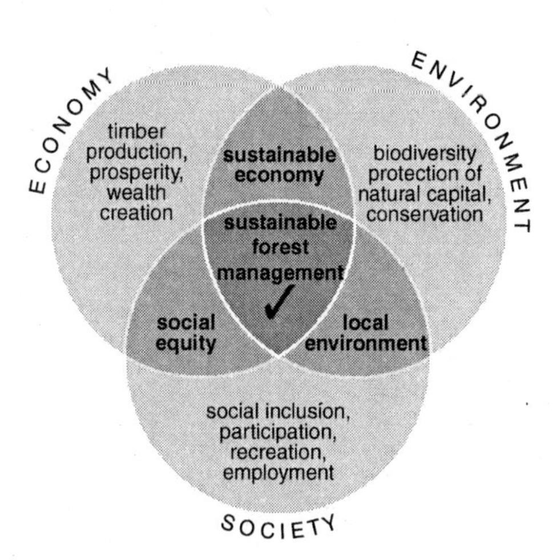


ments, carbon sequestration, landscape and aesthetic value are specifically important. However, it can be argued that all benefits fall within one of the three circles of the diagram. Strategies for achieving economic, environmental, and social benefits, and also an appropriate balance between the three, differ considerably across the various national forestry policies and programs (Buttoud et al., 2004). However, it is now widely recognized that in order to achieve sustainable forest management, all three areas need to be considered in detail. Failure to address any one of these three areas has resulted in problems in the past, and it is necessary to monitor forests to evaluate the extent to which sustainability is being achieved under current management. The requirement to monitor specific criteria within forests is necessary not only to achieve sustainable forest management, for which policy and practice are determined at the national level in Europe, but also to meet the commitments and reporting requirements of a number of European and international conventions such as the following:

- U.N. convention on Long-Range Transboundary Air Pollution 1979
  - Eight protocols and critical loads maps
  - International Cooperative Programme on Assessment and Monitoring of Air Pollution Effects on Forests (ICP Forests)
- U.N. Forest Forum and Statement of Forestry Principles

- European Union (EU) regulations on monitoring of forest condition:
  - European Economic Community Regulation No. 3528/86
  - European Economic Community No. 2158/92
  - Forest Focus European Council Regulation (EC 2152/2003)
- EU Habitats, Species, and Birds Directives
- EU Forest Reproductive Materials and Plant Health Directives (1990/105/EC)
- Ministerial Conferences on the Protection of Forests in Europe:
  - Strasbourg 1990
  - Helsinki 1993
  - Lisbon 1998
  - Vienna 2003
- U.N. Conference on Environment and Development: Rio de Janeiro 1992, Johannesburg 2002
  - Statement of Forest Principles and establishment of the Intergovernmental Panel on Forests and Intergovernmental Forum on Forests
  - Framework Convention on Climate Change and Kyoto Protocol
  - Convention of Biological Diversity

Many European forests are also certified by user-led accreditation schemes that require compliance with often quite complex requirements.

The EU Forestry Strategy provides a framework for forest-related actions in the EU. The strategy takes into account existing EU legislation in the forest sector and the commitments made by the EU and its member states in all relevant international processes, in particular the *United Nations Conference on Environment and Development in 1992 (UNCED)* and its follow-up, as well as the *Ministerial Conferences on the Protection of Forests in Europe (MCPFE)*. The strategy emphasizes the importance of the multifunctional role of forests and sustainable forest management (see Haynes, this volume), and focuses attention on key points (see Mason, this volume), as follows:

- Forest policy is a responsibility of the individual member states, but the EU can contribute to the implementation of sustainable forest management through common policies based on the principle of subsidiarity and the concept of shared responsibility.
- Implementation of international commitments, principles, and recommendations through national or subnational forest programs

developed by the member states and active participation in all forest-related international processes.
- The need to improve coordination, communication, and cooperation in all policy areas of relevance to the forest sector, both within the commission and the member states, and also among the member states.

Before detailed consideration of the various monitoring and reporting requirements and of the science/policy interface that becomes all important in this context, it is worth looking at a little of the history and at some of the recent and current environmental threats that confront European forestry today. Anthropogenic environmental change is of particular interest as forests have a significant role in both the mitigation of adverse impacts and in adaptation strategies. In this review, I do not specifically consider the economic and social factors identified in Figure 1; however, the integration of all three scientific disciplines is clearly essential to the achievement of sustainable forestry. An analysis of this type is not only of considerable interest in that we learn about how forests work biologically and physically, it also gives us the information we require to determine future priorities and to decide how they might be addressed.

## *THE ENVIRONMENTAL THREATS TO THE SUSTAINABILITY OF FORESTS*

### *Forest Decline*

On a geological timeframe, climate has been the principal factor in determining the extent and type of forest cover. To take a recent perspective only (of necessity), it is relevant that trees expanded from the southern European refugia in which they had survived the most recent ice age, extending their cover northward during the long retreat of the ice sheets between 18,000 and 7,500 BC. Some years ago we would have regarded climate as a natural factor responsible for influencing forest cover and condition, and would have listed it along with pests, fungal diseases, fire, wind, unseasonal frost and drought. However, today we have a different perspective and consider climate change along with the other known anthropogenic threats: pollutant deposition, overgrazing in pastoral woodland, insensitive harvesting, soil nutrient removal, physical damage to soil structure, the planting of trees with low genetic diversity, introduction of pest species, and alterations of hydrology.

Ciesla and Donaubauer (1994) have identified forest declines around the world that can be attributed to these factors and have also pointed out that forest dieback can be part of a normal ecological or cohort succession. Such successions can be driven by nutritional limitations, as in the case of *Metrosideros polymorpha* Gaud. in the Hawaiian Islands, or perhaps more commonly by fire or wind as natural disturbance factors. In managed forests, humans will have an influence even where the fundamental threat might be regarded as natural (e.g., the effect of species and genotype choice, cultivation method, or silvicultural system on susceptibility to windthrow).

The important conclusion is that natural factors including pests and diseases can have a very major influence on the condition and stability/sustainability of managed forests. I illustrate this important point by describing briefly the current threats posed by one entomological pest and one fungal pathogen in Europe today. I have chosen examples that pose a significant threat currently. Inevitably, their specific biology is of interest, but some more generic points also emerge.

### *Insect Pathogens*

Experience shows that systems made up of native tree hosts and native insects or fungal pathogens, that is systems that have co-evolved, are likely to be intrinsically more stable and ecologically better buffered than systems involving nonnatives. With an introduced tree species and a native pathogen or an introduced pathogen coming across a native tree species, co-evolution has not occurred and there is often a significant threat. Major pest and pathogen outbreaks have often occurred where either host or pathogen (invasive) or both are out of their natural range and ecological context (e.g., the natural enemies of insect pests or pathogens may be absent). The frequency of invasion by nonnative pests and pathogens has been increasing, perhaps as a result of increasing international/intercontinental movement of goods and people. The hosts that occur where co-evolution with new pathogens has not been possible may also be made more vulnerable by climate and other environmental change.

A current example of a risk of this type caused by an insect is the arrival of the pinewood nematode (*Bursapheleunchus* × *xylophilus*) in European forests (Evans et al., 2003). This nematode is a native of North America where it normally lives as a saprophyte on recently dead trees (Bergdahl, 1988) but can cause minor damage to introduced exotic tree species–notably Scots pine (*Pinus sylvestris* L.). The nematode is usually carried from tree to tree by longhorn beetles of the genus Monochamus. Over recent years *B. xylophilus* has spread from North America to Japan, China, and most recently to Portugal. Outside its native range and on tree species with which co-evolution has not occurred, the nematode is able to survive and breed in living trees by a combination of increased population density and the production of a toxin that causes rapid blockage of the xylem. Tree mortality occurs quickly and is characterized by a sharp decline in resin production followed by the reddening of foliage, wilt, and tree death. The nematode has killed millions of pine trees in Japan, China, and Korea, so its arrival in Portugal is a significant threat to European forests. The current area of forest infected in Portugal is 52,300 ha, and damage is to *Pinus pinaster* Soland., non Ait. (cluster pine) and *Pinus pinea* L. (umbrella pine).

A pest risk analysis for Europe concluded that the availability of susceptible tree species, particularly pines, is high. However, wilt disease is likely to be restricted to areas where the July or August isotherm is > 20°C and epidemic wilt occurs only at temperatures > 24°C (Evans et al., 1996). Spread is also dependent on the presence of its insect vector (*Monochamus*). Because of the dependence of disease symptoms on high temperature and greater susceptibility under conditions of moisture stress, the predicted climate change in Europe will extend the area of forestry that is at risk and will increase wilt expression. The concern is thus that although lower summer rainfall may reduce the incidence of some fungi, there will be other organisms, like pinewood nematode, for which the severity of outbreaks will increase. The combination of increased global transfer of exotic organisms associated with greater global trade and extended open markets, and climate change will result in more incidents of the type seen with pinewood nematode in Portugal.

### *Fungal Pathogens*

Globally, fungal pathogens are often cited as secondary or contributing factors to stand diebacks or forest decline (Ciesla and Donaubauer, 1994). However, there are also examples where detailed analysis shows that the introduction or evolution of new fungal pathogens is the principal cause of forest decline. Of the fungal pathogens that have caused problems in European forestry (Table 1) *Phytophthora* is undoubtedly the genera that causes the greatest concern to plant pathologists and foresters. The involvement of *Phytophthora cinnamomi* in the dieback of a

TABLE 1. Fungal pathogens considered to be the main causal agent or a major factor in forest declines and tree dieback in Europe

| Fungal pathogen (genus) | Disease | Location | Reference |
|---|---|---|---|
| *Ophiostoma* | Dutch elm disease | Europe and Southwest Asia | |
| *Phytophthora* | *Phytophthora* disease of alder (alder dieback) | Western Europe | Gibbs et al., 2003 |
| *Phytophthora* | Oak decline | Europe | Jung et al., 2000 |
| *Cryptococcus* and *Nectria* | Beech decline | Western and Central Europe | Schwerdtfeger, 1981 |
| *Mycosphaerella* | Redband needle blight (*Dothistroma* needle blight) | France and England | Villebonne and Maugard, 1999 |
| *Melampsora* | Poplar leaf rust<br>Chestnut blight | Europe | Newcombe et al., 1994 |

number of species of eucalyptus and of jarrah (*Eucalyptus marginata* Donn ex Sm.) in Australia is perhaps the most widely known tree decline caused by fungal pathogens. Many believe *P. cinnamomi* was introduced to Australia and that its occurrence and spread is certainly favoured by timber harvesting, road and power line construction, and other disturbances. Other fungal pathogens like *Heterobasidium*, the cause of conifer butt rot, are present relatively ubiquitously, and their control is part of normal management practice in some areas.

Table 1 lists examples of fungal pathogens that can be regarded as either the main causal agent or a major factor in forest declines in Europe. Of these examples, Dutch elm disease is probably the best known. This is because by the 1940s, the first wave of the disease, which was caused by *Ophiostoma ulmi*, resulted in the loss of 10 to 40% of elms in a number of European countries. A second and far more destructive outbreak began in the 1960s and was shown to be caused by a new, highly aggressive species of *Ophiostoma* (*O. novo-ulmi*) (Brasier, 2001). Although a scattering of trees escaped, this second fungus virtually eliminated all English elms (*Ulmus procera* Salisb.) from southern Britain. Dutch elm disease had a devastating effect on the English countryside, and although not a major crisis for forestry, these two major outbreaks illustrated the very major impact an introduced or "exotic" fungal pathogen can have. These examples also showed that outbreaks of fungal pathogens need to be approached through an understanding of their ecology and of the fungal population biology. Population studies of the two species, *O. ulmi* and *O. novo-ulmi*, raised the question of the rapid evolution of forest pathogens. It subsequently became clear that the virulent strain of alder *Phytophthora* was also the result of rapid evolution, in this case through hybridization of two species *P. cambivora* and *P. fragariae* (Brasier, 2001). Neither of these species was native to Europe, and both had probably been introduced through trade in plants or plant products.

Rapid evolution through hybridization between species has been a feature of rust on poplars, Dutch elm disease, and *Phytophthora* disease of alder (Brasier, 2000) and is also a factor in the current major concern over widespread oak mortality in California and Oregon. *P. ramorum*, the fungus responsible for the problems in California and Oregon, was found on trees in Europe in 2003 (Brasier et al., 2005). During surveys to establish the extent of the threat posed by *P. ramorum* in Britain, a previously unknown *Phytophthora* was isolated from a stem lesion on a mature European beech (*Fagus sylvatica* L.) in a woodland in southwest England. Like *P. ramorum*, the new fungus (named *P. kernoviae*) was causing widespread foliar necrosis and shoot dieback of dense understory rhododendrons. Although apparently not related to *P. ramorum*, *P. kenoviae* is associated with bark necrosis and bleeding stem lesions on *Fagus sylvatica*, *Quercus robur* L. and *Liriodendron tulipifera* L. and may well present a threat to forestry and natural ecosystems in Europe and the rest of the world (Brasier et al., 2005).

A series of introductions of invasive fungal pathogens and the apparent increasing occurrence of hybridization are probably both associated with increasing global trade in plants and plant products. This association needs careful consideration in the evaluation of risks to sustainability and design of plant health control policy.

### *Pollution*

The first recognition that emissions of the acidifying atmosphere pollutants sulphur dioxide ($SO_2$), nitrogen oxides ($NO_x$), and ammonia ($NH_3$) might damage both terrestrial and aquatic ecosystems on a regional and global scale came in the early 1970s (Graham et al., 1974). During the 1970s and 1980s scientific evidence for such damage accumulated, and the adverse impacts of lower pH in aquatic ecosystems became unequivocal. Scientific, public, and political alarm over the widespread impacts of air pollution resulted in the United Nations Economic Commission for Europe (UNECE) Convention on Long-Range Transboundary Air Pollution (CLTAP convention) in 1979. Since the establishment of the convention, national commitments set within the convention framework have attempted to reduce the emissions of acidifying pollutants across Europe through implementation of emission control protocols. In 1985, signatories agreed to reduce $SO_2$ emissions by 30% based on those in 1980 under the first sulphur protocol. Subsequent protocols have addressed $NO_3$, volatile organic compounds (VOCs), and ozone (O). The $NO_x$ protocol was the first to base targets on ecologically defined threshold depositions (critical loads and levels) although the critical loads approach had been developed in Scandinavia over some years prior to the 1990s.

A critical load is defined as "a quantitative estimate of the exposure to one or more pollutants below which significant harmful effects on specified sensitive elements of the environment do not occur according to present knowledge" (Nilsson and Grennfelt, 1988). Where pollutant inputs exceed the critical load, ecosystem damage may occur over an unspecified timescale. Comparison of critical loads for particular receptors with pollutant inputs on a spatial basis can be used to predict the risk and geographical extent of damage. The most recent protocols of the CLTAP were also based on the critical loads approach, e.g., the second sulphur protocol of 1994 and the Gothenburg protocol (1999), which targeted emission reductions for $SO_2$, $NO_x$ and $NH_3$ of 75%, 50% and 12% of the 1990 values by the year 2010. Clearly the development of the critical loads approach has dominated acidification research since the late 1980s and has achieved some notable successes in reducing pollutant emissions.

There has also been a major international effort to develop emission monitoring, modelling, and mapping of critical loads and their exceedence, and to monitor ecosystem response. Much of this effort has been coordinated under the CLTAP with the pollutant monitoring and mapping work being done by EMEP (Cooperative Programme for Monitoring and Evaluation of the Long-Range Transmission of Air Pollutants in Europe). Because of the widespread public concern over forest decline and the growing scientific evidence that pollutant depositions were part of the cause, the International Cooperative Programme on Assessment and Monitoring of Air Pollutant Effects on Forests (ICP Forests) was established in 1985.

### *Climate Change*

The Intergovernmental Panel on Climate Change (IPCC) was established by the United Nations and the World Meteorological Organisation in 1988. It was asked to assess the available information on climate change, assess the environmental and socioeconomic impacts, and formulate response strategies. The IPCC assessment reports have become standard references. The work of the first assessment report, and particularly the 1992 updates, were influential in the agreement on a framework convention on climate change at Rio in 1992. The few words in the second assessment report (IPCC, 1996)–"there is a discernible human influence on the climate"–represented a hugely important turning point in scientific consensus. Climate change represents not only a major challenge to science and society, but nearer to home, is a key reason for needing to develop understanding of, and policies to achieve, sustainable forest management.

The monitoring of carbon stocks in forest ecosystems and of the impacts of climate change on the distribution of forests, of species' composition within the different forest types, and of biodiversity in forests have become principal objectives of international forest monitoring and assessment programs (see below). The modelling of identified future climate scenarios and how they will influence the global distribution of dominant functional plant types (including types of forest) is just one illustration of the potential for some very major changes (Beerling and Woodward, 2001). A number of specific threats to forests are predicted to intensify with climate change. These include drought, storms (windthrow and damage), avalanches, snowbreak, and forest fires. Fire is a major problem for the sustainability of European forests, particularly in the Mediterranean region, and it is important for forest dynamics, and thus to some threatened species, in the boreal region. If consideration is extended into socioeconomic issues, then land use change (including deforestation), habitat fragmentation, and soil erosion become important. These are indeed major concerns currently, particularly in the tropics.

## *MONITORING AND EVALUATION OF FOREST SUSTAINABILITY*

Concern over the condition of the world's forests and, in particular, over the rate of deforestation worldwide, was the subject of intense debate at the Earth Summit on Sustainable Development in Rio in 1992.

The need to manage the world's forests sustainably was formally recognized through the signing of governments to a Statement of Forest Principles. Recognition of the multiple roles and values of forests has been a key influence in development of the concept of sustainable forest management with certification as a key instrument. At the European level this concept has influenced national forestry policies through the adoption of the guidelines for the sustainable management of forests at the second Ministerial Conference on the Protection of Forests in Europe in 1993. On a global scale, an important influence has been the United Nations Forum on Forests established in 2000 to promote the management, conservation, and sustainable development of all types of forests and to strengthen long-term political commitment to this end, based on the Rio Declaration of Forest Principles. Thus, international and many national forestry policies are now based firmly on the principles of sustainable development. Standards for sustainable forest management have been developed but do require maintenance, monitoring, and development as outlined in the MCPFC Pan-European criteria listed here:

1. Maintenance and appropriate enhancement of forest resources and their contribution to global carbon cycles.
2. Maintenance of forest ecosystem health and vitality.
3. Maintenance and encouragement of productive functions of forests (wood and nonwood).
4. Maintenance, conservation, and appropriate enhancement of biological diversity in forest ecosystems.
5. Maintenance and appropriate enhancement of protective functions in forest management (notably soil and water).
6. Maintenance of other socioeconomic functions and conditions.

The concept is dynamic, changing over time to reflect the situations in which it is applied and to adapt to social, economic, and environmental requirements.

### *Forest Area*

By Neolithic times (Neolithic culture spread across Europe between 6000 and 2000 BC), Europe had become almost entirely wooded (c. 80%). Rapid exploitation began in the Iron Age (c. 500 BC onward) and continued through the industrial revolution. In some European countries (e.g., Ireland and the United Kingdom) forest cover was as low as 2 to 4% in the early 20th century; in others the percentage of forest cover remained at c. 30% or more. In Europe as a whole, all forest cover probably sank as low as c. 20% and was at about this level in the early 1960s. However, as a result of active afforestation/reforestation policies, this has risen to a cover of c. 35% for the 25 EU (EU-25)countries today (27% in 1990). In the UK, for example, forest area has more than doubled since its low in 1919 (c. 4%) to 11.6% in 2003 (16.4% in Scotland, 13.8% in Wales and 8.4% in England). Sadly the situation is not the same globally, with forest cover continuing to decline according to the United Nations Food and Agriculture Organization's Global Forest Resource Assessment (FAO, 2000). World forest area has declined from c. 4.0 billion ha in 1946 (when the first assessment was initiated) to 3.9 billion ha in 2000. Between 1990 and 2000, the annual loss globally was c. 0.22% loss per year (an area the size of Portugal, 9.4 million ha). This loss is concentrated in tropical regions of America, Africa, and the Pacific Rim (see Figure 2). A degree of optimism can be taken from the reforestation in Europe and China, and from the decline in the rate of global forest loss that occurred during the 1980s and the 1990s.

It can be argued that for Europe and other developed countries, the condition or quality of wooded/forested areas is important, even perhaps more important than the area. At the extreme, a change from primary old-growth forest to secondary monoculture plantations might be seen as being as detrimental as deforestation. The monitoring of forest condition, quality, and biological diversity (perhaps all three together being at least indicative of biological sustainability) is thus important, and has been principally undertaken in Europe over the last 20 years as a result of the requirements of the U.N. Convention of Long-Range Transboundary Air Pollution and of a series of EU regulations that support the agreed actions (commitments) of the Ministerial Conventions on the Protection of Forests in Europe listed above.

Data on European forest cover, forest type (class), and potential or actual timber trends comes from the work of the UNECE Timber Committee and FAO European Forestry Committee. Early FAO forest assessments occurred in the late 1940s and early 1950s, and there have been 10 yearly assessments thereafter. The Timber Section of the joint FAO/ECE Agriculture and Timber Division has gradually developed more consistent systems of concepts and terms. Data are derived from national reports.

The EU regulations (Nos. 3528/86; 2158/92 and 2152/2003) were principally put into place as a result of the public concern that arose over forest decline (Waldsterben–literally meaning "death of the trees" in German) during the late 1980s and 1990s. A network of monitoring

FIGURE 2. Areas where forest cover decreased (red) and increased (green) at rates greater than 0.5% per year between 1990 and 2000. Reproduced with kind permission of FAO (source: FAO Global Forest Resources Assessment, 2000)

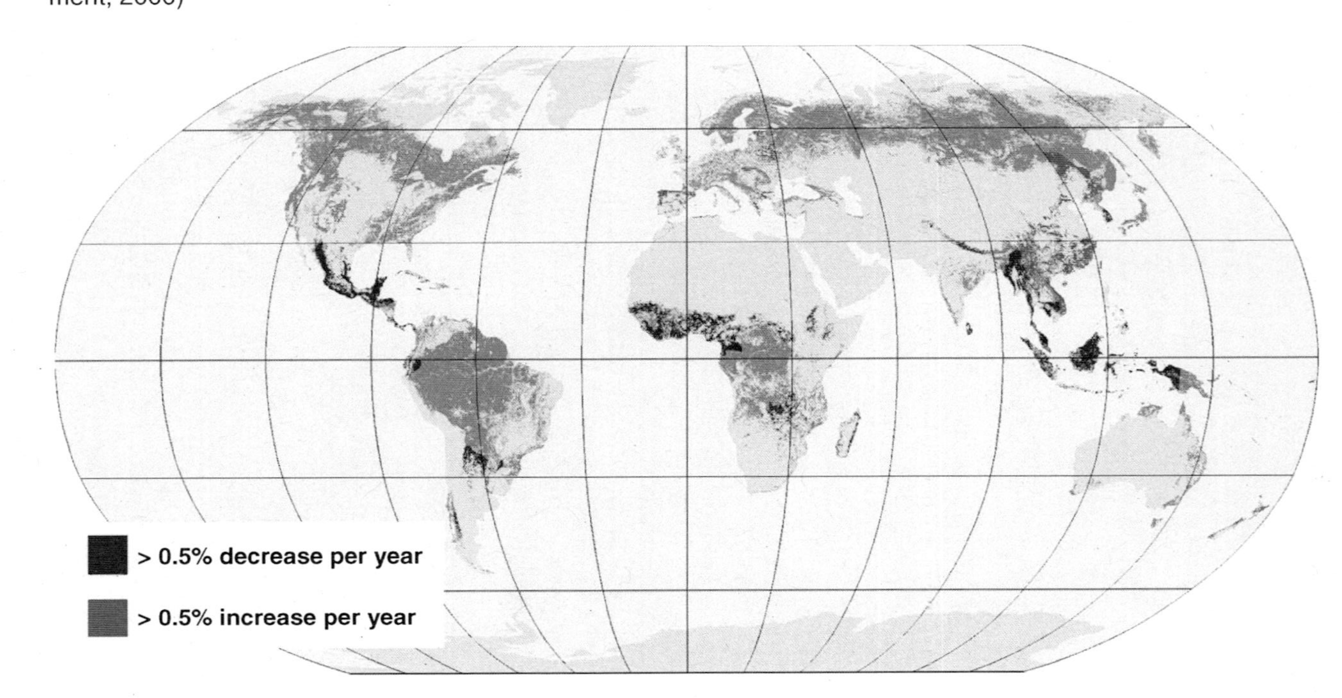

plots was established at two levels of measurement intensity (see Table 2). This monitoring scheme has been important scientifically but has perhaps had less impact at the science/policy interface than has the critical loads approach.

### *The EU Monitoring Scheme and National Forest Inventories*

The level 1 data have shown a rather extensive and sustained long-term decline of forest condition for some tree species (e.g., *Fagus sylvatica* and *Pinus pinaster*) as is indicated in Figure 3. Other species have been stable in condition over this wide geographical spread and time scale (e.g., *Quercus* spp. and *Pinus sylvestris*). Detailed analysis of the level 1 data has been undertaken in a number of studies (e.g., Schulze and Freer-Smith, 1991; Mather et al., 1995) and clearly indicates that ozone concentrations, sulphur depositions, and climatic conditions were important factors in determining trends of forest condition on regional and national geographic scales. At its simplest, it can be said that during the 1980s and early 1990s, forest condition declined more

TABLE 2. Parameters and frequency of measurement in the EU monitoring schemes under the Forest Focus Regulation

| Parameters surveyed | Level 1 | | Level 2 | |
|---|---|---|---|---|
| | **Frequency** | **Plots** | **Frequency** | **Plots** |
| Crown condition | Annual | All plots (6000) | At least annual | All plots |
| Foliar chemistry | Once | 1497 plots | Every 2 years | " " |
| Soil chemistry | Once | 5289 plots | Every 10 years | " " |
| Soil solution chem | | | Continuously | Part of all plots |
| Tree growth | | | Every 5 years | All plots |
| Ground vegetation | | | Every 5 years | All plots |
| Atmospheric deposition | | | Continuously | Part of all plots |
| Ambient air quality | | | Continuously | Part of all plots |
| Meteorology | | | Continuously | Part of all plots |
| Phenology | | | Several times/yr | Optional |
| Litterfall | | | Continuously | Part of all plots |
| Remote sensing | | | Preferable at plot installation | Optional |

FIGURE 3. Crown defoliation of six tree species between 1990 and 2004. Reproduced with kind permission of UNECE Data Coordination Centre, BFH Hamburg. (Source: UNECE, 2005)

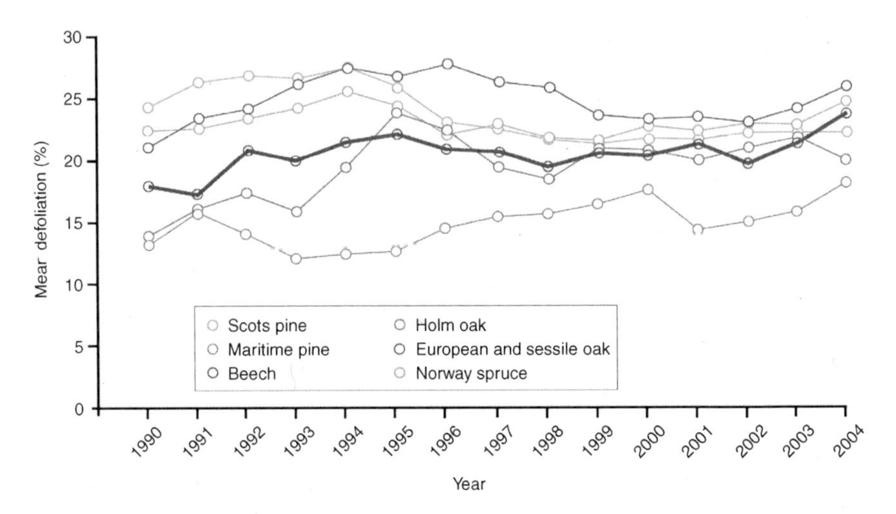


where sulphur depositions were greatest (Mylona, 1996). However studies like that of the impacts of the extreme drought in central Europe during 2003 have clearly indicated the importance of natural factors (insect pests, fungal pathogens) and particularly drought in determining forest condition (see Figure 4).

A more sensitive linkage between pollutant exposure or deposition and forest condition was achieved through application of the critical levels/loads approach. This approach originated from discussions between the USA and Canada, but for forest systems was developed particularly by Sverdrup in Sweden. Quantification of the terms in the simple mass balance equation (e.g., Freer-Smith and Kennedy, 2003) allowed critical load maps and critical load exceedance maps to be drawn. This approach is possible at national, European, and global scales (Hall et al., 1998).

There are a number of current initiatives to improve the survey and monitoring of Europe's forests. Many countries have recently increased the number of variables measured in their national inventories, in particular to extend the coverage of biodiversity and soil quality. There is increasing use of remote sensing and there is an initiative within the EU COST program to harmonize the national inventories in Europe (action E43–see http://cost.cordis.lu).

FIGURE 4. The deviation of mean plot defoliation of common beech in 2004 from the average defoliation 1997 to 2003, a strong indication of the impact of the 2003 European drought. See text for further details. Reproduced with kind permission of UNECE Data Coordination Centre, BFH Hamburg. (Source: UNECE, 2005)

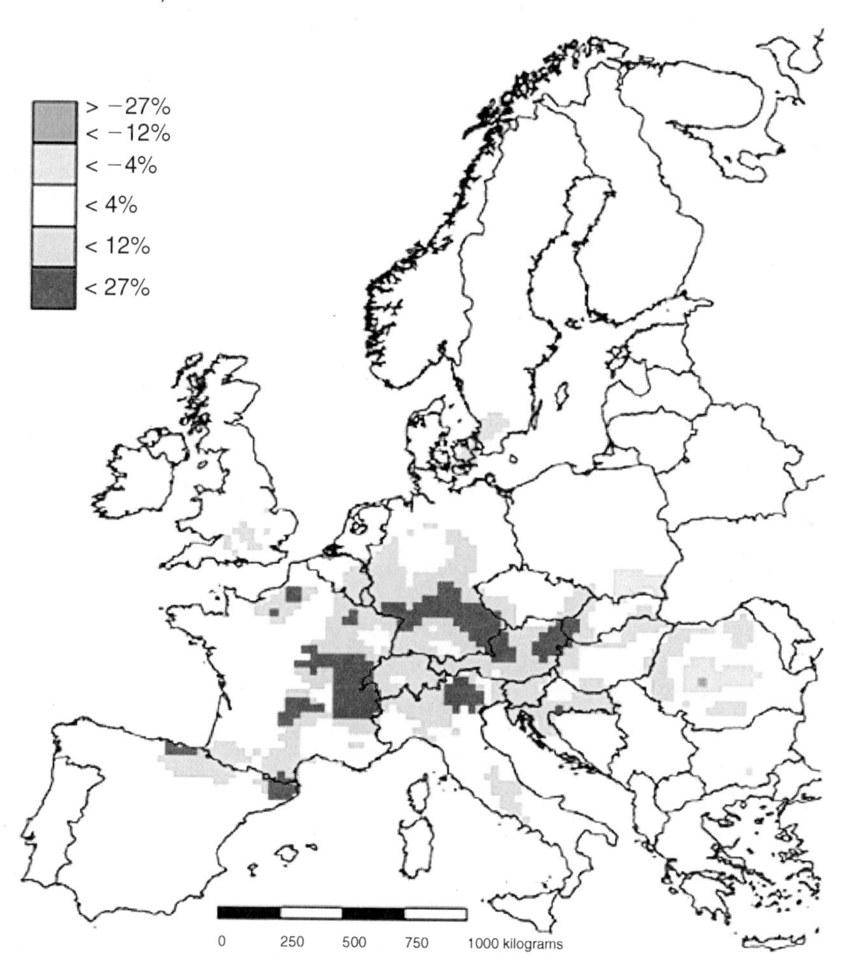


## *The Critical Loads and Ecosystem Approaches in Forestry*

The critical loads approach was first applied within the CLTAP at the time of negotiation of the second sulphur protocol (1994). Critical loads as defined in general terms by Nilsson and Grennfelt (1988) can be calculated for forest ecosystems by using the simple mass balance equation. The simple mass balance equation considers acid inputs, base-cation inputs, base-cation weathering (less uptake), nitrogen consumption, and acid leaching. If nitrogen consumption exceeds deposition (as in most forest systems), sulphur alone can be considered. Some countries who had not signed up to the first sulphur protocol did sign up to the second, and indeed to subsequent protocols on the convention (e.g., the 1998 heavy metals protocol, 1998 protocol on persistent organic pollutants, and 1999 eutrophication and ground-level ozone protocol).

The two sulphur protocols did lead to substantial declines in sulphur emissions in Europe, and responses have been observed in a range of parameters for both the aquatic and terrestrial environment. However, recent analyses by EMEP show that in Europe by 2010 there will still be significant areas of exceeding critical loads for acidification and even larger areas exceeding eutrophication levels as a result of nitrogen inputs (UNECE, 2005).

Foresters have been aware for some years that employment, agricultural, environmental, and now energy policies impact forest management. Within UNCED, the difficulty of taking a sectoral approach is identified in the convention on biological diversity (CBD) and has become known as the "ecosystem approach." This is presented as a strategy for the integrated management of land, water, and living resources to promote conservation and sustainable use. Within CBD, the concept of an ecosystem approach was set out in 5 guidelines and 12 principles in the conference of the parties meeting in Nairobi in 2000. A further challenge for the future is the integration of such ecosystem approaches with sustainable forest management (Sayer et al., 2003).

### *Monitoring of Sustainable Forest Management and Biodiversity Assessment in European Forests*

The Ministerial Conference on the Protection of Forests in Europe (MCPFE) was initiated at a conference in Strasbourg in 1990. The MCPFE is a high-level political initiative that has developed as a dynamic process concerned with the protection and sustainable management of European forests.

The political commitment involves 44 European countries and the European Community, and cooperation with other bodies including international organizations that participate as observers. Each ministerial conference has agreed to a series of commitments, and as a result of these, expert meetings are held to establish general principles, criteria for sustainability and forest management, and a coherent set of performance indicators for sustainable forest management (see list above and Table 3).

Particular outcomes include the production of general guidelines for the sustainable management of forest in Europe (resolution 1) at the Helsinki Conference in 1993. Following the Helsinki meeting, initiatives (workshops, etc.) were established to define criteria and performance indicators. At Lisbon in 1998, the conference decided to implement and to continue to review and improve the Pan-European indicators of

TABLE 3. Pan-European indicators for sustainable forest management under criterion 4

| **Indicator number** | **Indicator topic** | **Description** |
|---|---|---|
| Indicator 4.1 | Tree species composition | Area of forest and other wooded land, classified by number of tree species occurring and by forest type. |
| Indicator 4.2 | Regeneration | Area of regeneration within even-aged stands and uneven-aged stands, classified by regeneration type. |
| Indicator 4.3 | Naturalness | Area of forest and other wooded land, classified by "undisturbed by man," "semi-natural" or "plantations," each by forest type. |
| Indicator 4.4 | Introduced tree species | Area of forest and other wooded land dominated by introduced tree species. |
| Indicator 4.5 | Deadwood | Volume of standing deadwood and of lying deadwood on forest and other wooded land classified by forest type. |
| Indicator 4.6 | Genetic resources | Area managed for conservation and utilization of forest tree genetic resources (in situ and ex situ gene conservation) and area managed for seed production. |
| Indicator 4.7 | Landscape pattern | Landscape-level spatial pattern of forest cover. |
| Indicator 4.8 | Threatened forest species | Number of threatened forest species, classified according to International Union for the Conservation of Nature and Natural Resources Red List categories in relation to total number of forest species. |
| Indicator 4.9 | Protected forests | Area of forest and other wooded land protected to conserve biodiversity, landscapes, and specific natural elements, according to Ministerial Conferences on the Protection of Forests in Europe assessment guidelines. |

sustainable forest management (SFM). An MCPFE advisory group of scientists and technical experts from relevant organizations in Europe developed an improved set of Pan-European indicators for SFM based on the existing criteria and indicators. Four workshops were held during 2001 and 2002 that produced a system of 35 indicators grouped under six criteria. The use of these improved criteria and indicators was endorsed by ministers at Vienna in 2003 (see list above and Table 3). As an example of how these indicators are used to evaluate success against each main criteria, the detailed indicators used for criteria 4 on biological diversity are shown in Table 3. The first report against all these indicators for each of the six criteria was published in 2003 (MCPFE and UNECE, 2003). This report is perhaps the most comprehensive account of the state of European forests and presents in detail many of the data sets on which I have commented here.

## *CONCLUSIONS AND THE FUTURE*

Forests, which account for about 35% of the land area of the EU-25, are one of Europe's most important renewable resources and an extremely valuable component of European nature. The forests also make a considerable contribution to the economy of a number of member states, and their value for recreational purposes is having an ever-increasing effect on forest policy. Forest cover is still increasing in the EU (in contrast to the situation at the global scale) and the three EU candidate countries (Bulgaria, Romania, and Turkey) are also heavily wooded.

The extent to which understanding of biogeochemical cycling (sulphur, nitrogen, carbon, etc.) and of ecosystem responses (forest condition, freshwater chemistry, soil condition) have advanced over the last 20 years is remarkable. Much has changed and, in particular, the history of sulphur depositions and ecosystem responses to them (e.g., Moffat, 2004) gives cause for optimism. It is tempting to conclude that we now have highly effective institutional frameworks and have made excellent progress both scientifically and at the science/policy interface. However, concerns have changed. Significant pollutant problems remain, e.g., rising nitrogen emissions and rural ozone concentrations. Climate change has become of overwhelming importance, and new international conventions are the focus for science/policy interaction (see list above). The extent to which different factors interact to influence both forest system stability and long-term sustainability have not been fully explored. Such interactions are highly complex and have important consequences for species and management on the ground. Practical issues such as genetic quality (seed testing and registration of seed source), where to use natural regeneration, restoration of ancient and semi-natural woodlands, reduced chemical use, timber quality, continuous cover forestry, soil protection, and certification are all important to the forest manager. Traditional management practice, established species choice/forest type, and silvicultural systems that optimize economic returns may not maximize stability and sustainability in a changing environment. Progress on some of these more focused and applied issues of sustainable forest management is described in other papers of this session (e.g., Mason et al., 2005).

European countries have developed national forest programs, systems of certification of sustainable forest management, and also of associated forest product labelling. Forestry and forestry-based industries, with their wide economic, environmental, and social benefits, require a comprehensive policy framework, and in Europe this is being provided through the European Forestry Strategy and the MCPFE. On a global scale, the UNCED process and the United Nations Forum on Forests (UNFF) are to some extent equivalent processes. Signatory countries of the fourth MCPFE (Vienna 2003) now face the challenging task of developing effective implementation and monitoring of the improved set of criteria and indicators of sustainable forest management. In the wider-scale frameworks of the CLTAP and UNFF a coherent approach seems more difficult to achieve at the global scale. There is also an important question over the extent to which European and global policies and guidelines are affecting practice on the ground.

For European forestry, new requirements under the EU forestry strategy and forest monitoring activities are likely to be incorporated into a new LIFE+ instrument during 2007 (LIFE+ is a proposal from the Commission of the European Communities for providing financial support to implementing environmental policy in Europe). A newly devised forest monitoring scheme will need to be developed, taking into account and integrating the requirements for biological sustainability, climate change impacts and biodiversity monitoring in addition to the established criteria that relate principally to pollution impacts and potential production. This will not be an easy task, as requirements for forest monitoring are now in place at the national, European, and global scales, and these will need to be reconciled within an integrated monitoring scheme.

# REFERENCES

Beerling, D.J. and F.I. Woodward. 2001. Vegetation and the terrestrial carbon cycle. Cambridge University Press, Cambridge, UK.

Bergdahl, D.R. 1988. Impact of pinewood nematode in North America: present and future. *Journal of Nematology* 20: 260-265.

Brasier, C.M. 2000. The rise of the hybrid fugi. *Nature* 405:134-135.

Brasier, C.M. 2001. Rapid evolution of plant pathogens via interspecific hybridization. *Bioscience* 51:123-133.

Brasier, C.M., P.A. Beales, S.A. Kirk, S. Denman and J. Rose. 2005. *Phytophthora kernoviae* sp. nov., an invasive pathogen causing bleeding stem lesions on forest trees and foliar necrosis of ornamentals in Britain. *Mycological Research* 109(8): 853-859.

Buttoud, G., B. Solberg, I. Tikkanen, and B. Pajari (eds.). 2004. The evaluation of forest policies and programmes. European Forest Institute (EFI) Proceedings No. 52. European Forest Institute, Finland.

Ciesla, W.M. and E. Donaubauer. 1994. Decline and dieback of trees and forests. A global overview. FAO Forestry Paper No. 120. Food and Agriculture Organization, United Nations, Rome.

Evans, H.F., D.G. McNamara, H. Braasch, F. Chadoeuf and C. Magnusson. 1996. Pest risk analysis for the territories of the European Union on *Bursaphelenchus xylophilus* and its vectors in the genus *Monochamus*. *EPPO Bulletin* 26:199-249.

Evans H.F., N. Straw, and A. Watt. 2003. Climate change implications for insect pests. In Broadmeadow 2003 "Climate change: Impacts on UK Forests." Forestry Commission Bulletin 125.

Evans, J. 1994. Longterm experimentation in forestry and site change. In: pp. 83-94. R.A. Leigh and A.E. Johnston (eds.). Longterm experiments in agricultural and ecological sciences. CAB International, Wallingford, Oxfordshire, UK.

Food and Agriculture Organization of the United Nations [FAO]. 2000. Global forest resource assessment 2000, United Nations, Rome, Italy.

Forestry Commission. 2003. Sustainable forestry in the UK. The UK's National Forest Programme.

Freer-Smith, P.H. and F. Kennedy. 2003. Base cation removal in harvesting and biological limit terms for use in the simple mass balance equation to calculate critical loads for forest soils. *Water, Air, and Soil Pollution* 145:409-427.

Gibbs, J., C. Van Dijk, and J. Webber (eds.). 2003. *Phytophthora* disease of alder in Europe. Forestry Commission Bulletin 126. Forestry Commission, Edinburgh.

Graham, F., Persan, N. and Carson, R. 1974. Since silent springs. Fawcett Books.

Hall, J., K. Bull, I. Bradley, C. Curtis, P. Freer-Smith, M. Hornung, D. Howard, S. Langan, P. Loveland, B. Reynolds, J. Ullyett and T. Warr. 1998. Status of UK critical loads and exceedances January 1998. Part 1-Critical loads and critical loads maps. Report prepared under contract for the Natural Environment Research Council by the Institute of Terrestrial Ecology.

Intergovernmental Panel on Climate Change [IPCC]. 1996. The science of climate change. Contribution of working group 1 to the second assessment report of the In-

tergovernmental Panel on Climate Change. Cambridge University Press, Cambridge UK and New York, NY, USA.
Jung, T., H. Blaschke and W. Oßwald. 2000. Involvement of *Phytophthora ramorum. Mycological Research* 108:378-392.
Mather, A., P.H. Freer-Smith and P.S. Savill. 1995. Analysis of the changes in forest condition in Britain 1989 to 1992. Forestry Commission Bulletin 116. Her Majesty's Stationery Office, London.
Mason, W.L., Kerr, G., Humphrey, J. and Quine, C. 2005. Sustainable forest management in an era of declining timber prices: The British experience. *The International Forestry Review* 7(5):12.
Ministerial Conferences on the Protection of Forests in Europe Liaison Unit Vienna and UNECE/FAO [MCPFE and UNECE]. 2003. State of Europe's Forests 2003. MCPFE Liaison Unit, Vienna.
Moffat, A. 2004. Woodlands and the environment: hydrology, archaeology and environmental change. In: pp. 26-27. Forest Research Annual Report and Accounts 2003-2004. TSO (The Stationery Office), Norwich, UK.
Mylona, S. 1996. Sulphur dioxide emissions in Europe 1880-1991 and their effect on sulphur concentrations and depositions. *Tellus* 48B:662-689.
Newcombe, G., G.A. Chastagner, W. Schutte and B.J. Stanton. 1994. Mortality amongst hybrid poplar clones in a stool bed following leaf rust caused by *Melampsora meduscae f. sp. deltoides. Canadian Journal of Forestry Research* 24:1984-1987.
Nilsson, J. and P. Grennfelt (eds.). 1988. Critical loads for sulphur and nitrogen. Report from a workshop held at Skokloster, Sweden 19-24 March, 1988. *Miljørapport* 15:1-418.
Sayer, J., C. Elliott and S. Maginnis. 2003. Protect, manage and restore: conserving forests in multi-functional landscapes. Proceedings of XII IUFRO World Congress. Quebec, Canada.
Schulze, E.D. and P.H. Freer-Smith. 1991. An evaluation of forest decline based on field observations focussed on Norway spruce, *Picea abies*. In: pp. 155-168. Acidic deposition it's nature and impacts. Proceedings of the Royal Society of Edinburgh, 97B.
Schwerdtfeger, F. 1981. Die Waldkrankheiten. Verlag Paul Parey, Hamburg and Berlin.
Spiecker, H., K. Mielikäinen, M. Köhl and J.P. Skovsgaard (eds.). 1996. Growth trends of European forests: studies from 12 countries. European Forest Institute Research Report 5. Springer Verlag. 372 p.
U.N. Economic Commission for Europe [UNECE]. 2005. The condition of forests in Europe. Executive report of the convention on long-range transboundary air pollution. International Cooperative Programme on Assessment and Monitoring of Air Pollution Effects on Forests (ICP Forests), Bundesforschungsanstalt für Forst- und Holzwirtschaft (BFH) Hamburg.
Villebonne, D. de and F. Maugard. 1999. *Scirrhia pini*: un pathogène du feuillage en pleine expansion sur le pin laricio en France. In: pp. 30-32. Les Cahiers du DSF, 1-1999, La Santé des Forêts (France 1998). Ministère de l'Agriculture et de la Pêche (DERF), Paris.

doi:10.1300/J091v24n02_04

# Barriers to Sustainable Forestry in Central America and Promising Initiatives to Overcome Them

Glenn E. Galloway
Dietmar Stoian

**SUMMARY.** Over the past decades, Central America has suffered some of the highest deforestation rates worldwide. Vast tracts of forest have been converted to agriculture and pasture, encouraged by ill-designed government policies and perverse incentives. Recently, however, progress has been made toward more sustainable use of forest resources by adjusting forest policies, decentralizing forest administration, and providing conducive incentives through environmental service payments and forest certification. Valuable forest management experiences have been gained by indigenous and peasant communities. Community forestry in Central America is being increasingly recognized by national governments. Examples include the community concessions in Peten, Guatemala, and community-based forest operations in Honduras and Nicaragua. Stakeholder networks have been established that strengthen

Glenn E. Galloway is Director of the Education Program and Dean of the Graduate School at the Tropical Agricultural Research and Higher Education Center, Turrialba, Costa Rica (E-mail: galloway@catie.ac.cr).

Dietmar Stoian is Team Leader of the Center of Competitiveness of Ecoenterprises (CeCoEco) at the Tropical Agricultural Research and Higher Education Center, Turrialba, Costa Rica (E-mail: stoian@catie.ac.cr).

[Haworth co-indexing entry note]: "Barriers to Sustainable Forestry in Central America and Promising Initiatives to Overcome Them." Galloway, Glenn E., and Dietmar Stoian. Co-published simultaneously in *Journal of Sustainable Forestry* (Haworth Food & Agricultural Products Press, an imprint of The Haworth Press, Inc.) Vol. 24, No. 2/3, 2007, pp. 189-207; and: *Sustainable Forestry Management and Wood Production in a Global Economy* (ed: Robert L. Deal, Rachel White, and Gary L. Benson) Haworth Food & Agricultural Products Press, an imprint of The Haworth Press, Inc., 2007, pp. 189-207. Single or multiple copies of this article are available for a fee from The Haworth Document Delivery Service [1-800-HAWORTH, 9:00 a.m. - 5:00 p.m. (EST). E-mail address: docdelivery@haworthpress.com].

Available online at http://jsf.haworthpress.com

doi:10.1300/J091v24n02_05

horizontal and vertical alliances among wood producers and manufacturers and that help promote both community development and forest conservation. However, illegal logging, poor law enforcement, and lack of economic viability of forest management involving nontraditional species still provide barriers to the sustainable management of tropical broadleaved forests in the region. Future challenges include improved governance through decentralized forest administration, private sector involvement, and third-party certification. To improve traceability and value adding, development of integrated supply chains of forest and wood products is recommended. doi:10.1300/J091v24n02_05 

**KEYWORDS.** Sustainable forestry, community forestry, governance, Central America

## *INTRODUCTION*

Over the past decades, Central America has suffered some of the highest deforestation rates in the world. From 1990 to 2000 alone, the region has lost about 3.4 million ha or 19.1% of its forest cover (FAO, 2001) (Table 1). Vast tracts of tropical and subtropical forests, rich in biodiversity, have been converted to agriculture and pasture, often encouraged by ill-designed government policies and perverse incentives. Recently, however, some progress has been made toward more sustainable use of forest resources by adjusting forest policies, decentralizing forest administration, implementing capacity-building initiatives, and providing incentives through environmental service payments and forest certification.

Sustainable forestry in Central America, as elsewhere, is a complex task with social, technical, ecological, economic, cultural, institutional, and policy dimensions. Progress toward sustainable forestry requires that many barriers be overcome related to each of these dimensions. We will provide examples of promising initiatives that have sought to overcome these barriers, putting emphasis on community-based forest management, which plays an increasingly important role in the region's move toward sustainable forest management (SFM). It first focuses on institutional, technical, and business management barriers to SFM and

TABLE 1. Forest resources in Central America, 1990-2000

| Country/area | Land area | Forest area 2000 | | | | | Area change 1990-2000 (natural forest cover) | |
|---|---|---|---|---|---|---|---|---|
| | | Natural forest | Forest plantation | Total forest | | | | |
| | *Thousand ha* | | | | *% of land area* | *ha capita$^{-1}$* | *Thousand ha year$^{-1}$* | *% year$^{-1}$* |
| Belize | 2,280 | 1,348 | 3 | 1,351 | 59.2 | 4.9 | −36 | −2.32 |
| Costa Rica | 5,106 | 1,968 | 178 | 2,146 | 42.0 | 0.5 | −16 | −0.77 |
| El Salvador | 2,072 | 121 | 15 | 136 | 6.6 | 0.02 | −7 | −4.60 |
| Guatemala | 10,843 | 2,850 | 132 | 2,982 | 27.5 | 0.2 | −54 | −1.71 |
| Honduras | 11,189 | 5,383 | 48 | 5,431 | 48.5 | 0.8 | −59 | −1.03 |
| Nicaragua | 12,140 | 3,278 | 46 | 3,324 | 27.4 | 0.6 | −117 | −3.01 |
| Panama | 7,443 | 2,876 | 40 | 2,916 | 39.2 | 0.9 | −52 | −1.65 |
| Total Central America | 51,073 | 17,824 | 462 | 18,286 | 35.8 | 0.46 | −341 | −1.91 |

Note: Figures are recent estimates derived principally from FAO (2001); updated population figures are found in PRB (2004).

promising initiatives to overcome them. It then addresses the importance of appropriate services and an overall enabling environment conducive to SFM. We conclude by identifying knowledge gaps and research needed to further advance sustainable forest management in the region.

## *TOWARDS SUSTAINABLE FOREST MANAGEMENT IN CENTRAL AMERICA*

### *Lack of Legal Access to Forests*

In legal terms, most peasant and indigenous communities in Central America have long been excluded from participating in and benefiting from forest management. Lack of land tenure or long-term use rights has been and continues to be a common barrier to their participation. Since the 1990s, however, a growing number of rural communities have gained property rights and resource security in the region. As a result, community forestry in Central America is increasingly being recognized by national governments and the international community as a potentially effective strategy to slow forest loss and to contribute locally to the well-being of peasant and indigenous communities. Prominent examples include Guatemala, Honduras, and Nicaragua.

In Guatemala, the community concession process in the multiple-use zone of the Maya Biosphere Reserve (MBR) in Peten arguably is the most important example in Central America of community-based forest management contributing to sustainable forest management. When the MBR was created in 1990, several communities were enclosed within the newly established reserve without legal access to wood and other resources. Community groups outside its boundaries were also prohibited from harvesting timber within the reserve. This policy decision, instead of reducing timber extraction in the protected area, led to a marked increase in illegal logging and processing of timber with chainsaws within the reserve as well as a disordered advance of the agricultural frontier (CONAP, 2002). Furthermore, restrictions prohibiting access to forest resources generated great friction between personnel of the National Protected Areas Council (CONAP) and representatives of local community groups (Carrera and Prins, 2002). Indeed, the situation became so tense that CONAP personnel were no longer able to enter the reserve because of security risks.

In the early 1990s, the Tropical Agricultural Research and Higher Education Center (CATIE) promoted the concept of sustainable development in Central America through conservation minded initiatives. A conceptual pillar of these initiatives was that conservation and production objectives can be compatible if adequate planning and operational controls are put into place. After much debate–sometimes heated–the decision was made in 1994 to grant the first community concession in the multiple-use zone of the MBR to the community San Miguel La Palotada. This community had worked closely with CATIE's Project for the Sustainable Development of Central America (OLAFO), which received Norwegian, Swedish, and Danish financial support. They were granted a concession of 7,039 ha for a period of 25 years (Carrera and Prins, 2002). This early concession experienced a number of inherent weaknesses unsurprising in an incipient program (for example, deficient technical support, excessive bureaucratic red tape, conflicts of interest, and lack of experience in business management practices and in the commercialization of forest products). On the other hand, this concession pointed the way for the expansion of this approach, which, in the last decade, has placed the primary responsibility for sustainable management of the forests in the MBR into the hands of community-based groups. The scope of the community concession approach is now quite ambitious, with nearly 500,000 hectares under certified management (FSC, 2004).

Although the community concession process is still quite young, a number of tangible successes can be cited: reduction of forest fires during the dry season (Carrera and Prins, 2002), control of the expansion of the agricultural frontier within the MBR multiple-use zone and economic benefits to participating community-based groups (Mollinedo, 2000; Carrera et al., 2006). Indeed, dramatic satellite images of the community concessions and adjacent national parks have shown that the incidence of fire is much greater in the national parks, where fire suppression is supposed to be absolute (see CEMEC and CONAP, 2000; Carrera and Prins, 2002).

In Honduras, although community groups have been granted usufruct contracts in broadleaf forests in the northern provinces, the overall success of the approach has been less successful than in the case of Guatemala (see, for example, Fermi, 2005; Nygren, 2005). There are several considerations that help explain this apparent discrepancy:

- The areas under contract are much smaller, with low volumes of species most valued on local markets. Access to international markets has been essentially blocked by a regulation requiring that tropical timber undergo secondary processing prior to exportation as community groups and local timber processing facilities do not generate products of sufficient quality for export markets. The community concessions in Guatemala, in contrast, are permitted to export sawnwood to international markets and have obtained attractive prices for this product.
- The abrupt topography limits technological options. Wood must be processed into rough-cut blocks in the forest and then be carried considerable distances by mules to access roads.
- The granting of usufruct contracts has not been complemented with the development of agile bureaucratic procedures. Arbitrary limits have also been placed on the amount of wood that can be harvested by each community group, even when inventory data indicates a greater availability under the principles of sustainable management.
- Scant institutional presence in most areas has led to rife illegal logging. Recent estimates indicate that over 70% of the tropical hardwoods processed in Honduras come from illegal sources (Richards et al., 2003). This influx of illegal wood into the market leads to lower prices, and hence, unfair competition.

These problems have led to a marked reduction in the number of persons participating in community-based forest management groups. Indeed, evidence suggests that many forest workers have returned to illegal logging to avoid onerous bureaucratic procedures. The payment of taxes and other transaction costs are avoided by simply cutting trees and inserting them into markets through clandestine channels. Even with these constraints and serious problems, a number of community-based management operations under the umbrella of the Regional Agro-forestry Cooperative, Colon, Atlántida, Honduras Ltd. (COATLAHL) have been certified by the Forest Stewardship Council (FSC) as well-managed forest operations. In these communities, local producers strive to control illegal logging and have been effective in reducing the expansion of the agricultural frontier.

Few would question that land tenure or long-term use rights over forested areas is an important condition for community-based SFM. Access to forest resources, however, must be complemented by a host of other conditions and services in order to be successful. Their absence often represents barriers to SFM, some of which will be discussed below.

### *Lack of Technical Capacity to Carry Out Forest Management Operations*

Sustainable forest management requires technical competence in a considerable number of tasks, some quite complex (Table 2). The promotion of community-based forest management implies that required skills are gradually acquired by representatives of community groups, a large proportion of which have had few (if any) opportunities for formal education. These persons, however, often possess vast practical experience and knowledge about local forest resources (Zamora, 2000). Informal community-based groups and individuals throughout Central America have struggled to make a living from timber exploitation over the last few decades, many through their participation in illegal logging. For example, representatives of some community-based groups that have been granted community concessions in Guatemala openly admit their past involvement in this illicit activity. Consequently, in some cases, community-based forest management has converted persons involved in forest degradation into allies of conservation. This conversion necessitates the creation of local capacity to carry out forest management activities in an appropriate manner.

TABLE 2. Partial list of technical skills required for forest management that be developed with community-based groups

| | |
|---|---|
| Quality management of wood and nonwood products | • Selection and grading<br>• Local processing<br>• Stacking and drying<br>• Transportation |
| Management of tropical broadleaf forests | • Ecological and economic considerations<br>• Forest inventories<br>• Management plans<br>• Silvicultural treatments<br>• Cost-benefit analysis<br>• Organization for forest management<br>• Fire control<br>• Control of illegal logging |
| Low-impact harvesting techniques | • Roads and skidder trails<br>• Directional felling<br>• Prevention of accidents<br>• Technical evaluation of harvesting operations<br>• Equipment operation and maintenance |

The knowledge and skills required to carry out the activities listed in Table 2 make imperative the realization of an effective capacity-building strategy. Over the last two decades, a number of initiatives have been financed by the international donor community to foster the development of local capacity in forest management. CATIE has participated in various research and development initiatives. These activities have taken place in Costa Rica, Guatemala, Honduras and Nicaragua with support from the Swiss Agency for Development and Cooperation and the U.S. Agency for International Development (USAID), in conjunction with national institutions and other partners. The Canadian International Development Agency supported the long-term Broadleaf Forest Development Project in the northern tropical forests of Honduras, implemented by the Honduran Forest Development Corporation (COHDEFOR). The Department of International Development of the United Kingdom supported the Natural Forest Integrated Management Project in Costa Rica, leading to the formation of the forest management service of the Forest Development Commission of San Carlos (CODEFORSA). These and other initiatives involved the development of low-impact harvesting techniques applicable to conditions in Central America. Long-term research plots have been established and monitored to determine the impact of timber harvesting on forest composition, regeneration, and growth. Experiences and information generated was used to develop simplified forest management plans. A major component of all these initiatives has been the implementation of capacity-building strategies including courses and workshops and technical assistance activities in which tens of thousands of person-days of training and technical assistance have been given.

An important result of these efforts has been a great increase in technical capacity for forest management in the Central American region. Depending on the skills and knowledge required, capacity development has targeted different end users of the information. For example, capacity development in low impact harvesting techniques has normally been directed to forest workers participating in community-based groups. Training in the processing and interpretation of forest inventory data, on the other hand, generally targets forest technicians and professional foresters with some analytical skills. Over time, a growing number of people in Central American countries possess the ability and knowledge to take on leadership roles in locally organized capacity-building programs and serve as capable instructors.

Exceptional, highly skilled forest workers from Costa Rica were the first to become trainers to other community-based groups in Guatemala, Honduras, and Nicaragua on reduced-impact logging techniques. Professional forest technicians from Switzerland and Costa Rica dedicated several years to building up this capacity. Now local trainers in low-impact harvesting can also be found in Guatemala, Honduras, and Nicaragua. Local professionals throughout Central America today provide technical services to community-based forest management operations, sometimes with technical support from CATIE, international organizations, and national universities. Downsizing of the national forest services has led to an increased reliance on professional services from persons outside the public sector. This growing role of local service providers and consultants has not progressed without problems. The erratic availability of contractual opportunities has often made difficult the consolidation of these local service providers. Forest workers who also provide services as instructors in low-impact harvesting often lack a formal certificate vouching for their hard-earned capabilities. Finally, although much progress has been made in generating greater local capacity in technical skills required for community-based forest management, less support is available to communities in business development and marketing of forest products. These deficiencies represent the next barriers addressed in this paper.

## *Lack of Business Management Skills Required for Rural Enterprise Development*

Community-based sustainable forestry faces barriers common to the development of many rural enterprises: deficient business planning and administration; lacking demand orientation; and inability to provide sufficient volumes, meet quality standards, and ensure on-time delivery. Until the recent past, most efforts to promote community-based forest management have concentrated primarily on technical concerns. Technical competence, however, is only one set of skills required to consolidate a successful rural enterprise based on forest resources (Table 3). Gradually, more emphasis is being placed on strengthening business management skills and marketing of forest products in community organizations and small and medium forest enterprises (SMFE) in Central America (Galloway et al., 2005). A number of strategies are being used to achieve this depending on local conditions: organizing at the intra- and inter-business levels, increasing access to forest resources, orienting to niche markets, strengthening the capabilities of

TABLE 3. Partial list of skills required for the management of forest enterprises

| | |
|---|---|
| Organization of forest enterprises | • Intra-business organization<br>• Inter-business organization<br>• Vertical integration vs. vertical alliances<br>• Trust relationships with service providers and other actors along the supply chain<br>• Institutional arrangements ensuring information flow and more equitable distribution of benefits |
| Administration | • Strategic and operational planning<br>• Financial planning and management<br>• Personnel management<br>• Market analysis and marketing management<br>• Product development<br>• Quality control<br>• Procurement and purchases<br>• Auditing<br>• Reporting and documentation |
| Certification | • Individual or group certification<br>• Opportunities<br>• Requirements<br>• Costs and benefits<br>• Chain of custody<br>• Documentation |

SMFE managers, and providing improved embedded and externally-sourced business development services. For example, in Costa Rica individual private forests are too small to generate an adequate supply of products to access international markets, and it is not realistic that small owners develop autonomous expertise in forest management. Several organizations have been formed (for example the *Fundación de Desarrollo para la Cordillera Volcáncia Centra*–FUNDECOR–and the aforementioned CODEFORSA) that provide both technical and, more recently, business development services.

Over time, a growing proportion of technical assistance being provided to community concessions in the MBR has also concentrated on the development of business management skills and marketing. Much of this support is being channeled through a local umbrella organization, the Association of Forest Communities in the Peten (ACOFOP). With technical and financial support from Chemonics through a project funded by the USAID, ACOFOP has been instrumental in creating an association of SMFE called FORESCOM. One objective of FORESCOM is to coordinate the joint commercialization of products from the community concessions to access more attractive markets. Efforts are also being made to create market opportunities for lesser-known species, the most highly valued species being mahogany (*Swietenia macrophylla* King) and Spanish cedar (*Cedrela odorata* L.).

The Network for the Management of Broadleaf Tropical Forests in Honduras (REMBLAH) has promoted the formation of alliances among community-based groups in the northern mountainous region of the country and the eastern lowlands (La Mosquitia) to improve market opportunities. In its strategic plan, the REMBLAH's Commission of Industry and Commerce seeks to improve access to national and international markets for goods and services generated by network members. Specific objectives include improving market options for lesser known species, participating in trade shows, and strengthening business management skills of network members, among others. Because REMBLAH's membership includes both primary producers and small forest industries involved in the processing of timber, both horizontal and vertical alliances are promoted within the network to improve economic returns to participants along tropical timber supply chains. This strategy is viewed as necessary to promote both community development and forest conservation.

Many community-based forest management initiatives in Central America seek to take advantage of opportunities created by the certification of management operations and chains of custody (Molnar, 2003), especially in Guatemala and Honduras. Although certification has provided producers with periodic independent, unbiased evaluations of their operations, it has rarely resulted in price premiums for certified products and tangible economic benefits. One of the few well-documented cases where forest certification resulted in higher wood prices is the Peten, Guatemala, where the certification process brought new intermediaries on the scene. Consequently, prices have risen because of increased competition for raw material and semi-elaborated products (Carrera et al., 2006). The general picture, though, is that many certified community operations have been facing unfavorable cost-benefit ratios with respect to certification, requiring subsidies from donor agencies and local nongovernmental organizations (NGOs) to maintain their certificates. This practice is unsustainable, and many SMFE run the risk of losing certification once external funding is removed (Thornber and Markopoulos, 2000). Even in cases where external funding is available for recertification, for example COATLAHL in Honduras, drop-out rates from community-based forest organizations are significant and caused in part by high compliance costs of meeting certification standards (Fermi, 2005).

One way to address the cost problem is group certification, which helps dilute the cost of certification by bringing together various small forest owners or community operations under the umbrella of a sole certificate. Certified groups may still find it difficult to comply with all the standards established by the FSC as the sole forest certification scheme in place in Latin America (Nussbaum et al., 2001; Nussbaum and Simula, 2004). In addition, certification cannot solve problems related to insufficient volumes, lack of quality, or untimely delivery. Moreover, the vast majority of the established distribution channels of forest products from the region do not require certification. If certification is to be successful for a given community-based enterprise or group of enterprises, alternative distribution channels need to be identified, which in turn may impose further quality or volume requirements in addition to certification.

Several initiatives have been established in the region that seek to address these challenges. The CATIE-based Center for the Competitiveness of Ecoenterprises (CeCoEco), for example, has installed a web-based bilingual market intelligence system, *EcoNegocios Forestales–Forest EcoBusiness* (see www.catie.ac.cr/econegociosforestales) to promote trade in certified forest products. Several interactive business development tools and services, including B2B and B2C[1] market places, help reduce transaction costs of enterprises offering or buyers seeking certified forest products. In addition, CeCoEco provides technical assistance and capacity building services to SMFE with the aim of fostering the development of integrated forest- and wood-product supply chains. Toward the same end, the World Wildlife Fund launched JagWood initiative (see www.jagwood.org/) seeks to link buyers and sellers of certified forest products from Central America and the Caribbean. Similarly, a project funded through the Multilateral Investment Fund of the Interamerican Development Bank and executed by CATIE in Guatemala, Honduras, and Nicaragua, has helped create favorable business environments by strengthening the relationships between the providers of raw materials and primary and secondary wood manufacturers along the supply chains of wood products originating from sound forest management. All of these initiatives share the common vision that sustainable forest management in the region requires major efforts to ensure its economic viability, and in turn calls for improved business development services to generate and strengthen the respective entrepreneurial capacities.

### *Fragmented Support for Community-Based Forest Management*

In the introduction of this paper, the complexity of SFM was pointed out, indicating that it involves social, technical, economic, cultural, institutional, ecological, and policy dimensions. Few, if any, organizations possess the capacity to provide community groups with required support in all these dimensions. Often, entities implementing initiatives seeking to promote SFM have done so in an isolated fashion, resulting in community-based groups with perhaps enhanced technical capacity, but with considerable deficiencies in business management skills (for example, administration, accounting, and marketing). These groups are vulnerable to bureaucratic inefficiencies and perverse forest policies. In these and in many other cases, community-based forest management initiatives are less than successful because of the support provided often reflects the professional background of service providers such as NGO and government staff, rather than the needs of community-based forest enterprises.

To better address the complexity of SFM and to achieve a larger, more sustained impact on tropical forest conservation and reduction of rural poverty, a considerable number of entities have joined together in multistakeholder platforms in Central America. In the case of pilot experiences in Honduras and Nicaragua, these platforms are deemed operational networks of horizontal cooperation (for example, REMBLAH–

see www.remblah.org/). They bring together entities from the pubic and private sector, producer cooperatives, universities, wood processing associations, projects, local NGOs, and research organizations in an effort to strengthen horizontal cooperation among diverse actors involved in tropical forest management and conservation. These networks have pioneered a number of innovative processes.

Networks were initially structured into "working groups" and eventually commissions to provide more direction to cooperative efforts. For example, the REMBLAH network has formed the following four commissions: community development; forest and environment; industry and commerce; and policy and legal concerns. Training, technical assistance, dissemination of information, and research are viewed as cross-cutting activities applicable to all commissions. Once the networks were structured, shared strategic planning was carried out to define the mission of the networks, the shared visions within each commission, objectives, and indicators for monitoring progress. These strategic plans have served as platforms for operational planning.

The networks have also implemented cooperative capacity-building programs in topics relevant to the aforementioned commissions. Annual, shared planning exercises have served to identify capacity-building priorities, which have been observed to evolve over time. One important benefit of cooperative capacity-building initiatives has been the gradual development of uniform approaches ("school of thought") to the planning and execution of forest management among diverse stakeholders.

Gradually, more emphasis is being placed on concerns related to the business administration of community-based forest enterprises and to the commercialization of forest products. Cooperative courses have been given on the topics listed in Table 3. Networks in several instances have shared costs of market studies and applied research into the wood properties and processing of lesser-known species. Within REMBLAH, alliances have been formed among producer organizations and between producer organizations and cooperatives involved in the processing of timber to enhance economic benefits along the supply chain.

Finally, the operational networks have become increasingly involved in policy issues in related to community-based forest management–for example, illegal logging, fiscal policy, incentive programs, and streamlining bureaucratic procedures. The barriers related to policy and governance issues are addressed in the following section.

## *Lack of an Enabling Environment that Favors Community-Based SFM*

Even where the aforementioned initiatives and approaches have yielded a certain degree of progress, barriers of a legal and institutional nature often limit advances toward SFM. An enabling environment that facilitates participation in and success of forest management initiatives is essential. Unfortunately, illegal logging, deficient law enforcement, and the lack of basic services in remote forest areas still make sustainable forest management a difficult prospect in many regions of Central America. Often times policies meant to contribute to forest conservation end up stimulating illegality. For example, many indigenous communities are not permitted to take part in forest management, even when they have occupied traditional lands for centuries (Contreras-Hermosilla and Global Witness, 2003). In some cases, the need for complex permits to harvest a few trees has led to farmers eliminating young trees from their lands in order to avoid legal problems in the future. In many countries, difficult-to-obtain permits are required to transport timber, but not agricultural products. When small producers perceive the incapacity of officials from the forest administration to enforce regulations, the tendency is to ignore the regulations altogether (Contreras-Hermosilla and Global Witness, 2003). Experience has shown that one of the best ways to promote conservation from a policy perspective is to encourage the responsible participation of community-based groups in sustainable practices and the commercialization of forest products.

A number of governance issues must also be resolved to consolidate SFM in the long term. As has been indicated, some political-legal barriers are products of perverse policies and regulations. Other constraints are more local in origin and could be construed as problems of local governance. A survey and analysis of these types of problems was carried out within communities in the region. Examples of problems detected include the following:

- Unscrupulous buyers underestimating wood quality and volume.
- Reduction of agreed-on prices at the moment of commercial transactions.
- Lack of transparency and adequate participation within the communities in commercial negotiations.
- Buyers exploiting lack of knowledge of market opportunities.
- Payments made with checks without funds.
- Imbalances in bargaining power between buyers and sellers.

- Assault and robbery in rural areas, especially when large sums of money are being carried. Often a general lawlessness predominates in remote rural areas, sometimes as a sequel to political conflict.
- Committing the sale of forest products to two (or more) buyers and accepting related advance payments without having the capacity to deliver all the products committed.
- Poor utilization or theft of funds within community organizations.

These problems greatly reduce the benefits perceived from forest management, undermining the interest and commitment of rural communities to participate in forest development activities. It is paramount that community-based groups be aware of these types of threats to their forest enterprises to avoid them whenever possible.

Finally, in several countries of Central America, budgetary crises and the lack of effectiveness of centralized institutions in charge of administering and regulating the forestry sector have given rise to efforts to delegate responsibilities for forestry sector development to municipalities and other local stakeholders. The effectiveness of local governments and organizations sometimes suffers from the same problems affecting the centralized ones: lack of technical expertise, budgetary constraints, and in some cases corruption and illegality (Ferroukhi, 2003). Successful decentralization of forest governance and a more equitable distribution of powers and benefits therefore depend upon the integration of potentially conflicting resource interests, institutional democratization, and political accountability of forest authorities and community representatives to local populations (Nygren, 2005). Democratizing the use and conservation of natural resources also requires a process of reconceptualizing who counts as a political actor, allowing for new social movements that mobilize around alternative notions of politics, citizenship, identity, and nature (Sundberg, 2003). Despite these challenges, the overall impression is that local governments and communities will take on greater responsibilities in the future, although there is nothing intrinsic in decentralization that ensures improved forest management and more equitable distribution of benefits. Local governments need to strengthen their capacities as resource managers, along with providing incentives that ensure long-term commitment (Larson, 2002). The existing challenge is how best to provide them with the resources, expertise, and authority to respond to their growing mandate. Given the inherent heterogeneity of rural realities, improving the efficiency of rural-sector policies requires the promulgation of both democratic decentralization and community-driven development programs (Ruben and

Pender, 2004). These, however, will only be successful if rural infrastructures and basic services such as health care, education, communication, and transportation are significantly improved.

### *Gaps in Available Knowledge for SFM*

Although much has been learned about the technical, ecological, and even commercial dimensions of tropical forest management in Central America, important gaps continue to exist in available knowledge. Periodic monitoring of forests under management and well-directed research projects are necessary to provide information to adjust management prescriptions over time in a continuous feedback process sometimes termed "adaptive management." Related research requires broad, transdisciplinary approaches and a diverse set of methods for data collection, without privileging certain factors a priori when seeking explanations for environmental or behavioral changes of local populations (Vayda, 1996).

Recognizing the deficiencies in available information, a number of projects have established and maintained permanent sample plots in Central America to monitor forest dynamics and responses to logging and silvicultural treatments. Information generated from these permanent sample plots have indicated necessary adjustments in cutting cycles, in acceptable harvesting intensity, and in lower diameter limits for harvesting of different species. But knowledge gaps exist beyond the silvicultural aspects of forest management, for example, regarding the evolution of "indigenous" forest management systems vis-à-vis externally developed systems; ways of how best to communicate research results to forest managers; and the influence of spatial and ecological factors, as well as those pertaining to the market and political-legal frameworks (Walters et al., 2005). Even though enough knowledge is available to make good forest management technically feasible, rarely is it found in practice because of socioeconomic and political constraints (Finegan et al. 1993). Thus monitoring and research must also include carefully selected social, economic, and institutional variables (or indicators) in order to make informed adjustments in community-based forest management programs in a timely manner. After years of largely unsuccessful efforts to promote SFM in Rio San Juan, Nicaragua, studies showed that large discrepancies existed between conservation organizations and small landowners in their perceptions regarding sustainability and forest use (Altamirano et al., 2004). Similar divergences were found in the MBR in Guatemala (Sundberg, 2003). More effort is required to understand the social, cultural, and economic context that largely determines the viability of SFM in the region.

Finally, although policy initiatives exist that seek to foster SFM in remote forested regions, they have often been ineffective in stemming environmental degradation or the worsening poverty that characterizes them. The failure of policies to generate desired results has several causes. On the one hand, well-intended policies often generate outcomes quite different than those envisioned by policy makers. On the other hand, there is often a large gap between policy formulation and its application, especially in remote regions such as those found on the agricultural frontier where the State and its institutions are largely absent. Decentralization of forest administration has not helped to reduce this gap considerably. These two problems indicate the need for innovative strategies both to favor policy implementation and to monitor results of this implementation to ensure that desired impacts are generated.

## *FINAL COMMENTS*

As has been pointed out, SFM is a complex endeavor requiring adequate progress and acceptable conditions in social, technical, ecological, economic, cultural, institutional, and policy dimensions. Efforts to attain technical competence are for naught should a forest management operation prove to be economically unviable. Capacity-building initiatives in aspects related to rural enterprise development are irrelevant if a community operation is lacking basic technical skills for forest management, or if transaction costs are too high for legally constituting the enterprise and its operations. Indeed, the Central American experience has shown that a resolute commitment is needed to create an enabling environment for forest management and rural enterprise development based on it. Clear economic and nonmonetary incentives are needed to reward responsible stewardship. As no institution or organization possesses all the capacities and strengths required for sustainable forestry and the development of integrated forest- and wood-product supply chains, cooperation within multistakeholder platforms (for example, networks and business roundtables) appears to be an important condition for advancing this complex endeavor. Finally, the monitoring of progress is essential to make necessary adjustments in the proposed strategies and methods applied to achieve SFM. Only continuous monitoring and the flexibility to make adjustments in existing proposals will result in marked improvements in efforts to encourage community-based SFM over time.

## NOTE

1. Business to business (B2B) and business-to-consumer (B2C) market places are the most common platforms for e-commerce.

## REFERENCES


Carrera, F. and C. Prins. 2002. Desarrollo de la política en concesiones forestales comunitarias en Petén, Guatemala: el aporte de la investigación y experiencia sistematizada del CATIE. *Revista Forestal Centroamericana* 37:33-40.

Carrera, F., D. Stoian, J.J. Campos, J. Morales and G. Pinelo. 2006. Forest certification in Guatemala. In: Cashore, B., F. Gale, E. Meidinger and D. Newsom (eds.). Confronting sustainability: forest certification in developing and transitioning countries. Yale School of Forestry and Environmental Studies Press, New Haven, CT, USA, pp. 363-406.

CEMEC and CONAP. 2000. Map of forest fires in the Department of Peten in 1998. Center for Monitoring and Evaluation of the National Council of Protected Areas (CEMEC/CONAP), Guatemala.

CONAP [National Protected Areas Council]. 2002. Política marco de concesiones para el manejo integral de recurso naturales en Áreas Protegidas de Petén. CONAP, Guatemala.

Contreras-Hermosilla, A. and Global Witness. 2003. Emerging best practices for combating illegal activities in the forest sector. DFID-World Bank-CIDA 2003. United Kingdom.

Fermi, V. 2005. Efectos de la certificación forestal sobre los medios de vida de los integrantes de grupos de aserrío del Departamento de Atlántida, Honduras. Unpublished M.Sc. Thesis. CATIE, Turrialba, Costa Rica.

Ferroukhi, L. (ed.) 2003. La gestión forestal municipal en América Latina. CIFOR/IDRC, Bogor, Indonesia.

Finegan, B., C. Sabogal, C. Reiche and I. Hutchinson. 1993. Los bosques húmedos tropicales de América central: su manejo sostenible es posible y rentable. *Revista Forestal Centroamericana* 6:17-27.

Food and Agriculture Organization of the United Nations [FAO]. 2001. Global forest resources assessment 2000 (online). FAO, Rome. Retrieved October 6, 2004, from http://www.fao.org/forestry/fo/fra/index.jspe.

Forest Stewardship Council [FSC]. 2004. Forest management report by continents up to August 2, 2004. (Accessed 07 Oct 04).

Galloway, G., Kengen, S., Louman, B., Stoian, D., Carrera, F., Gonzalez, L., Trevin, J. 2005. Changing paradigms in the forestry sector of Latin America. In: pp. 243-263. G. Mery, R, Alfaro, M. Kanninen, M. Lobovikov (eds.). Forests in the Global Balance–Changing Paradigms. IUFRO World Series Volume 17. Helsinki. 318 p.

Larson, A.M. 2002. Natural resources and decentralization in Nicaragua: Are local governments up to the job? *World Development* 30(1):17-31.

Mollinedo, P. del C. 2000. Social benefits and financial profitability of sustainable forest management in two communities of the Mayan Biosphere Reserve, Peten, Guatemala. Unpublished M.Sc. Thesis. CATIE, Turrialba, Costa Rica.

Molnar, A. 2003. Forest certification and communities: looking forward to the next decade. Forest Trends, Washington, DC.

Nussbaum, R. and M. Simula. 2004. Forest certification: a review of impacts and assessment frameworks. TFD Publication 1. The Forests Dialogue (TFD), Yale School of Forestry and Environmental Studies, New Haven, CT, USA.

Nussbaum, R., M. Garforth, H. Scrase, and M. Wenban-Smith. 2001. Getting small forest enterprises into certification: An analysis of current FSC accreditation, certification and standard-setting procedures identifying elements which create constraints for small forest owners. PROFOREST, Oxford, UK.

Nygren, A. 2005. Community-based forest management within the context of institutional decentralization in Honduras. *World Development* 33(4):639-655.

PRB–Population Reference Bureau 2004. World Population Data Sheet Retrieved October 6, 2004, from http://www.prb.org/

Richards, M., F. del Gatto and G. Alócer López. 2003. The cost of illegal logging in Central America. How much are the Honduran and Nicaraguan governments losing. ODI Forestry Briefing. ODI, London.

Ruben, R. and J. Pender. 2004. Rural diversity and heterogeneity in less-favoured areas: the quest for policy targeting. *Food Policy* 29(4):303-320.

Sundberg, J. 2003. Conservation and democratization: constituting citizenship in the Maya Biosphere Reserve, Guatemala. *Political Geography* 22(7):715-740.

Thornber, K. and M. Markopoulos. 2000. Certification: its impacts and prospects for community forests, stakeholders and markets. International Institute for Environment and Development (IIED), London.

Vayda, A.P. 1996. Methods and explanations in the study of human actions and their environmental effects. CIFOR/WWF Special Publication. Center for International Forestry Research, Jakarta.

Walters, B.B., C. Sabogal, L.K. Snook and E. de Almeida. 2005. Constraints and opportunities for better silvicultural practice in tropical forestry: an interdisciplinary approach. *Forest Ecology and Management* 209(1-2):3-18.

Zamora, N. 2000. Árboles de la Mosquitia Hondureña. CATIE. Serie Técnica, Manual Técnico No. 43. CATIE, Turrialba, Costa Rica.

doi:10.1300/J091v24n02_05

# A United States View on Changes in Land Use and Land Values Affecting Sustainable Forest Management

Ralph J. Alig

**SUMMARY.** With increasing opportunity costs of keeping land in forests because of increasing values for other land uses, such as for developed uses, forest ownership may become less attractive for some landowners and the return on investment less viable for both private and public landowners. This raises the question of what will become of the forests and the resources the forest supports, such as water and wildlife, if owners of the forests find it too costly to manage the forest. Land markets provide evidence on revealed behavior about what people are willing to actually pay for a bundle of rights necessary to gain access to land that can provide forest-based goods and services into perpetuity. However, owing to market failures and the nature of some forest land values, markets do not always reveal true forest land values, as discussed

---

Ralph J. Alig is Team Leader and Research Forester, USDA Forest Service, PNW Research Station, Land Use and Land Cover Dynamics Team, 3200 SW Jefferson Way, Corvallis, OR 97331 USA (E-mail: ralig@fs.fed.us). Team web site: www.fsl.orst.edu/lulcd.

Earlier research was funded by the PNW Station Focused Science Delivery Program's Sustainable Wood Production Initiative, and Human and Natural Resources Interactions Program; the Environmental Protection Agency; and the USDA Forest Service Strategic Planning and Resource Assessment unit in the Washington Office.

[Haworth co-indexing entry note]: "A United States View on Changes in Land Use and Land Values Affecting Sustainable Forest Management." Alig, Ralph J. Co-published simultaneously in *Journal of Sustainable Forestry* (Haworth Food & Agricultural Products Press, an imprint of The Haworth Press, Inc.) Vol. 24, No. 2/3, 2007, pp. 209-227; and: *Sustainable Forestry Management and Wood Production in a Global Economy* (ed: Robert L. Deal, Rachel White, and Gary L. Benson) Haworth Food & Agricultural Products Press, an imprint of The Haworth Press, Inc., 2007, pp. 209-227. Single or multiple copies of this article are available for a fee from The Haworth Document Delivery Service [1-800-HAWORTH, 9:00 a.m. - 5:00 p.m. (EST). E-mail address: docdelivery@haworthpress.com].

Available online at http://jsf.haworthpress.com
doi:10.1300/J091v24n02_06

below. Allocation of land by use and cover types is a key determinant in sustainable forest management, with changes in land values providing important signals to land managers. Land valuation differs under market-oriented economies, emerging values in transition economies, and administered values in countries with command economies and is influenced by interactions between the environment and humans, including land ownership, use, and management. doi:10.1300/J091v24n02_06 *[Article copies available for a fee from The Haworth Document Delivery Service: 1-800-HAWORTH. E-mail address: <docdelivery@haworthpress.com> Website: <http://www.HaworthPress.com>.]*

**KEYWORDS.** Land values, land allocation, externalities, deforestation, forest benefits

## *INTRODUCTION*

Increasing opportunity costs to keep land in forests because of increasing values for other land uses can cause forest ownership to become less attractive for landowners and the return on investment less viable for both private and public landowners. Actions that affect values for forest uses, such as zoning, taxation, and regulation, can provide incentives for the landowner to seek a higher valued alternative, which can mean shifting to a nonforest use, such as a developed use. On the other hand, forestland values can be enhanced by wise management, for example, increased vigor of heavily stocked forest stands or re-establishing previous conditions of certain forest ecosystems (Adams and Latta, 2004), and recognition of all the values that a forest produces. Changes in forest land values can result from forest management practices, timber demand, and nonmarket influences, such as forest practices regulation. Emerging issues include management for nontraditional forest-based goods and services, such as forest carbon. A key challenge is maintaining options for forest sustainability and a robust suite of benefits in the face of continuing population growth. The dynamics of forests and society, along with complex economic-environmental-social relationships, complicate efforts to attain sustainability. Potentially, many paths lead to a sustainable future, but each is associated with tradeoffs–such as different economic benefits for consumers, amounts and quality of habitat for wildlife species, and living and recreational space–that affect the quality of life for people.

Examining changes in land uses and land values can provide valuable information for forest sustainability deliberations. Forest land prices capture a wealth of information regarding the current as well as the anticipated uses of land and resources contained on it. They reflect the valuation of current uses and incorporate information regarding productivity, standing timber capital, and effects of taxes and regulations that apply to the land and production. Land prices are also speculative, so that they also incorporate information regarding anticipated future uses of the land and resources. Use of land value information has been limited in forest sustainability analyses to date, and this paper discusses usefulness of land use and value information beyond the typical boundaries of forestry-related indicators. This includes indicators pertaining to the ability of forestry as a land use to compete among other major land uses, a fundamental necessary condition for having forests, much less having long-term sustainable forests and associated resources. In assessing such potential, this paper also looks back and helps to uncover important relationships as a basis for looking forward, such as projecting changes in forest area for use in large-scale studies of forest sustainability.

## *OWNERSHIP OF FORESTS*

Land ownership can be an important determinant of how forestland is managed and the levels of investments in different practices, e.g., Alig et al., 1999. The majority of the world's forests are publicly owned, with most administered by governments. In contrast, the majority of U.S. forests are privately owned. The relative proportions of U.S. private and public timberland have remained fairly stable since 1953, with about 29% of U.S. timberland in public ownership. Within the private timberland group, the proportion of nonindustrial private forest (NIPF) ownership dropped slightly, from 84% to 82% of total private between 1953 and 2002 (Smith et al., 2004). Family forests are a large component of the NIPF ownership class, and the number of family forest owners increased from 9.3 million in 1993 to 10.3 million in 2003; this diverse set of owners controls 42% of the nation's forest land (Butler and Leatherberry, 2004).

Although forest policy debates often focus on whether to manage public forestlands for timber, ecological, or recreation purposes, almost three-fifths of U.S. forests are privately owned. Although public forestlands will be with us for the foreseeable future, private forestlands have the potential to be lost to developed uses. Nationally, about 400,000 ha of U.S. forestland were lost to development annually from 1992 to 1997. More than another 10 million ha could be lost by 2030 (Alig et al., 2003). Global forest area was reduced by 9.4 million ha annually between 1990 and 2000 (UN FAO, 2001).

Per capita amount of forestland has been declining in the United States (Figure 1). Recent per capita amounts are about half of 1950 levels. Per capita volume of growing stock on timberland in the United States has also declined since 1950, given the growth in population relative to the smaller increase in aggregate growing stock volume. As with per capita forest area, projections are for that trend to largely continue.

Causes for development of U.S. forests are market forces resulting from population and personal income growth (Alig et al., 2004). Population, income, and economic growth combine to increase demands for land in residential, commercial, and industrial uses, and public infrastructure. Demands also increase with people's lifestyle choices when, for example, people relocate to rural areas or desire second homes in scenic forest settings. When demands for developed land uses increase,

FIGURE 1. Amount of forest area (hectares) per resident in the United States and the Pacific Northwest region, 1952-1997 (Smith et al. 2004), with projections to 2050.

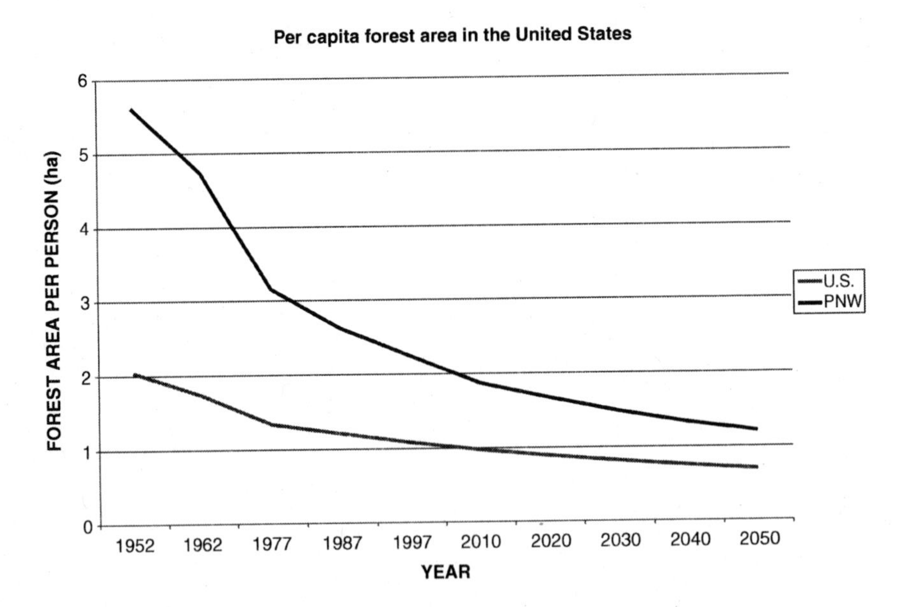

so do the financial incentives some forestland owners have to sell land for development. The incentive is the revenue they can earn from selling land over and above what they can earn from maintaining land in forest. When these market forces are at play, some forestland development is inevitable. For forestland retention, market forces dictate the need for acceptable financial returns for forest ownership, in view of opportunity costs associated with alternative land uses.

Policy-related concerns about forestland development include (1) How does development affect our ability nationally and globally to produce sufficient forest commodities? (2) How does development affect the many ecological, water resource, recreation, and scenic values we also desire from forests as open space? Removal of trees leads to a spectrum of forestland conditions, spanning forest clearing to partial removal. In addition, forest clearing can lead to different spatial patterns of forest fragmentation (Butler et al., 2004; Alig et al., 2005). Further, ownership changes can lead to forest parcelization–the breaking up of large forest parcels into smaller parcels–which can affect the economic feasibility of commercial forest management. Regarding open space values, there is concern about loss of fish and wildlife habitat owing to forest fragmentation and changes in forest structure that accompany development, concern about the adverse effects development can have on water quality and the timing of supply, and concern about loss of access to many private forestlands for recreation and aesthetic enjoyment.

Given the expected U.S. and world population increases and changes in economic activity, a key question is how society can make positive progress toward "forest sustainability" in the face of needing more developed land to serve more people in the future. Forests currently cover about one-third of the United States, and long-term assessments of their condition and relationships to changes in demographics and other socioeconomic factors can aid in defining key policy-relevant questions and developing any beneficial policy actions. With robust economic activity and strong population growth, the United States in the 1990s lost more than 1,000 hectares of forest each day, with more being impacted by sales, fragmentation, or parcelization. Human-related pressures on land condition are likely to increase as population continues to increase. Knowledge of the drivers behind land use change is useful for informing policy discussions concerning society and the environment. One important characteristic of policy formation is that it is concerned with the future. It is the role of policy makers to anticipate future conditions and estimate the effects that proposed policies are likely to produce. Next, we look at examples of information contained in land values as one tool to help in anticipating future uses of land and how the overall forestland base may change.

## *VALUE OF FORESTS*

Valuation of forests varies widely around the world. In Europe, Japan, and the United States, well over half of the forest and wooded land is privately owned. In contrast, a large majority in Canada and Australia is publicly owned. In many of the Central and Eastern European countries with economies in transition, the forest ownership pattern is undergoing substantial change as forestland is restituted to its former owners or privatized. Experience in centrally planned economies of the world, where forests are usually owned by the government, shows that forestland value is rarely an issue. Forests are viewed as common property with rights of use dispensed by the government, either arbitrarily by dictatorships, or by some combination of forest law and policy in more representative forms of government. Forest productivity differences are recognized and valued implicitly, but it is rare to find someone who has a sense of the market value per hectare of a given tract (Beuter and Alig, 2004).

Forestland values provide clues to what amounts and types of forest land are likely in the future and prospects for the provision of mixes of land-based goods and services. Land prices embody information on relative valuations by different sectors of the economy. The ability to retain land in forest cover is dependent on how competitive the uses and values provided by forests are with uses and values produced by other uses of the land, such as agriculture (Alig et al., 1998) or residential/commercial development (Alig et al., 2004). The land values relate to the return on investment in forest enterprises or activities compared to investments in other sectors and to investments in forest sectors in other countries. With growth in population, demand for land for developed uses is increasing at the same time that demand for some forest products is increasing. Increases in population and income tend to increase the demand both for wood and developed land, and conceivably can have different short-term (increased timber harvest) vs. long-term (smaller timberland base) effects when they are viewed from a forestry perspective. With the likelihood of substantial population growth and some of that resulting in more development in forested areas, one policy question is how to mitigate the impacts of housing developments on forests and their environmental services, while allowing for positive contributions for the economy. Housing development not only has a large impact on forests, but also plays a very large role in the U.S. economy–housing development along with consumer spending accounted for a large proportion of U.S. Gross Domestic Product (GDP) growth over the past 4 years (Economist, 2005a). Awareness of the large impacts could elicit some interest in more sustainable systems for housing development, such as "green home-building" guidelines for mitigating adverse impacts of housing development. Alongside this, recognition of macro-economic forces on housing markets and wood products employment includes increased media coverage of a possible housing bubble. Never before have real house prices risen so fast, for so long, in so many countries (Economist, 2005b). A housing bubble bust could have severe consequences for the banking system, heavily invested mutual funds and pension funds, hundreds of thousands of housing-related jobs, and the typical American consumer who is debt-heavy and savings-light. It is also an example of a possible episodic event that may be part of cycles and short-term perturbations that are not typically considered in forest sustainability analyses.

Potential loss of industrial infrastructure pertaining to wood products is another concern with continued forest land loss. Public land management may be affected if timber harvest and forest products processing capacity shrinks with the timberland base. This may hamper ecological restoration activities. Many millions of hectares of national forest land in the United States are highest priority for treatments to reduce fire risk. In addition, without those treatments, the public may lose valuable ecosystem goods and services such as disturbance regulation, water purification, and wildlife habitat.

Valuation of land currently in forest uses in some areas is strongly influenced by trends in developed areas (e.g., Wear and Newman, 2004). Land values for developed uses can exceed those for forest uses by a substantial amount (Alig and Plantinga, 2004). Although agricultural land values are usually second to developed uses in potential value, they often exceed forestland values, although that can be altered by government programs. Trends over time show an increase in forest values but not as large as for developed uses. Developed use values have mostly risen steadily, in contrast to some cyclic movements in agricultural values, related in part to influence of government programs. The potential for switching land uses can give land value a speculative component, so that both forest and agricultural land values are often influenced by development potential.

Forestland values can differ in a variety of geographic, biological, regulatory, economic, and social situations. Given that complex of factors, forest use valuation is increasingly becoming more complicated, as is our economy, by overlays of land use zoning, environmental laws, forest practices regulation, site-specific environmental considerations, and recognition of forest resource values other than timber.

People differ in the values that they place on environmental, economic, and social aspects of forests. This affects the social valuation; this is in contrast to the cost of providing goods and services that others may value from private forestland. An example is that many forestlands and open spaces comprise social values–ecological, scenic, recreation, and resource protection values–which are typically not reflected in market prices for land. When these social values are present, more forestland will be developed than is optimal for society (Kline et al., 2004). For open space policy, one needs to understand social values in the context of forestland market values and the economic rationale and impetus for public and private efforts to protect forestland as open space. Land values reveal what people are actually willing to pay for a bundle of rights necessary to gain access to land that can provide goods and services for a certain period. Changing perceptions about forestland mirror those in U.S. farmland preservation. National interest in preserving farmland arose in the 1970s from concerns about rapid loss of farmland to development, and the supposed threat to food security and agricultural viability. These concerns led to the gradual nearly nationwide implementation of local, state, and federal farmland preservation programs. More recently, recognition has grown for the environmental amenities and the social values of farmland and the role they play in motivating public support for preserving farmland.

## *LAND USE AND LAND COVER CHANGES: 1953 TO 2002*

Examining historical trends can provide helpful guidance in identifying key factors that are likely to influence forestland condition and associated natural resources in future years. As mentioned above, land values capture much information about the current and anticipated uses of land; now we will look at past trends to see how land allocation provides a foundation for projections of sustainability and its implications. In this paper, several broad measures are examined, ranging from total forest area measures to classification schemes such as the rural-urban continuum, that reflect the changing mosaic of land uses within which forestry operates.

### *Total Forest Area*

A key indicator at the national level is total forest area. Between 1953 and 2002, the net change in U.S. forest area was a reduction of about 3 million ha or 1%. Timberland area was reduced by a similar amount. For the largest forest ownership aggregate in the country–NIPF owners, timberland area was reduced by 5.6 million ha, or 5%. The largest concentration of NIPF owners is in the U.S. South, a key timber supply region. Most forestland development occurs on land owned by NIPF owners; NIPF owners control the most U.S. timberland–58% (118 million ha) of the total. In the Pacific Northwest, NIPF land is often critical to threatened and endangered species, especially in lowlands or riparian areas (Bettinger and Alig, 1996).

The declining trend for NIPF timberland area in the South also holds for the Pacific Northwest and Pacific Southwest, two other regions also experiencing above-average growth rates in population and increases in developed area. In the Pacific Northwest, NIPF timberland area dropped by 1.8 million ha or 34% between 1953 and 2002, while the corresponding reduction in California (Pacific Southwest) has been 0.6 million ha or 25%. The trend of a shrinking forestland base has contributed to a decline in per capita forest area (Figure 1), in the face of growth in population.

Recent findings in key U.S. timber supply regions are consistent with broader national trends. Findings for the states of Washington and North Carolina show an accelerated loss of forest land, in line with national findings for the 1990s in Natural Resource Conservation Service (USDA NRCS 2001) surveys. For Washington, the amount of timberland declined by about 5% between 1990 and 2000 (Gray et al., 2005). This was an acceleration from the 1980-90 period, when timberland area declined by about 3%. Most of the losses in timberland came from conversion to urban and developed land uses, mainly on nonindustrial private lands. North Carolina is part of the U.S. South, which now has more timber harvest than any other country in the world. However, even some southern states have had net forest losses in recent decades, including Florida, Louisiana, and Texas. North Carolina recently lost 5% of its timberland in a decade, mostly to urban development, according to forest surveys (Brown, 2004).

Investment in remaining forestland (Alig et al., 1999) has allowed total forest volume to increase in the face of a shrinking forestland base. In the United States, the prime case is that of the area of planted pine in the South increasing more than tenfold since 1953, mostly on private lands. Intensive forestry on private timberland has generally reduced rotation lengths, which leads to more frequent regeneration opportunities and increases the probabilities of more forest cover changes (Alig and Butler, 2004).

### *Forests in the Rural-Urban Continuum*

Forestland development brings more people living closer to remaining (standing) forestlands, in view of growing cities and other urban areas. Identification of forest lands by rural-urban continuum classes indicates that 13% of U.S. forestland now is in major metropolitan counties and 17% in intermediate and small metropolitan counties and large towns, together composing 30% of all U.S. forestland (Smith et al., 2004). Between 1997 and 2002, the area of standing forests in major metro counties increased by 5% or more than 2 million ha (Smith et al. 2002, 2004), as the developed areas in the U.S. expanded considerably.

### *Spatial Changes*

Change in the spatial arrangement of forestlands includes forest fragmentation, considered to be a primary threat to terrestrial biodiversity, with most forested parcels in fragmented landscapes. Fragmentation of U.S. private forestlands is likely to spread in the future (Stein et al., 2005). Positively correlated with forest fragmentation are population density and the percentage of agricultural land (Alig et al., 2005). Forest fragmentation complicates designing multiple-objective policies and land conservation and ecosystem services objectives, for example, forest carbon sequestration.

With rural land uses subject to increasing conversion pressure from a growing population, open space concerns have heightened. The earliest significant open space preservation efforts in the United States (U.S.) involved preserving and restoring publicly owned forests and parks at national and state levels. These efforts were inspired by public concern for rapid loss of forests to agriculture and logging in the later 19th century, and the desire to protect timber and water resources and lands of extraordinary beauty and uniqueness. Since then, public concern for land use change has evolved to recognize the contribution of open space to our day-to-day quality of life–its recreation, aesthetic, ecological, and resource protection benefits.

A privately-owned optimal landscape can depart from a socially optimal landscape, the latter which reflects society's preferences for public goods associated with interior forest parcels. Policy-related research can examine land use shifts for parcels in a way that is optimal for reducing forest fragmentation. However, spatial configuration considerations make this complex, in that benefits of converting (or retaining) a parcel will depend on the land uses on the neighboring and other parcels affected by the policy. Such spatial arrangements affect the landscape conditions, and at larger scales, the regional and national forestland conditions.

## *FORESTLAND CONDITIONS*

The type of removal of trees has large implications for resulting land conditions. For example, when a forest is converted to a developed use, the loss of ecosystem services is direct and immediate, with permanent habitat loss. Some trees will remain in certain developed areas, but the mix of wildlife often changes drastically. Some wildlife species, such as elk or grizzly, require large areas. Such wide-roaming species are often associated with the national forests, and they can use entire forested landscapes, including privately owned bottomlands at certain times of the year. As these private lands are developed, the wildlife disappears, no matter how good the habitat remains on adjacent public land.

An example of a national effort to monitor and project forestland conditions is the periodic national assessments in the United States. Periodic U.S. natural resource assessments, mandated by the National Forest and Rangeland Renewable Resources Planning Act (RPA) of 1974, support USDA Forest Service (e.g., USDA FS, 2001) strategic planning and policy analyses, including the current situation and prospective area changes over the next 50 years (Alig et al., 2003, Alig and Butler, 2004). The RPA assessments interface with international assessments and regional assessments, e.g., USDA Forest Service (1988). With each succeeding RPA assessment, more details have been added in characterizing current and projecting future forestland conditions. Diversity indices that combine information on both age class and forest types exhibit limited change over the projection period for the United States as a whole (Haynes, 2003). Based on broad-scale measures of forest resource conditions, the RPA assessment does not project large or dramatic changes in U.S. forest conditions over the next 50 years, even as timber harvest levels rise. This does not suggest that deliberations of forest protection and stewardship are moot, rather that related discussions do not need to proceed in an atmosphere based on large-scale ecological change on existing forestland.

Consideration of mixed ownerships is important for U.S. forest sustainability discussions. Within a holistic view of forestland conditions, maintaining or enhancing benefits from forest ecosystems will be affected by the level of forest investment and current institutional arrangements. For public lands, institutional arrangements work best where a government has the resources to deliver ecosystem services as public goods. However, it may be another matter on private land. In the United States, private timberlands are quite important, because about 90% of U.S. timber production is from private lands. Across the United States, forestland conditions are altered by timber harvesting, fire management practices, conversion to other land uses, recreation, and climate change. For example, wood use has increased by 40% since 1960 and is expected to rise by about 30% in the next four decades, which has implications for domestic timber harvest levels (Haynes, 2003). Some people suggest that sustainable forest management in the private sector currently depends heavily on sustained commercial outputs of forest products such as timber. However, opportunity costs of keeping private land in forests are rising as the relative value of such lands for residential or urban development increase compared to forest use. This, in turn, may place additional pressures on public forests to provide diverse land-based ecosystems containing a variety of habitats for wildlife, to help cleanse the air and water, to serve as places for recreation, and to provide other goods and environmental services.

Deforestation has important impacts on forestland conditions. In recent years, most U.S. deforestation has been for conversion to developed uses. Urbanization does not just result in direct conversion of forestland but can also involve forest fragmentation (Alig, 2000; Butler et al., 2004), forest parcelization, and ownership changes. Development pressures can also add to uncertainty about how forestland will be managed, if owners anticipate higher financial returns in an alternative use. Because forestland prices capture information regarding current as well as anticipated uses of land, land prices anticipate future development of forestland near urbanizing areas, casting a speculative shadow over timberland values (Wear and Newman, 2004). With anticipated population and income growth, such dynamics could hold important implications for conditions of forestland and environmental benefits.

Projections suggest continued urban expansion over the next 25 years, with the magnitude of increase differing by region (Alig et al., 2004). The U.S. developed area is projected to increase by 79%, raising the proportion of the total land base that is developed from 5.2% to 9.2%. Because much of the growth is expected in areas relatively stressed with respect to human-environment interactions, such as some coastal counties, implications for landscape and urban planning include potential impacts on sensitive watersheds, riparian areas, wildlife habitat, and water supplies. The projected developed and built-up area of about 71 million ha in 2025 represents an area equal to 38% of the current U.S. cropland base, or 23% of the current U.S. forestland base.

When examining land use dynamics, we find that the many different pathways by which land use can change warrant examining both net and gross area changes for major land uses. The total or gross area shifts involving U.S. forests are relatively large compared to net estimates. Gross area changes involving U.S. forests totaled about 20 million ha between 1982 and 1997, an order of magnitude greater than the net change of 1.6 million ha. Movement of land between forestry and agriculture in the last two decades resulted in net gains to forestry that have offset forest conversion to urban and developed uses in area terms. However, the conditions of forested areas entering and exiting the forestland base can be quite different; entering areas may be bare ground or have young trees, while exiting areas often contain large trees before conversion to developed uses. Concern about the attributes of exiting or entering forested areas was heightened in the 1990s when the rate of development increased, with about 400,000 ha of forests converted to developed uses per year (USDA NRCS, 2001).

Forest environmental services are in increasing demand by society. Treatment of related elements of the landscape has progressed from new forestry to ecosystem management to broader awareness that economic and ecological systems are significantly linked. Joint economic and ecological thinking can help when considering environmental services. Key questions include: What are current motivations for using markets to finance conservation? Who buys ecosystem services and who benefits? What are the opportunities and risks for forest owners and producers? This includes consideration of different types of markets and payment schemes, the mechanics of new market solutions, and insights from business models around the globe.

An example of a forest environmental service with increasing demand is forest carbon. Sale of carbon sequestration credits is slowly proceeding. Many investors are wary of investing in decades-long carbon sequestration projects, which are seen as too risky. A portfolio approach may help, as rules are still being defined. Market intervention in the form of government programs for agriculture has traditionally dwarfed that for forestry (Ingram et al., 2005), in part leading to more land being allocated to agriculture compared to other enterprises than is likely without subsidies for agricultural production. Spending trends in farm programs affect land prices, amount of production in agriculture and forestry, allocation of land between the forestry and agricultural sectors, and the stewardship of land. If agricultural prices are driven down by farm subsidies, this can affect farmers around the globe, some who may export in efforts to boost relatively low incomes. Equity questions arise at both the national and international levels, as huge sums of taxpayers' money go to relatively few producers.

Intervention in markets can affect resource allocation, in contrast to price signals from market-based decisions of private landowners. Prices signal the relative scarcity of land for different uses, and also point to the attractiveness of substitutes for land in certain production practices, such as the intensification of forest management on remaining forest if other uses are causing a reduction in total forestland. Intervention also affects the attractiveness of investing in forest stewardship, in the general context of promoting a more sustainable forest sector.

Investments in conservation practices have been proposed as one means for the transition from commodity-dominated programs to conservation ones. Past studies have shown high retention rates for conservation practices, such as for tree planting on surplus agricultural land. Overall, investment in forestry as measured by tree planting has been increasing (Alig et al., 2003), but not to the degree that spending in agricultural programs has.

## *STATE OF THE LAND*

Sustainability assessments have received increasing attention in recent years, including at the 2005 World Congress of the International Union of Forest Research Organizations, but major questions still remain about what is to be sustained and how linkages across major uses of the land and sectors are to be recognized within any single-sector sustainability effort, for example, forest sustainability. Use of land is shared across sectors, but no well-established comprehensive set of indicators exists to describe the state of the land. The decreasing amount of per capita forest area and the increasing fragmentation of remaining forest, largely owing to socioeconomic forces, are examples where coordinated indicator information and policy discussions could usefully extend beyond traditional forestry boundaries. Enhancement opportunities and threats to forests (e.g., invasive species, loss of open space, unmanaged recreation, and fire and fuels management) are increasingly interlinked, along with more stresses from a growing population that on average has higher incomes and more mobility than earlier generations. Improved indicators of sustainability will need to progress with societal changes, and next I list some attributes of such an assessment system related to the state of the land:

- Uses data and methods to describe indicator conditions that are firmly based on a scientific process, as in the RPA assessments, to support long-range regional and national projections of future supply and demand for agricultural crops, animal products, forest products, recreation land, wildlife habitat, water use, and other landscape and environmental measures (see, for example, USDA FS, 1988, 2001).
- Provides a "big picture" over both time and space that would involve a balanced set of biophysical/ecological, economic, and social components and the key interrelationships. Dynamics of natural resources need to be reflected, as well as socioeconomic realities and the status of legal and economic systems conducive to sustainable land management.
- Reflects that forestland conditions are affected by what happens in other sectors, and examines cross-sector linkages (Ingram et al., 2005) while recognizing that the complexities of such linkages, identification, and reporting of state-of-the-land measures is an ongoing challenge; approach sustainable development from the perspective of connecting concerns about farms, forests, and communities through community- and watershed-based approaches and other means, e.g., *U.S. National Report on Sustainable Forests* (USDA FS, 2004).
- Increasingly links forest resource data to socioeconomic data, such as characteristics of who owns the forest land; for example, trends in population density warrant further study for different classes of rural and urban forestland (Alig, 2000). Demographic changes increase the size of the wildland-urban interface, whose expansion has exacerbated wildfire threats to structures and people. This includes more people on the landscape near public lands, to pursue attractive recreational land and aesthetic amenities, often involv-

ing forests, and related to concerns about changes in quality of life.

- Links scales in a general way. Transforming information obtained into summary statistics, models, or any other format needed for use in sustainability assessments involves the challenges of information quality, frequency, or scale, e.g., local, state, regional, national, global. Linkages among scales are influenced by relevant policy questions, e.g., the concentration of the population and development within coastal zones may be impacted by global climate changes as well as local land markets affected by expectations regarding future land uses.
- Reports value information where possible. Current markets typically value land for development (e.g., residential house site) and for forest products, such as timber, but current valuation is typically imperfect for providing water, maintaining open space, sequestering carbon, and supporting wildlife. The value of such ecosystem services can supplement those for traditional forest products, and, as with carbon, some new markets are emerging.
- Involves multiple disciplines in policy deliberations that encompass the natural, socioeconomic, and cultural dimensions of sustainability and stewardship issues. For example, monitoring forest health or forest quality issues has several facets beyond biophysical or ecological characteristics.
- Recognizes private and social values and why they may differ in some cases. This can be part of designing incentive systems for owners to promote conservation objectives (e.g., opportunities in the next U.S. Farm Bill for more investment in conservation practices, in the context of long-run stewardship) and other social values, while recognizing property rights.
- Uses a broad approach to land and ecosystem valuation to help land managers provide for a diverse array of societal needs, including ecological (e.g., biodiversity), economic, and social ones, such as the growing interest in spiritual values associated with forests.
- Has useful sustainability guidelines supported by measurable indicators that fundamentally reflect the *long-term* ecological, economic, and social well-being, with a more integrated approach for describing the complex interplay between human activity and the environment.

Such recommendations for improving state-of-the-land sustainability analyses are offered with the caveat that any measurement and assessment system will need periodic evaluation for updating and enhancement. Requirement for a dynamic approach is consistent with how natural resources and society change. Many forestry-related challenges cannot be solved by working solely within the forest sector; instead, we must plan more for working across sectors and political boundaries. Selecting indicators involves value judgments about what aspects of forestland are most important to track over space and time and what data are needed to address policy questions that arise with increased attention to sustainability and activities associated with the environment, economy, and societal institutions. In global dialogues on sustainable development, broad views across sectors are increasingly warranted by human alteration of global conditions, such as the impacts of climate change on forests, e.g., Alig et al. (2002). As shown by the Kyoto Protocol process, multiple jurisdictions and scales across the globe can complicate efforts to manage and conserve land and natural resources. Addressing the problems of sustaining ecologically viable and socioeconomically feasible landscapes is a complex task for which neither easy, inexpensive, nor perfect solutions exist.

## REFERENCES


Adams, D. and G. Latta. 2004. Effects of a forest health thinning program on land and timber values in eastern Oregon. *Journal of Forestry* 102(8):9-13.

Alig, R., 2000. Where do we go from here? Preliminary scoping of research needs. In: pp. 371-372. N. Sampson and L. DeCoster (eds.). Proceedings, Forest Fragmentation 2000, September 17-20, 2000, Annapolis, Maryland. American Forests, Washington, DC.

Alig, R., D. Adams, J. Chmelik and P. Bettinger. 1999. Private forest investment and long run sustainable harvest volumes. *New Forests* 17:307-327.

Alig, R., D. Adams and B. McCarl. 1998. Impacts of incorporating land exchanges between forestry and agriculture in sector models. *Journal of Agricultural and Applied Economics* 30(2):389-401.

Alig, R., D. Adams and B. McCarl. 2002. Projecting impacts of global change on the U.S. forest and agricultural sectors and carbon budgets. *Forest Ecology and Management* 169:3-14.

Alig, R. and B. Butler. 2004. Area changes for forest cover types in the United States, 1952 to 1997, with projections to 2050. Gen. Tech. Rep. PNW-GTR-613. U.S. Department of Agriculture, Forest Service, Pacific Northwest Research Station, Portland, OR. 106 p.

Alig, R., J. Kline and M. Lichtenstein. 2004. Urbanization on the US landscape: looking ahead in the 21st century. *Landscape and Urban Planning* 69(2-3):219-234.

Alig, R., D. Lewis and J. Swenson. 2005. Is forest fragmentation driven by the spatial configuration of land quality? The case of western Oregon. *Forest Management and Ecology* 217(2-3):266-274.

Alig, R. and A. Plantinga. 2004. Future forestland area: impacts from population growth and other factors that affect land values. *Journal of Forestry* 102 (8):19-24.

Alig, R., A. Plantinga, S. Ahn and J. Kline. 2003. Land use changes involving forestry for the United States: 1952 to 1997, with projections to 2050. Gen. Tech. Rep. PNW-GTR-587. U.S. Department of Agriculture, Forest Service, Pacific Northwest Research Station, Portland, OR. 92 p.

Bettinger, P. and R. Alig. 1996. Timber availability on non-federal land in western Washington: implications based on physical characteristics of the timberland base. *Forest Products Journal* 46:30-38.

Beuter, J. and R. Alig. 2004. Forestland values. *Journal of Forestry* 102 (8):4-8.

Brown, M.J. 2004. Forest statistics for North Carolina, 2002. Resour. Bull. SRS-88. U.S. Department of Agriculture, Forest Service, Southern Research Station, Asheville, NC.

Butler, B. and E. Leatherberry. 2004. America's family forest owners. *Journal of Forestry* 102 (7):4-9.

Butler, B., J. Swenson and R. Alig. 2004. Forest fragmentation in the Pacific Northwest: quantification and correlations. *Forest Management and Ecology* 189: 363-373.

Economist. 2005a. Consumer spending trends. Retrieved September 9, 2005 from www.economist.com.

Economist. 2005b. The giant housing boom: In come the waves. Retrieved September 10, 2005 from www.economist.com.

Gray, A., C.F. Veneklase, R.D. Rhoads and D. Robert. 2005. Timber resource statistics for nonnational forest land in western Washington, 2001. Resour. Bull. PNW-RB-246. U.S. Department of Agriculture, Forest Service, Pacific Northwest Research Station, Portland, OR. 117 p.

Haynes, R. (tech. coord.). 2003. An analysis of the timber situation in the United States: 1952 to 2050. Gen. Tech. Rep. PNW-GTR-560. U.S. Department of Agriculture, Forest Service, Pacific Northwest Research Station, Portland, OR. 254 p.

Ingram, D., R. Alig, and K. Lee. 2005. Analysis of cross-sectoral impacts: The U.S. forest sector. IUFRO 2005 World Congress, Brisbane, Australia.

Kline, J., R. Alig and B. Garber-Yonts. 2004. Forestland social values and open space preservation. *Journal of Forestry* 102(8): 39-45.

Smith, W., P. Miles, J. Vissage and S. Pugh. 2004. Forest resources of the United States, 2002. Gen. Tech. Rep. NC-241, U.S. Department of Agriculture, Forest Service, North Central Research Station, St. Paul, MN.

Smith, W., J. Vissage, D. Darr and R. Sheffield. 2002. Forest resources of the United States, 1997. Gen. Tech. Rep. NC-219, U.S. Department of Agriculture, Forest Service, North Central Research Station, St. Paul, MN.

Stein, S.M., R.E. McRoberts, R.J. Alig [et al.]. 2005. Forest on the edge: housing development on America's private forests. Gen. Tech. Rep. PNW-GTR-636. Portland, OR: U.S. Department of Agriculture, Forest Service, Pacific Northwest Research Station. 16 p.

U.S. Department of Agriculture, Forest Service [USDA FS]. 1988. The South's fourth forest: alternatives for the future. Forest Resource Rep. 24. Washington, DC.

U.S. Department of Agriculture, Forest Service [USDA FS]. 2001. 1997 Resources Planning Act (RPA) Assessment. Washington, DC.

U.S. Department of Agriculture, Forest Service [USDA FS]. 2004. National report on sustainable forests–2003. Report 766. Washington, DC.

U.S. Department of Agriculture, Natural Resource Conservation Service. [USDA NRCS]. 2001. Summary report: 1997 national resources inventory (revised December 2001). Washington, DC.

Wear, D. and D. Newman. 2004. The speculative shadow over timberland values in the U.S. South. *Journal of Forestry* 102(8):25-31.

doi:10.1300/J091v24n02_06

# Landowner-Driven Sustainable Forest Management and Value-Added Processing: A Case Study in Massachusetts, USA

David T. Damery

**SUMMARY.** The Massachusetts Woodlands Cooperative, LLC (MWC) is working to help members conduct sustainable forestry of the highest standards while increasing financial returns from harvest activities. The forests of Massachusetts, the third most densely populated state in the United States, are threatened. Decades of high grading and the threat of forest conversion to alternative use present challenges for maintaining a forested landscape. Despite being 60% forested, Massachusetts imports approximately 98% of the wood fiber that its citizens consume.

The Massachusetts Woodlands Cooperative is a forest management, processing, and marketing cooperative organized by and on behalf of forest landowners in western Massachusetts. An umbrella group certification protocol was developed to provide cost-effective forestland management certification. Members benefit from cooperative management of harvest operations, above-market stumpage payments, and value-added processing and production including marketing traditionally low-value

David T. Damery is Lecturer, Department of Natural Resources Conservation, University of Massachusetts-Amherst, Amherst, MA 01003 USA (E-mail: ddamery@forwild.umass.edu).

[Haworth co-indexing entry note]: "Landowner-Driven Sustainable Forest Management and Value-Added Processing: A Case Study in Massachusetts, USA." Damery, David T. Co-published simultaneously in *Journal of Sustainable Forestry* (Haworth Food & Agricultural Products Press, an imprint of The Haworth Press, Inc.) Vol. 24, No. 2/3, 2007, pp. 229-243; and: *Sustainable Forestry Management and Wood Production in a Global Economy* (ed: Robert L. Deal, Rachel White, and Gary L. Benson) Haworth Food & Agricultural Products Press, an imprint of The Haworth Press, Inc., 2007, pp. 229-243. Single or multiple copies of this article are available for a fee from The Haworth Document Delivery Service [1-800-HAWORTH, 9:00 a.m. - 5:00 p.m. (EST). E-mail address: docdelivery@haworthpress.com].

Available online at http://jsf.haworthpress.com

doi:10.1300/J091v24n02_07

and small-diameter material. The added revenue from developing these new markets is used to fund timber, wildlife habitat, recreation, and other sustainable forest management activities. The cooperative works in partnership with local wood-processing businesses to spur community economic development. This study on cooperatives may be a successful example of sustainable forest management that can be applied in other regions with private land ownerships. doi:10.1300/J091v24n02_07 *[Article copies available for a fee from The Haworth Document Delivery Service: 1-800-HAWORTH. E-mail address: <docdelivery@haworthpress.com> Website: <http://www.HaworthPress.com>* 

**KEYWORDS.** Forest certification, cooperative, cooperation, NIPF, private forest, marketing

## *INTRODUCTION*

Private forests are important in addressing the issue of sustainability both in the United States (National Research Council, 1998) and around the globe (Vilkriste, 1998). The amount of forestland controlled by private forest landowners is large, and privatization of the world's forests is expanding (FAO, 2001). In some countries, including Finland, Norway, Sweden, and the United States, nonindustrial private forest (NIPF) landowners own the majority of forestland (Lindstad, 2002). In the United States, for example, 71% of all timberland area is privately owned (Smith et al., 2004). Nonindustrial private forest landowners own 118 million hectares of timberland, and in 2001 they accounted for 63% of the volume of trees removed in the United States. A continuing challenge is how to encourage private forest stewardship that ensures the provision of both public and private services now and in the future (Best and Wayburn, 2001).

This paper presents a case study of a landowner-driven initiative that supports the twin goals of sustainable forest management and increased revenue from forest management activities. The organization profiled is located in the state of Massachusetts in the northeastern United States. Significant forest characteristics, economic circumstances, and social and cultural differences exist across private forest ownerships around the globe. However, portions of the forest management model presented here can be applied in many areas with concentrations of NIPF ownership.

Massachusetts is the third most densely populated state in the United States and home to 6.2 million residents. The forest has been altered dramatically in concert with changes in land use over the past 300 years. With the arrival of European colonists and the expansion of an agrarian society, land was cleared for farming over the 18th and much of the 19th centuries. In the 1870s, the reduction in forest cover owing to conversion to agriculture slowed. More productive agricultural land was beginning to contribute to the young nation's economy as the population expanded westward (Barten et al., 2001). Breunig (2003) reported that agricultural land covered 50% of the Massachusetts land area in the late 19th century and has since declined to just 7% in 2003. A significant portion of the formerly agricultural land has now returned to forest. Forest area increased steadily from a low of about 30% of land area in 1870 to a high of nearly 75% in 1960. The most recent, 1998, forest inventory analysis yielded area estimates for forestland (62%), nonforest (33%) and agriculture (5%) (Alerich, 2000). Since 1960, development pressures, including subdivision for homebuilding, have led to a new erosion of the forest base. Forest loss averaged 5,900 hectares (14,600 acres) annually over the period 1985-1999 (Breunig, 2003). The spread of suburbanization and conversion of forestland to other uses poses a growing concern in Massachusetts.

Even though forest area is losing ground in Massachusetts, the volume of trees continues upward. The growing-stock volume of trees increased by 17% between 1985 and 1998 (Alerich, 2000). A potential tool to help stem conversion may lie in the increased volume of trees. If forest landowners can derive higher levels of revenue from the growing forest resource, then the economic incentive for converting to other uses may be lessened.

Ownership changes in the 20th century have caused many of the regenerated forests of Massachusetts to be managed for reasons other than timber production or neglected entirely (Beattie et al., 1993). Forest stands are often too densely stocked and contain a high proportion of low-grade trees, a result of the practice of high grading (Barten et al., 2001). Land ownership patterns have decreased the average tenure of forest properties (Birch, 1996). In our increasingly mobile society, property changes ownership more frequently. New landowners may not have had a close relationship to the land prior to their forestland acquisition. They often lack knowledge of forest management principles and practices. Many forest landowners are beginning to awake to the fact that actively managing the forest has the potential to significantly increase the provision of both timber and nontimber values.

The forests of Massachusetts are in a transitional area composed of both central and northern hardwood and softwood types (Barten et al., 2001). Although tree volume continues to increase overall, the increase is not uniform across species. The rate of growth in volume across species reflects, in part, the relative market values and merchantability of the predominant species. In recent decades, northern red oak (*Quercus rubra* L.) and eastern white pine (*Pinus strobes* Englem.) have commanded the highest stumpage prices. New markets need to be developed for the less-valued species of red maple (*Acer rubrum* L.) and hemlock (*Tsuga canadensis* (L.) Carr.) that are the second and fourth most abundant species by growing-stock volume. Table 1 shows the percentage of growing-stock volume for all species in Massachusetts from the most recent, 1998, forest inventory analysis (Alerich, 2000).

Berlik et al. (2002) highlighted the inconsistency of U.S. consumers' desire for global environmental protection and the reality of their current level of forest products consumption. Wood consumption in Massachusetts is estimated at over 13 million cubic meters annually, but harvests from state timberland amounts to only 300,000 cubic meters. Massachusetts currently produces only 2% of the wood fiber that it consumes (Berlik et al., 2002: p. 13). This stems from a combination of in-

TABLE 1. Volume of growing stock by species in Massachusetts, USA, 1998

| Species | Total growing-stock volume |
|---|---|
| | *Percent* |
| White pine (*Pinus strobes* Engelm.) | 25 |
| Eastern hemlock (*Tsuga canadensis* (L.) Carr.) | 10 |
| Other softwoods | 3 |
| Red maple (*Acer rubrum* L.) | 18 |
| Sugar maple (*Acer saccharum* Marsh.) | 4 |
| Birch (*Betula papyrifera* Marsh., *B. lenta* L., *B. alleghaniensis* Britt.) | 7 |
| Beech (*Fagus grandifolia* Ehrh.) | 3 |
| Ash (*Fraxinus americana* L.) | 4 |
| Black cherry (*Prunus serotina* Ehrh.) | 3 |
| White oak (*Quercus alba* L.) | 2 |
| Red oak (*Quercus rubra* L.) | 17 |
| Other hardwoods | 4 |

Source: Alerich, 2000.

creasing population and demand for wood products coupled with a shrinking sawmill industry (Damery and Boyce, 2003). The number of sawmills operating in the state has fallen by 55% from 1971 to 2001. Berlik et al. recommend reducing consumption and increasing harvest levels to improve the level of self-sufficiency. Current rates of forest growth exceed harvest rates by a significant margin. An argument can be made for higher levels of sustainable production of forest products within Massachusetts. Private landowners own almost 80% of the state's forestland. Coordinating this diverse group through outreach and education activities presents a major challenge to achieving higher levels of sustainable production (Clawson, 1979).

## *LANDOWNER GOALS*

Although we may recognize, from a policy perspective, a need for higher levels of local production, the decision to actively manage Massachusetts' forests still lies with individual landowners. Apart from the "macro" perspective on demand and supply, individual landowners have varying perspectives and goals. Private forest landowners have many values and ownership objectives beyond revenue from timber harvesting (Young et al., 1985).

To gain insight into the values of private forest landowners, Damery (2001) surveyed 232 western Massachusetts forest landowners who had active forest management plans. When asked to rate their level of interest in eight forest management goals, landowner income came in fifth. Wildlife habitat, tree and plant quality, ecosystem health, and water quality all ranked higher. Rankings for all options are shown in Table 2.

Clearly, these landowners desire to maintain and improve their forestlands. The results from this survey led to the adoption of forest certification as a method of achieving recognized standards of sustainable forest management. The survey also showed that timber management is only one goal among many. Achieving the various landowner goals is costly. Timber harvesting is often looked on as the means to finance other activities. A confounding factor is the need to improve the overall stand quality for long-term management. This often requires thinning or selective harvests. Small-diameter, crooked-stem, or low-value species that might be targeted for removal often bring little or no value in the marketplace. This problem of finding markets for this material has been identified as a key goal by the landowners of the MWC.

Traditional market factors also stand in the way of a solution regarding the financing of thinning cuts. Local markets for small-diameter

TABLE 2. Landowner rankings of forest management goals

| Issue | Respondents who ranked issue as "strong" or "moderate" interest |
|---|---|
| | *Percent* |
| Wildlife habitat | 58 |
| Tree and plant quality | 56 |
| Ecosystem health | 52 |
| Water quality | 38 |
| Landowner income | 36 |
| Aesthetics | 16 |
| Recreational opportunity | 14 |
| Local economy | 10 |

n = 232, Source: Damery, 2001.

wood are thin (Clawson, 1979). That is, there may be few, or no buyers for this material within an economically feasible trucking distance of the property. This has sometimes led to the practice of high grading, where the largest and best-formed trees are regularly harvested leaving an inventory of poorer quality trees (Beattie et al., 1993). Recent outreach and forestry education activities are raising the consciousness of both foresters and landowners, and this is serving to lower the level of high-grading activity.

Landowners are sometimes unaware of the benefits they can gain by obtaining professional assistance, such as that offered by consulting foresters. Landowners who undertake harvest activities without professional assistance have been shown to receive lower returns from their sales and were less satisfied than those who contracted with a consulting forester (Clark et al., 1992). Owners of smaller forest parcels are less likely to seek professional assistance and may be at a competitive disadvantage in negotiating stumpage prices with buyers (Clawson, 1979; Birch et al., 1998). In summary, the individual forest landowner is faced with both educational and economic challenges in achieving their desired management goals.

## *COOPERATION AS A SOLUTION*

A group of forest landowners in western Massachusetts began meeting in 1999 to share their experiences and address some of the issues discussed above. Two needs emerged from the initial meetings; a desire to improve their knowledge of sustainable forest management and the desire to improve the economic return from their harvest activities. Cooperation is one method with the potential to address these needs. The potential benefits of cooperation can stem from both economies of scale and from economies of scope (Baumol et al., 1982).

Economies of scale provide the potential to develop better markets for the small-diameter and lesser valued materials. Coordinated harvest activities can lower the unit costs of harvest through more efficient use of consultants and loggers. By combining their management efforts, a group of landowners are able to offer a steady stream of forest products, over time, from a much larger combined forest area. Larger harvest volumes and more reliable supply have the potential to generate higher timber sales revenues (Clawson, 1979; Simon and Scoville, 1982).

Economies of scope affect the number of different types of activities that are enabled by coordinating efforts. The ability to achieve landscape-level ecosystem objectives is enhanced through cooperation (Belin et al., 2005). Nontimber management activities including management for wildlife habitat, ecosystem health, and recreational activities were identified as examples of activities enhanced by cooperation in the 2000 landowner survey (Damery, 2001). Other examples of the potential benefits of cooperation include identifying capabilities and quality of work histories of service providers such as consultants and loggers.

## *THE COOPERATIVE FORMATION PROCESS*

The mission of the MWC contains three primary objectives. The first is a desire to perform forest management in an environmentally responsible fashion. The second is to coordinate value-added operations in order to improve the financial returns to the landowner. The third is the desire to conduct business operations with local partners to foster local community economic development. The MWC is a for-profit organization and membership is by invitation. The organizing process leading to incorporation took 2 years. The Limited Liability Corporation (LLC) form of organization was selected based on ease of formation and operating flexibility. Although formed as an LLC, the MWC operates as a cooperative where each member has one vote, and profits are returned to the members based on their share of the value of the wood that the cooperative processes. In the case where profits are available to distribute back to members, they are apportioned on the basis of the stumpage value that each member contributed to the cooperative operations that year. In addition to the landowner founders, a group of resource professionals was assembled to advise and inform the membership regarding forestry and business operations. Professional advice was sought from forest products producers, lawyers, accountants, university faculty, state forestry professionals, consulting foresters, and nonprofit groups interested in cooperative formation and economic development. The group of resource professionals provided specific knowledge in the areas of sustainable forest management, incorporation, business management, accounting, manufacturing, drying, and marketing.

## *COST-EFFECTIVE FOREST CERTIFICATION*

Members of MWC have a strong desire to conduct forestry activities consistent with the world's highest standards. The northeastern United States regional certification standard of the Forest Stewardship Council (FSC), a nonprofit group that provides an environmentally appropriate and economically viable certification system for forest products, was chosen at the outset as a recognized measure for achieving this goal. With the help of grant funding from the U.S. Department of Agriculture, Forest Service, the MWC developed a protocol for certifying members' forestlands. Working with the University of Massachusetts-Amherst faculty and students, and state forestry personnel, an "umbrella" protocol for group certification was developed. This was designed to provide a cost-effective way for small landowners to have their forests certified as being sustainably managed. The MWC FSC group certification was approved in 2003.

One requirement for MWC's group certification was that each member must have an approved forest management plan that addresses the nine guiding principles of FSC. The majority of landowners joining MWC are already covered under a forest management plan. Typically these plans are part of a Massachusetts forest property tax reduction program known as Chapter 61. This program requires a 10-year commitment to keep the land in its forested state and to follow the management activities that are described in the plan. Chapter 61 participants are eligible for a property tax reduction of up to 95% of the assessed value of the forestland; penalties are applied for any early withdrawal from the plan. The required forest management plan is comprehensive–addressing both timber and nontimber objectives. Landowners typically contract with a consulting forester to survey their property and work with them to identify specific landowner goals that are written into the plan. Members with these types of plans need only slight modifications, at a modest cost, to upgrade their Chapter 61 plans so that they meet the broader FSC guidelines. The annual audit fee is approximately $1,500 USD, which is covered by the cooperative's general operating expenses. Apportioned across more than 40 members, this represents roughly $40 USD per member annually.

To keep audit costs to a minimum, MWC acts as an internal auditor to ensure compliance with FSC guidelines. Staff, members, and volunteers, including forestry faculty from the University, survey each member's property annually. Harvest activities performed on member properties are reviewed by MWC staff. Smartwood, part of the Rainforest Alliance, has been the FSC auditor. Having the internal review team allows Smartwood to select a random sample of all member properties to review during their annual audit. By selecting a subset of all member properties, the time and expense of the outside auditor can be kept to a reasonable minimum. Documentation of MWC's internal audit team visits are also reviewed as part of the annual outside audit. The process developed provides a lower-cost method of forest certification than individual members could have achieved on their own.

## *OPERATIONS AND MARKETING*

Three part-time staff people currently manage MWC operations. Two co-executive directors manage membership, harvest activities, value-added production, and sales and marketing. An office manager handles day-to-day clerical operations for the cooperative. The cooperative contracts a part-time bookkeeper and retains accounting and legal help as needed.

The staff reports to a board of directors and the general membership. Staff is presently funded through a major U.S. Department of Agriculture, Rural Business-Cooperative Services Grant. These funds, awarded in 2004, were designed to provide start-up working capital that would enable the MWC to be financially self-sustaining by 2007. Membership fees and product sales generate other income. An annual membership fee of $85 USD has been assessed each member for each of the years 2004 and 2005. The MWC business plan was developed by using data from pilot projects where trees were harvested from member properties and then managed through a variety of value-added processing steps. The business plan included operating, marketing, and financial plans, and projections. A break-even production rate was estimated and is projected to be 1,200 cubic meters (350,000 board feet) of processed timber annually. The MWC staff coordinates steps in the value-added process. Staff consults with landowner members and their consulting foresters to purchase stumpage at above-market rates. In addition, staff contracts with independent loggers for harvesting, sawmills for primary processing, dry kilns, and molders for secondary processing into flooring and other products.

Manufactured products include rough-sawn lumber, finished tongue-and-groove hardwood flooring, timbers, and logs. These products are marketed to local homeowners, building contractors, and architects for inclusion in residential and commercial construction. Several pilot-project case studies were conducted to produce hardwood flooring, timbers, poles, and lumber. Data were gathered from each of these projects including yields, costs, prices, and time involved. Data from the pilot projects and the field experience gained were used to develop a business plan.

Pilot projects included production of hardwood flooring and the harvest and production of black locust (*Robinia pseudoacacia* L.) for lumber, posts, railings, and firewood. These projects required the organization of landowners, consulting foresters, and loggers, and included coordination of trucking, sawmilling, drying, and value-added processing (moulder). Yields were higher than expected, but at the added cost of additional labor time and management expense (Campbell, 2004).

The flooring project involved processing 30 logs, mostly of black cherry (*Prunus serotina* Ehrh.), that were left over from a harvest and lumber milling project. This material was of relatively small diameter and low to medium low in quality. This provided an opportunity to experiment with the production of strip hardwood flooring. Strip flooring can be sold in relatively narrow widths and short lengths. The material was trucked to a local circular sawmill, owned by a cooperative landowner member where it was milled and dried. The dried material was trucked to a custom moulder for production of tongue-and-groove flooring in a variety of widths. Costs were tracked throughout the process, and the material was marketed at a competitive price. Two factors should be noted in this project. The first is that the species, black cherry, commanded a premium price in the marketplace. Marketing the less popular species, red maple, yielded lower margins. Secondly, the flooring material was marketed locally as a "natural" or "character grade" flooring, and was not separated into traditional grades of flooring. This "run of the log" grading process requires education of the consumer in the marketplace.

A second pilot project involved the harvest of a small stand 0.4 hectare (1 acre) of black locust located close to a cooperative member's home. As black locust is not a common species in the region, this project involved finding a potential buyer for the harvested and processed material prior to starting the project. One of the best uses of black locust is in exterior applications, as black locust is considered to have very high decay resistance (USDA FS, 1999). A buyer was identified who specializes in the production of outdoor walkways. The MWC co-director coordinated the activities of the landowner, forester, logger (who also purchased the end material), and sawmiller. Details of the particular end-uses that the buyer was interested in were obtained. This information was used by the logger and sawmiller to produce a much greater volume of material than either the initial forest inventory, or subsequent log-tally indicated. The overall sawlog volume recovery was 2.2 times what was expected from a conventional harvest. The estimated value of products recovered was even greater. Value at the log landing was estimated at $3,547 USD under the cooperative scenario vs. an estimated $1,158 USD that might have been received in a conventional stumpage sale (Campbell, 2004).

The MWC is developing value-added markets for traditionally under-valued and smaller diameter wood. The production of red maple flooring is an example. In recent decades red maple has attracted lower stumpage prices than many other species. One result has been rapidly growing volume of this species. In a comparison of the two most recent forest inventory results, 1985 and 1998, growth of red maple was shown to have exceeded removals by a factor of 6.5 times (Alerich, 2000). Product markets like flooring can use smaller diameter logs, and short cuttings. The pilot projects helped establish the economic feasibility for this market. Further market development may provide members with higher returns for red maple that can lead to increased harvest activity.

If the cooperative is to succeed on an expanded level, markets must be found to absorb a much higher level of production. Two primary markets have been identified as holding the best potential. The western Massachusetts region has a reputation for cultivating "buy-local" markets. A small but significant portion of the consuming public prefers to buy local material over possibly less expensive products imported from outside of the state. On a more global scale, mechanisms such as the low-energy and environmental design (LEED) architectural standards promote the purchase of locally produced materials. Marketing materials and promotion campaigns for MWC-produced products will promote this buy-local message.

The second major market opportunity is associated with sustainable management and green certification. A different, but perhaps overlapping, consuming public appreciates the notion of sustainable forest management. To support the expansion of green-certified forest products, the MWC has embarked on a project to certify local forest products businesses. A group "chain-of-custody" umbrella protocol has been developed and several local businesses are in the process of applying for certification. The chain-of-custody process ensures that businesses can document their material purchases and manufacture of products that use green-certified wood. This provides the consumer with an audited level of assurance that the product they are purchasing can indeed be traced back to material that came from a sustainably managed forest.

## *MEMBERSHIP GROWTH AND THE FUTURE*

Growth of the MWC has been incremental. Conservative governance has focused on developing partnerships with existing manufacturers rather than purchasing new plant and equipment. This has helped to avoid large capital investment, borrowing, and debt load. Challenges remain in expanding the membership to enable balancing the harvest activities of the membership with the market demand for MWC value-added products. Educating potential new members of the costs and benefits associated with the cooperative is time consuming. Current members of MWC include 42 individuals, families, and organizations with 1,900 hectares (4,600 acres) of forestland. The average property size is approximately 45 hectares (110 acres).

The economies of scale and scope provided by the cooperative are dependent on the identification of higher value markets for these less-valued materials. If successful, more timber stand improvement work can be conducted on member properties. These management activities have the potential to increase timber growth rates and stumpage values. Nontimber management goals, such as recreation, wildlife habitat, and ecosystem management, will also be less costly to achieve if higher values are received from associated harvest activities.

If MWC can achieve its management and marketing goals, additional revenues will be generated for the member landowners that should provide an incentive to keep their property in forestry. Improved forestry returns combined with practicing the highest recognized standards of sustainable forestry would enable members to help preserve the forested landscape of western Massachusetts for future generations.

The MWC has worked continuously to broaden its own membership and to assist other groups with similar goals. Publications, conference presentations, and a recent grant aimed at promoting this sustainable forestry model to farmers are some of the methods that have been used to disseminate the knowledge that has been gained. Pilot projects of MWC are featured in the book, *Profiles from Working Woodlands* by Susan M. Campbell (2004). Various projects undertaken by MWC have been presented at forestry-related conferences in the United States, Canada, and Australia. The U.S. Department of Agriculture, Sustainable Agriculture Research and Education program awarded MWC a 2-year grant, begun in 2005, to encourage farmers with woodlots to adopt sustainable forestry practices. As knowledge is gained and problems are solved, MWC will continue to publish their findings with the goal of promoting sustainable forestry practices worldwide.

## *CONCLUSIONS*

The MWC has shown that small private forest landowners can successfully collaborate to achieve sustainable forestry goals. They have developed a cost-effective model for obtaining FSC land certification. They are producing value-added products for "buy local" and "green certified" markets and are providing the landowner members with higher rates of return for their forest management activities. Recommendations for landowners in other regions who wish to achieve similar results are:

- Obtain grant financing to assist with start-up costs.
- Obtain technical assistance from forestry and business professionals.
- Minimize capital investments, if possible, to limit borrowing and its associated debt load.
- Partner with existing value-added service providers (sawmills, dry kilns, moulders) to produce value-added products.
- Identify and sell into niche markets where a competitive advantage exists.

When landowners are able to cooperate, they may be able to achieve recognized standards of sustainable forestry at a lower cost. Through careful forest management and business planning, they may be able to increase returns from management activities and improve the level of both public and private services from the world's private forests.

## REFERENCES

Alerich, C.L. 2000. Forest statistics for Massachusetts: 1985 and 1998. Resour. Bull. NE-148. U.S. Dept. of Agriculture, Forest Service, Northeastern Research Station, Newtown Square, PA.

Barten, P.K., D. Damery, P. Catanzaro, J. Fish, S. Campbell, A. Fabos and L. Fish. 2001. Massachusetts family forests: birth of a landowner cooperative. *Journal of Forestry* 99(3):23-30.

Baumol, W.J., J.C. Panzar and R.D. Willig. 1982. Contestable markets and the theory of industry structure. Harcourt Brace Jovanovich, Inc., New York.

Beattie, M., C. Thompson and L. Levine. 1993. Working with your woodland: a landowner's guide. University Press of New England, Hanover, NH.

Belin, D.L., D.B. Kittredge, T.H. Stevens, D.C. Dennis, C.M. Schweik and B.J. Morzuch. 2005. Assessing private forest owner attitudes toward ecosystem-based management. *Journal of Forestry* 103(1):28-35.

Berlik, M.M., D.B. Kittredge and D.R. Foster. 2002. The illusion of preservation: a global environmental argument for the local production of natural resources. Harvard For. Pap. No. 26. Harvard Forest, Harvard University, Petersham, MA.

Best, C. and L.A. Wayburn. 2001. America's private forests. Island Press, Washington, DC.

Birch, T.W. 1996. Private forest landowners of the Northern United States, 1994. Resour. Bull. NE-136. U.S. Dept. of Agriculture, Forest Service, Northeastern Forest Experiment Station, Radnor, PA.

Birch,T.W., S.S. Hodge and M.T. Thompson. 1998. Characterizing Virginia's private forest owners and their forest lands. Research Paper NE-707. U.S. Dept. of Agriculture, Forest Service, Northeastern Forest Experiment Station, Radnor, PA.

Breunig, K. 2003. Losing ground: at what cost? Summary report. Mass Audubon, Lincoln, MA.

Campbell, S.M. 2004. Profiles from working woodlands. Massachusetts Woodlands Institute, Montague, MA.

Clark, B.J., T.E. Howard and R.G. Parker. 1992. Professional forestry assistance in New Hampshire timber sales. *Northern Journal of Applied Forestry* 9(1):14-18.

Clawson, M. 1979. The Economics of U.S. nonindustrial private forests. Res. Pap. R-14. Resources for the Future, Washington, DC.

Damery, D.T. 2001. Report on the western Massachusetts forest landowner interest survey. White paper, University of Massachusetts-Amherst, Dept. of Natural Resources Conservation, Amherst, MA.

Damery, D.T. and G. Boyce. 2003. Massachusetts directory of sawmills & dry kilns–2003. Dept. of Conservation and Recreation, Boston, MA.

Food and Agriculture Organization of the United Nations [FAO]. 2001. Global forest resources assessment 2000. FAO For. Pap. 140. FAO, Rome, Italy.

Lindstad, B.H. 2002. A comparative study of forestry in Finland, Norway, Sweden and the United States. Gen. Tech. Rep. PNW-GTR-538, U.S. Deptartment of Agriculture, Forest Service, Pacific Northwest Research Station, Portland, OR.

National Research Council. 1998. Forested landscapes in perspective: prospects and opportunities for sustainable management of America's nonfederal forests. Committee on Prospects and Opportunities for Sustainable Management of America's Nonfederal Forests, Board of Agriculture, National Research Council, National Academy Press, Washington, DC. 249 p.

Simon, D.M. and O.J. Scoville. 1982 Forestry cooperatives: organization and performance. ACS Res. Rep. 25. U.S. Department of Agriculture, Agricultural Cooperative Service, Washington, DC.

Smith, W.B., P.D. Miles, J.S. Vissage and S.A. Pugh. 2004. Forest resources of the United States, 2002. Gen. Tech. Rep. NC-241. U.S. Department of Agriculture, Forest Service, North Central Research Station, St. Paul, MN.

U.S. Department of Agriculture, Forest Service [USDA FS]. 1999. The wood handbook: wood as an engineering material. Gen. Tech. Rep. 113. Forest Products Laboratory, Madison, WI.

Vilkriste, L. 1998. Sustainability and private forestry in Latvia. In: pp. 123-130. M. Hytönen (ed.). Social sustainability of forestry in the Baltic Sea region. Res. Pap. 704. Finnish Forest Research Institute, Helsinki, Finland.

Young, R.A., M.R. Reichenbach and F.H. Perkuhn. 1985. PNIF management: a social-psychological study of owners in Illinois. *Northern Journal of Applied Forestry* 2:91-4.

doi:10.1300/J091v24n02_07

# Monitoring Sustainable Forest Management in the Pacific Rim Region

Gordon M. Hickey
Craig R. Nitschke

**SUMMARY.** The Pacific Rim is rich in forest resources. It contains the world's largest contiguous forest areas, high levels of biodiversity, millions of forest-dependent people, and the world's leading wood-product exporting and importing nations. However, because of a range of issues, the Pacific Rim region is also experiencing high rates of deforestation and forest degradation. An important step in addressing these issues and moving toward sustainable forest management is improved monitoring and information reporting at the local, national, and international levels. A number of criteria and indicators initiatives have been developed throughout the countries of the Pacific Rim. These have ranged from international processes to local initiatives such as forest certification. Although there is considerable variability in the issues facing forest policy makers in the countries of the Pacific Rim, it is often expected that criteria and indicators will reflect a level of comparability. This paper presents the results of a comparative analysis designed to identify similarities and differences in sustainable forest management criteria and indicators

---

Gordon M. Hickey is Post-Doctoral Research Fellow at the University of British Columbia, Vancouver, Canada (E-mail: ghickey@interchange.ubc.ca).

Craig R. Nitschke is a PhD candidate in the Sustainable Forest Management Laboratory, Department of Forest Resources Management, University of British Columbia, 2045, 2424 Main Mall, Vancouver, Canada, V6T 1Z4.

[Haworth co-indexing entry note]: "Monitoring Sustainable Forest Management in the Pacific Rim Region." Hickey, Gordon M., and Craig R. Nitschke. Co-published simultaneously in *Journal of Sustainable Forestry* (Haworth Food & Agricultural Products Press, an imprint of The Haworth Press, Inc.) Vol. 24, No. 2/3, 2007, pp. 245-278; and: *Sustainable Forestry Management and Wood Production in a Global Economy* (ed: Robert L. Deal, Rachel White, and Gary L. Benson) Haworth Food & Agricultural Products Press, an imprint of The Haworth Press, Inc., 2007, pp. 245-278. Single or multiple copies of this article are available for a fee from The Haworth Document Delivery Service [1-800-HAWORTH, 9:00 a.m. - 5:00 p.m. (EST). E-mail address: docdelivery@haworthpress.com].

Available online at http://jsf.haworthpress.com

doi:10.1300/J091v24n02_08

initiatives in the Pacific Rim region. When considered in the context of globalization, the research findings support international efforts to encourage comparability in sustainable forest management-related monitoring and information reporting. doi:10.1300/J091v24n02_08 *[Article copies available for a fee from The Haworth Document Delivery Service: 1-800-HAWORTH. E-mail address: <docdelivery@haworthpress.com> Website: <http://www.HaworthPress.com>* 

**KEYWORDS.** Criteria and indicators, forest policy, forest certification, constant comparison

## *INTRODUCTION*

The countries of the Pacific Rim (see Figure 1) are rich in forest resources. They contain the world's largest contiguous forest areas, high levels of biodiversity, millions of forest-dependent people, and the world's leading wood-product exporting and importing nations (Forest Trends, 2000). However, because of a range of issues associated with shifting cultivation, illegal logging and trade, policy and market failures, and excess demands, the Pacific Rim is also experiencing high rates of deforestation and forest degradation (Geist and Lambin, 2001; Contreras-Hermosilla, 2001). Forest managers in the Pacific Rim region face many challenges for achieving sustainability. Hickey (2004b) identified that improved monitoring and information reporting is a necessary step for forest managers to take, at any scale, if they are going to move successfully toward sustainable forest management (SFM). Our research sought to identify consistencies and differences in the SFM indicators being applied in different jurisdictions. We then considered the implications of our findings for sustainable forest management in the Pacific Rim region.

## *BACKGROUND: FORESTRY IN THE PACIFIC RIM REGION*

### *Forest Ecology and Management*

The countries of the Pacific Rim contain four broad forest types: tropical, subtropical, temperate, and boreal (FAO, 2000). This can be

FIGURE 1. Jurisdictions in the Pacific Rim region.

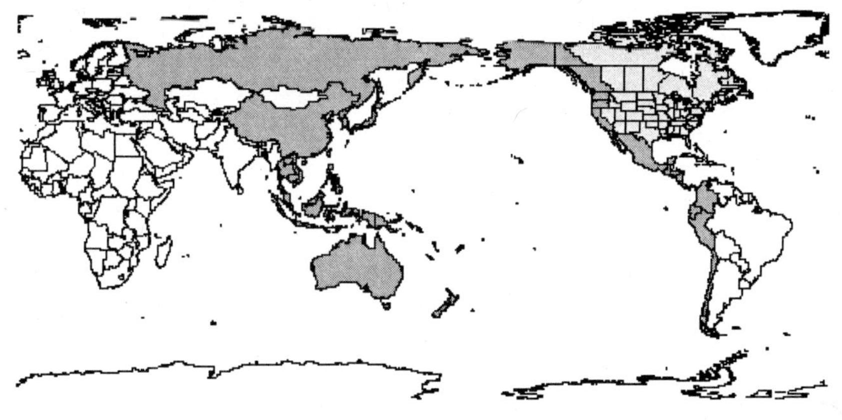

Source: ESRI data (2004).

attributed to the large geographical diversity that exists in countries in the region. Table 1 provides a summary of the forest area, rate of change, total number of species, and number of endangered and endemic species by Pacific Rim country.

Because of the diversity of ecosystems present in the region, dramatic differences exist in the species abundance of fauna and flora. As a consequence, biodiversity differs significantly between countries, as does the potential impact of forest management on these species. Not surprisingly, the countries that contain tropical and subtropical forests also contain a greater number of known species (Myers et al., 2000; FAO, 2000). It follows that the greatest impact of forest management on forest biodiversity is occurring in the countries that have tropical and subtropical forest or are island nations (see Figure 2). Comparing Table 1 to Figure 2, we observe that countries that have lost forest cover between 1990 and 2000 also tend to have the highest number of endangered forest species. There are, however, exceptions (i.e., Chile and El Salvador), which could reflect their forest management paradigm or the current issues that impact those paradigms (see Kimmins, 1995).

Forest ownership in the region also differs widely, with countries such as Indonesia and Russia having 100% of their forests classified as publicly owned, and countries such as Papua New Guinea, having almost all of their forest estate (97%) under community/indigenous ownership (White and Martin, 2002). The Food and Agriculture Organization

TABLE 1. Forest area and species characteristics for the countries of the Pacific Rim

| Country | Forest area statistics | | | | | Total species | Endangered | | Endangered forest occurring species | | | | | | | |
|---|---|---|---|---|---|---|---|---|---|---|---|---|---|---|---|---|
| | Total forest area | Area per capita | Forest cover | Plantation area | Annual change (1990-2000) | | Total | Endemic | Amphibians | Birds | Ferns | Mammals | Palms | Reptiles | Trees | Total |
| | *Thousand ha* | *ha* | *%* | *%* | *%* | *n* | *n* | *n* | *n* | *n* | *n* | *n* | *n* | *n* | *n* | *n* |
| Australia*† | 154,539 | 8.3 | 20.1 | 0.9 | −0.2 | 2439 | 300 | 184 | 17 | 19 | 48 | 17 | 26 | 11 | 17 | 155 |
| Brunei Darussalam* | 442 | 1.4 | 83.9 | 0.7 | −0.2 | 623 | 119 | 14 | - | - | - | - | 1 | - | 10 | 11 |
| Cambodia | 9,335 | 0.9 | 52.9 | 1.0 | −0.6 | 1071 | 80 | 2 | - | - | - | - | - | - | 2 | 2 |
| Canada*† | 244,571 | 7.9 | 26.5 | 0.0 | - | 1013 | 43 | 2 | - | - | 3 | - | - | - | - | 3 |
| Chile*† | 15,536 | 1.0 | 20.7 | 13.0 | −0.1 | 798 | 131 | 42 | - | 2 | 27 | 1 | 3 | - | 14 | 47 |
| China*† | 163,480 | 0.1 | 17.5 | 27.6 | 1.2 | 4310 | 402 | 137 | - | 16 | 10 | 8 | 4 | 1 | 69 | 108 |
| Chinese Taipei* | - | - | - | - | - | - | - | - | - | - | - | - | - | - | - | - |
| Colombia† | 49,601 | 1.2 | 47.8 | 0.3 | −0.4 | 5133 | 435 | 216 | - | 27 | 9 | 3 | 23 | - | 78 | 140 |
| Costa Rica† | 1,968 | 0.5 | 38.5 | 9.0 | −0.8 | 2629 | 214 | 56 | - | 5 | 23 | 3 | 10 | - | 36 | 77 |
| North Korea | 8,210 | 0.3 | 68.2 | 0.0 | - | 579 | 29 | 1 | - | - | - | - | - | - | 1 | 1 |
| Ecuador† | 10,557 | 0.9 | 38.1 | 1.6 | −1.2 | 4031 | 303 | 157 | - | 6 | 6 | 1 | 5 | - | 127 | 145 |
| El Salvador† | 121 | - | 5.8 | 11.6 | −4.6 | 1028 | 46 | 5 | - | - | - | - | - | - | 4 | 4 |
| Fiji | 815 | 1.0 | 44.6 | 11.9 | −0.2 | 455 | 110 | 95 | - | 6 | 3 | 1 | 19 | - | 54 | 83 |
| Guatemala† | 2,850 | 0.3 | 26.3 | 4.7 | −1.7 | 2283 | 130 | 19 | - | - | - | 1 | 6 | - | 9 | 16 |
| Honduras† | 5,383 | 0.9 | 48.1 | 0.9 | −1.0 | 1849 | 145 | 52 | - | 1 | - | 2 | 1 | - | 36 | 40 |
| Indonesia*† | 104,986 | 0.5 | 58.0 | 9.4 | −1.2 | 5952 | 762 | 370 | - | 55 | 38 | 50 | 27 | - | 125 | 295 |
| Japan*† | 24,081 | 0.2 | 64.0 | 44.4 | - | 1351 | 353 | 41 | - | 5 | 53 | 6 | 2 | - | n/a | 66 |
| Kiribati | 28 | 0.3 | 38.4 | 0.0 | - | 69 | 6 | - | - | - | n/a | - | - | - | - | - |
| Malaysia*† | 19,292 | 0.9 | 58.7 | 9.1 | −1.2 | 3121 | 966 | 534 | - | 1 | 5 | 10 | 107 | - | 358 | 481 |
| Marshall Islands | - | n/a | n.s. | n.s. | - | 82 | 3 | - | - | - | n/a | - | - | - | - | - |
| Mexico*† | 55,205 | 0.6 | 28.9 | 0.5 | −1.1 | 4033 | 362 | 185 | - | 15 | 11 | 13 | 17 | - | 65 | 121 |
| Micronesia | 15 | 0.1 | 21.7 | 0.0 | −4.5 | 135 | 20 | 11 | - | 5 | | 2 | 1 | - | 2 | 10 |
| Nauru | - | n/a | n.s. | n.s. | - | 22 | 2 | 1 | - | - | n/a | - | - | - | - | - |
| New Zealand*† | 7,946 | 2.1 | 29.7 | 19.4 | 0.5 | 553 | 90 | 55 | - | 10 | 7 | 1 | 2 | - | 13 | 33 |
| Nicaragua† | 3,278 | 0.7 | 27.0 | 1.4 | −3.0 | 1832 | 63 | 3 | - | - | 1 | - | - | - | 2 | 3 |
| Palau | 35 | 1.8 | 76.1 | 0.0 | - | 161 | 10 | 1 | - | - | n/a | - | - | - | 1 | 1 |
| Panama† | 2,876 | 1.0 | 38.6 | 1.4 | −1.6 | 2645 | 297 | 129 | - | 2 | 6 | 5 | 12 | - | 106 | 131 |
| Papua New Guinea*† | 30,601 | 6.5 | 67.6 | 0.3 | −0.4 | 3372 | 323 | 166 | - | 11 | 45 | 20 | 2 | - | 103 | 181 |

| Country | Forest area statistics | | | | | | En-dangered | | Endangered forest occurring species | | | | | | | |
|---|---|---|---|---|---|---|---|---|---|---|---|---|---|---|---|---|
| | Total forest area | Area per capita | Forest cover | Plantation area | Annual change (1990-2000) | Total species | Total | Endemic | Amphibians | Birds | Ferns | Mammals | Palms | Reptiles | Trees | Total |
| | *Thousand ha* | *ha* | *%* | *%* | *%* | *n* | *n* | *n* | *n* | *n* | *n* | *n* | *n* | *n* | *n* | *n* |
| Peru*† | 65,215 | 2.6 | 50.9 | 1.0 | −0.4 | 4247 | 462 | 277 | - | 13 | 10 | 6 | 4 | - | 182 | 215 |
| Philippines*† | 5,789 | 0.1 | 19.4 | 13.0 | −1.4 | 2097 | 447 | 317 | - | 70 | 36 | 37 | 40 | 1 | 131 | 315 |
| Republic of Korea*† | 6,248 | 0.1 | 63.3 | 0.0 | −0.1 | 551 | 25 | - | - | - | - | - | - | - | n/a | - |
| Russian Federation*† | 851,392 | 5.8 | 50.4 | 2.0 | - | 138 | 77 | 3 | - | - | - | - | - | - | - | - |
| Samoa | 105 | 0.6 | 37.2 | 4.8 | −2.1 | 10 | 7 | 6 | n/a | n/a | n/a | n/a | 6 | n/a | - | 6 |
| Singapore* | 2 | - | 3.3 | 0.0 | - | 567 | 87 | 3 | - | - | - | - | - | - | 5 | 5 |
| Solomon Islands | 2,536 | 5.9 | 88.8 | 2.0 | −0.2 | 798 | 77 | 41 | - | 6 | 3 | 6 | 13 | - | 6 | 34 |
| Thailand*† | 14,762 | 0.2 | 28.9 | 33.3 | −0.7 | 2293 | 195 | 17 | - | | | 1 | 2 | - | 8 | 11 |
| Tonga | 4 | - | 5.5 | 25.0 | - | 141 | 7 | 2 | - | 1 | - | - | - | - | 1 | 2 |
| United States*† | 225,993 | 0.8 | 24.7 | 7.2 | 0.2 | 1283 | 436 | 292 | 13 | 20 | 13 | 4 | 18 | - | 147 | 215 |
| Vanuatu | 447 | 2.4 | 36.7 | 0.7 | 0.1 | 269 | 31 | 21 | - | 5 | - | - | 9 | - | 5 | 19 |
| Viet Nam* | 9,819 | 0.1 | 30.2 | 17.4 | 0.5 | 1999 | 258 | 85 | 1 | 7 | 8 | 4 | 17 | - | 32 | 69 |

* Asia-Pacific Economic Cooperation member country (no FAO data for Chinese Taipei).
† Signatory to at least one international sustainable forest management initiative (i.e., national-level criteria and indicators of sustainable forest management).
\- No data available.
n/a Not applicable.

Source: Global Forest Resources Assessment, FAO (2000)

(FAO) (2000) stated that forest management throughout many of the Pacific island nations was complicated by the customary land tenure system. Although this land cannot usually be sold, the forest resources can be. This has led to difficulties in establishing sustainable management practices and has contributed to a loss of forest cover (see Table 1).

Plantation forestry is increasing in a number of Pacific Rim nations. In many cases, this has resulted from a desire to reduce the anthropogenic pressure on natural forests. Of the countries in the Pacific Rim, Chile, China, Indonesia, Thailand, Malaysia, and Viet Nam have undertaken massive plantation expansion to rehabilitate areas of natural forests that were previously exploited and exhausted (FAO, 2000) (see Table 1). This has allowed the expansion of their forest industries. Whereas many of the Pacific Rim countries are focussed on establishing

FIGURE 2. Proportion of endangered species that occur in the Pacific Rim countries.

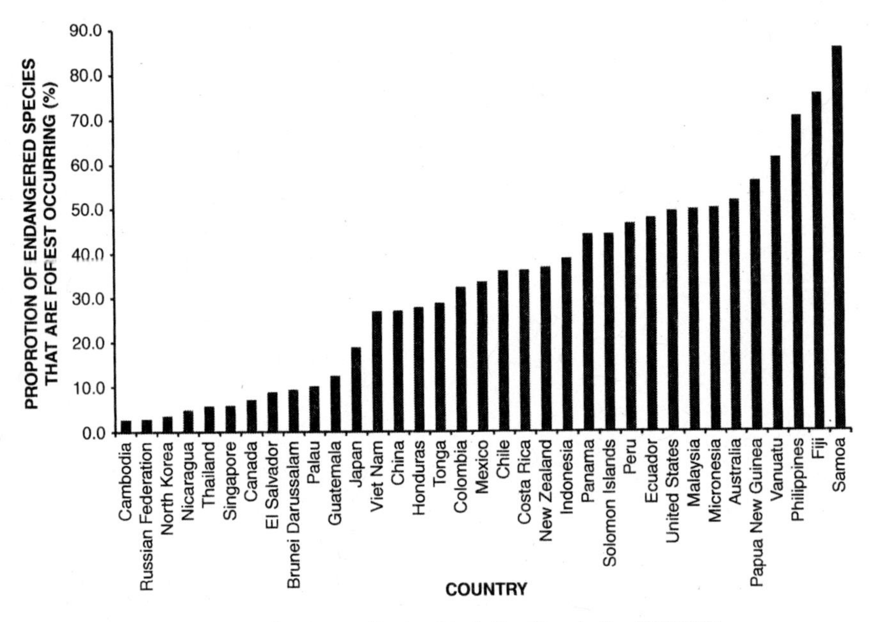


Source: Global Forest Resources Assessment, Food and Agriculture Organization (FAO 2000).

plantations, four of the largest countries, Russia, Canada, United States, and Australia, are still focussing on managing natural forests. These countries have large areas of forest per capita in comparison to many of the countries that have moved toward plantation forestry.

## *Forest Products and Trade*

Over the last decade, a number of factors have been affecting the trade of timber products in the Pacific Rim region. As post-materialist consumers have become more sophisticated, new markets have emerged for high-quality and competitively priced products that are derived from sustainably managed forests. There has also been a shift toward greater value-added processing, particularly in North America's sawn softwood industry. This has been driven, in part, by the need to compete with the rising timber volumes being produced by Russia and the plantation-based resources in the Southern Hemisphere (i.e., Chile, New Zealand, and Australia). Owing to the negative impacts of deforestation, and the increasing areas of protected forest in the Pacific Rim region, the area of high-quality natural forest available for timber harvesting is decreasing. For many countries, this has caused a shift toward intensive forest management and plantations (e.g., Chile [Clapp, 2001] and China [Zhang et al., 2000; Zhang, 2005]).

A study completed by the World Trade Organization (WTO) identified a positive relationship between the removal of trade restrictions and improved environmental quality. Specifically, the report suggested that enhanced competition would lead to more efficient consumption patterns, reduce poverty, and increase the availability of environment-friendly goods and services (WTO, 1997). This was supported by Thiele (1995) and Tumaneng-Diete et al. (2005) who identified that trade restrictions are ecologically ineffective and do not contribute to achieving SFM objectives. In the Pacific Rim region, the Asia-Pacific Economic Cooperation (APEC) forum is encouraging trade and investment liberalization. The APEC forum was established in 1989 and has a membership of 21 economic jurisdictions (see Table 2).

These countries represent a population of over 2.5 billion people and a combined gross domestic product of over 19 trillion USD. This accounts for approximately 47% of world trade. The APEC economies also account for a large proportion of the world's trade in forest products (Forest Research, 1999). In 1995, exports of forest products between APEC economies were valued at $100 billion USD. Furthermore, for many members, the APEC region is the destination for 95 to 99% of their forest product exports. Owing to the vastly different operating conditions, exchange rates and trade policies present in the Pacific Rim region, cost asymmetries have been an important issue for timber producers. Although APEC is a voluntary organization with no binding implications, it does have a policy mandate for cooperative work on sustainable development.

However, Southgate et al. (2000) found that liberalization did not necessarily result in improved environmental quality or investment in Ecuador owing to weak property rights, corruption, and dated legislation. Similarly, Clapp (2001) found that trade liberalization contributed to the overexploitation of Chilean native forests owing to the boom and bust of export patterns (Vincent, 1992). Trade liberalization is also related to the concept of shadow ecology (Dauvergne, 2001), a situation where one country's economy has environmental impacts on resource management in another. The "ecological shadow" that international market forces cast upon the local resources in forest-rich countries can have an influence on the ability of a jurisdiction to promote SFM on the

TABLE 2. Key economic indicators for Asia-Pacific Economic Cooperation member economies (December 2005)

| Member Economy and Year Joined | Area | Population | GDP | GDP per capita | Exports | Imports |
|---|---|---|---|---|---|---|
| | *(Thousand km$^2$)* | *(Million)* | *(Billion USD)* | *(USD)* | *(Million USD)* | *(Million USD)* |
| Australia (1989) | 7,692 | 20.2 | 692.4 | 33,629 | 86,551 | 103,863 |
| Brunei Darussalam (1989) | 6 | 0.4 | 5.7 | 15,764 | 4,713 | 1,638 |
| Canada (1989) | 9,971 | 32.0 | 1,084.1 | 33,648 | 315,858 | 271,869 |
| Chile (1994) | 757 | 15.4 | 105.8 | 6,807 | 32,548 | 24,769 |
| China (1991) | 9,561 | 1,299.8 | 1,851.2 | 1,416 | 593,647 | 560,811 |
| Hong Kong, China (1991) | 1 | 6.9 | 174.0 | 25,006 | 265,763 | 273,361 |
| Indonesia (1989) | 1,905 | 223.8 | 280.9 | 1,237 | 71,585 | 46,525 |
| Japan (1989) | 378 | 127.3 | 4,694.3 | 36,841 | 566,191 | 455,661 |
| Korea (1989) | 99 | 48.2 | 819.2 | 16,897 | 253,845 | 224,463 |
| Malaysia (1989) | 330 | 25.5 | 129.4 | 4,989 | 125,857 | 105,297 |
| Mexico (1993) | 1,958 | 105.0 | 734.9 | 6,920 | 177,095 | 171,714 |
| New Zealand (1989) | 271 | 4.1 | 108.7 | 26,373 | 20,334 | 21,716 |
| Papua New Guinea (1993) | 463 | 5.9 | 3.5 | 585 | 4,321 | 1,463 |
| Peru (1998) | 1,285 | 27.5 | 78.2 | 2,798 | 12,111 | 8,872 |
| Philippines (1989) | 300 | 86.2 | 95.6 | 1,088 | 39,588 | 40,297 |
| Russia (1998) | 17,075 | 144.0 | 719.2 | 5,015 | 171,431 | 86,593 |
| Singapore (1989) | 1 | 4.2 | 116.3 | 27,180 | 179,755 | 163,982 |
| Chinese Taipei (1991) | 36 | 22.5 | 335.2 | 14,857 | 174,350 | 168,715 |
| Thailand (1989) | 513 | 64.6 | 178.1 | 2,736 | 97,098 | 95,197 |
| United States (1989) | 9,364 | 293.0 | 12,365.9 | 41,815 | 818,775 | 1,469,704 |
| Viet Nam (1998) | 332 | 82.6 | 51.0 | 610 | 26,061 | 32,734 |

GDP = gross domestic product.

Source: APEC (2005).

ground. This situation has driven many governments, forest companies, and wood product retailers to promote comparable SFM initiatives to meet the growing demand for better forest management.

## *Sustainable Forest Management Initiatives in the Pacific Rim*

One of the main issues associated with ensuring SFM is to define what is meant by sustainability and then determine progress toward this goal (Hickey and Innes, 2005). This is the reason for criteria and indicators (C&I). Prabhu et al. (2002) defined a "criterion" as "a standard that a thing is judged by" while an "indicator" is "any variable . . . used to infer performance."

Criteria and indicator initiatives present in the Pacific Rim include the Pan-European criteria and indicators for SFM (also referred to as the Helsinki Process [1993]); the criteria and indicators for SFM in temperate and boreal forests (referred to as the Montreal Process [1995]); the International Tropical Timber Organization (ITTO) (1992); the Amazon Cooperation Treaty (the Tarapoto Agreement) (1995); the Lepaterique Process for Central America (1999) and the Bhopal-India Process (1999). The Centre for International Forest Research (CIFOR) has also developed a generic C&I standard for use in any jurisdiction. Some of the countries in the Pacific Rim region have also identified national criteria and indicators based on the relevant international processes (e.g., the Commonwealth of Australia [1998] and the Canadian Council of Forest Ministers [CCFM] [1995] have developed national indicators based on the Montreal Process).

Forest certification initiatives (see Rametsteiner, 1999; Cashore et al., 2003; Hickey, 2004b) are another important driver for SFM-related indicator development in the region. These standards are often more specialized and designed for application within a particular jurisdiction. Forest certification standards currently operating in the region include the Forest Stewardship Council (FSC), the Sustainable Forestry Initiative (SFI), CERTFOR Chile, the Australian Forestry Standard (AFS) and the Indonesian Ecolabelling Institute (LEI) (see Innes and Hickey, 2005).

Process-based certification is also prominent in the region. Whilst this form of certification does not specify performance indicators for SFM per se, many organizations now use their Environmental Management System (EMS) to facilitate monitoring and reporting on SFM indicators. According to the International Organization for Standardization (ISO) (2004), Japan, China, USA, and the Republic of Korea led the list of Pacific Rim nations by number of ISO 14001 certificated organizations in 2003.

## *METHODOLOGY*

This study used grounded theory (Glaser and Strauss, 1967) to compare C&I standards in the Pacific Rim region. The analytical techniques used for the analyses were constant comparison and content analysis (Creswell, 1997; Babbie, 2001). Constant comparison was conducted according to six broad SFM criteria. This facilitated the cross-case analysis of monitoring requirements. Content analysis was then used to code the qualitative data extracted from the C&I standards into a standardized form. This resulted in a presence/absence data matrix, where the rows represented indicators, and the columns represented the C&I standards. Using these data, we assigned cell values of 1 (presence) or 0 (absence) to each of the indicators. A comparative analysis was then conducted by using the distribution of the indicators associated with each C&I standard.

Hierarchical cluster analysis (HCA) was used to identify which C&I standards were most likely to be similar to each other. A cluster analysis is an empirical technique that classifies objects without prior assumptions about their population (Punj and Stewart, 1983; Archer et al., 2005). In this analysis, the "Manhattan distance" and the "maximum distance between cluster centroid" techniques were used (Manly, 1994).

Following HCA, a single-factor analysis of variance (ANOVA) ($\alpha = 0.05$) was used to determine if the indicators in each C&I standard were drawn from different sampling distributions of means (Tabachnick and Fidell, 2001). Each C&I standard was used as an independent treatment with the combined indicators of all the standards used to represent the total sample size from which the grand mean for all the scores over all the treatments was derived (Tabachnick and Fidell, 2001). To determine which treatments were significantly different from each other, a Bonferroni test was conducted (Manly et al., 1993; Genov, 1999). If the variation between means exceeded the Bonferroni critical distance, they were considered to be significantly different. The ANOVA and Bonferroni test were conducted on the entire dataset and on the data associated with each broad SFM criterion. The Bonferroni tests identify which standards are statistically different and which ones are statistically similar. This allows the clusters of standards that are grouped together by the HCA to be validated and allows standards to be grouped into clusters that are statistically similar to each other.

Both the HCA and the Bonferroni analysis allow standards to be grouped into clusters based on the number and distribution of indicators. However, they accomplish this in much different ways. The HCA is an exploratory technique that searches for patterns of similarities whereas the Bonferroni test is a statistical test that searches for similarities in the means of a variable. Nevertheless, both tests base their interpretations of homogeneity on distance; the closer they are, the more similar they are. For this reason, both tests were used to better validate the results.

## *Jurisdictions and Documentation*

The study concentrated on jurisdictions located in the Pacific Rim region that are signatories to at least one international C&I initiative. This resulted in 21 C&I standards related to forest management being analyzed (see Table 3 and Figure 3).

Our analysis considered all of the indicators required or implied to implement the standard. As noted by Holvoet and Muys (2004), the terminology used to describe forest-related subjects in C&I documents often differ. This variability did not pose a problem for our analysis, which relied on the presence of elements in the C&I standards rather than focussing on their exact wording (see also Holvoet and Muys, 2004; Hickey et al., 2006).

TABLE 3. Criteria and indicators standards and the associated acronyms

| Criteria and indicators standards | Acronym |
|---|---|
| Asia Initiative: Bhopol, Criteria & Indicators (1999) | AIB |
| Australian Forestry Standard Criteria & Indicators (2003) | AFS |
| Canadian Council of Forest Ministers Criteria & Indicators (2003) | CCFM |
| Center for International Forestry Research Criteria & Indicators (1998) | CIFOR |
| CERTFOR Chile Criteria & Indicators (2004) | CERTF |
| Commonwealth of Australia Criteria & Indicators (1998) | COM |
| FSC Boreal Criteria & Indicators (2004) | FSC_BO |
| FSC British Columbia Criteria & Indicators (2003) | FSC_BC |
| FSC Colombia Criteria & Indicators (2003) | FSC_CO |
| FSC Pacific Coast Criteria & Indicators (2003) | FSC_PC |
| FSC Peru Criteria & Indicators (2001) | FSC_PE |
| Helsinki Process Criteria & Indicators (1993) | HEL |
| Indonesian Ecolabelling Institute Certification Standard Criteria & Indicators (1998) | LEI |
| International Tropical Timber Organisation Criteria & Indicators (1992) | ITTO |
| Lepaterique Process of Central America Criteria & Indicators (1997) | LEP |
| Malaysian Criteria & Indicators (2002) | MCI |
| Montreal Process Criteria & Indicators (1995) | MON |
| Natural Resources Canada Local-Level Criteria & Indicators (2000) | NRC |
| Sustainable Forestry Initiative Criteria & Indicators (2003) | SFI |
| Tarapoto Proposal, Amazon Forest Criteria & Indicators (1995) | TPA |
| USDA Forest Service Local Unit Criteria & Indicators Development (2003) | LUCID |

FIGURE 3. Criteria and indicator initiatives selected for analysis.

a. International Standards

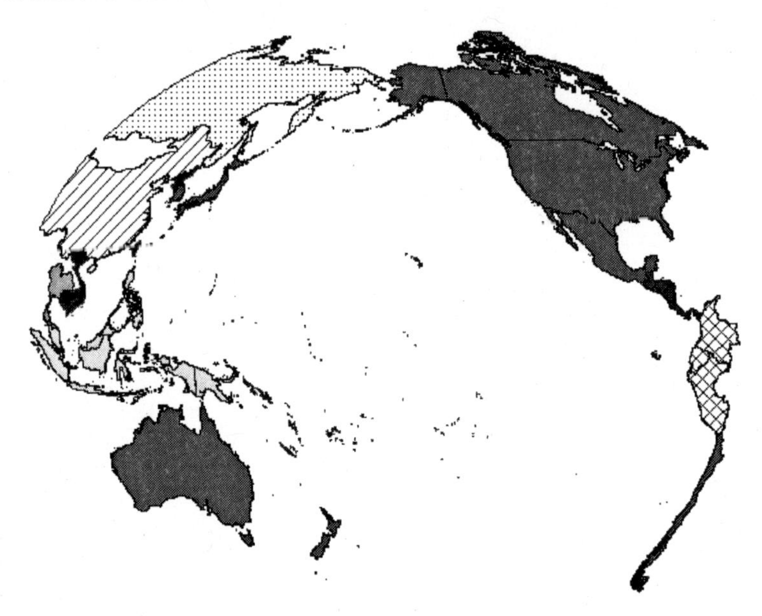

**Legend**

**International Standards**

Montreal Process

Montreal Process and Helsinki Process

Montreal Process and Bhopal Asia Initiative

Bhopal Asia Initiative

International Tropical Timber Organisation

Lepaterique Process

Tarapoto Proposal

No International Standard

b. National and Local-Level Standards

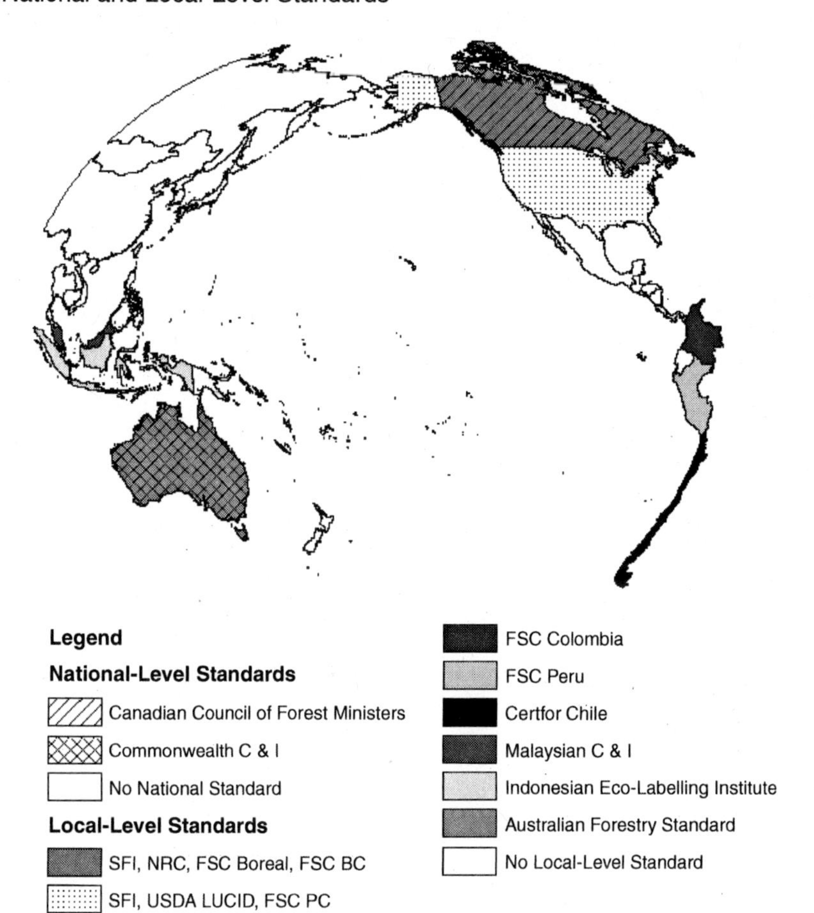


Source: ESRI data (2004).

## *Assumptions and Limitations*

The study was based upon the assumption that the indicators documented in various SFM standards could provide insight into SFM phenomena in different jurisdictions. As not all indicators are of the same significance to SFM, it is important to note that our analysis was not designed to demonstrate "better" or "worse" standards, but rather to describe the broad nature of the standards.

There are a number of limitations associated with a study of this kind. The geographical and cultural diversity within the Pacific Rim region limited our ability to verify the application of many of the C&I standards. However, we can, under the assumption that they are applied as documented, gain an understanding of the importance of each broad SFM criterion within the region and whether any similarities exist across different scales. Another issue related to our research was the quality of the data used in the analysis. To maximize the accuracy of our qualitative data, we employed the matrix analysis technique to identify unexpected phenomena. This allowed our data to be verified as the analysis proceeded through recognizing patterns and regularities, searching for meanings and explanations, and describing propositions and apparent causal flow prior to conducting any statistical analysis (see Miles and Huberman, 1984).

## *RESULTS*

### *Constant Comparison and Content Analysis*

From the constant comparison and content analysis, 866 separate indicators were identified among the 21 standards. These indicators were, subsequently, coded according to one of six broad SFM criteria. Table 4 shows the number of indicators in each standard and the breakdown of indicators by criterion. The LEI standard (Indonesia) was found to have the least number of indicators, and the Natural Resources Canada (NRC) local-level standard (Canada) was found to have the most. With the exception of the ITTO standard, all of the international standards were found to have between 46 and 62 indicators while the number of indicators documented within national-level standards ranged from 65 to 128 indicators. The local-level standards were found to have the widest range of variation (i.e., from 30 to 251 SFM indicators).

In terms of each individual criterion, "economics" had the greatest number of indicators (251), followed by "biodiversity" (203). "Global cycles" was found to have the fewest indicators (35), with seven standards having no apparent indicators related to this criterion. Of the local-level standards, FSC_BC, FSC_CO, FSC_PE, MCI, LUCID, and NRC (see Table 3 for definitions) were found to have the greatest number of indicators associated with "economics," and SFI had the fewest. The CIFOR standard had the greatest number of indicators related to "biodiversity" (73), and the AFS standard had the fewest (5). Based on

TABLE 4. The number of indicators within each broad sustainable forest management criterion and the overall number for each standard

| C&I Standard | Biodiversity | Productivity and health | Soil and water | Global cycles | Economics | Society | Total**** |
|---|---|---|---|---|---|---|---|
| AFS*** | 5 | 23 | 9 | 2 | 16 | 17 | 72 |
| AIB* | 10 | 8 | 7 | 3 | 13 | 5 | 46 |
| CCFM** | 11 | 5 | 4 | 5 | 20 | 20 | 65 |
| CERTF*** | 7 | 12 | 12 | 0 | 29 | 24 | 84 |
| CIFOR* | 73 | 3 | 8 | 0 | 32 | 12 | 128 |
| COM** | 13 | 11 | 8 | 7 | 31 | 30 | 100 |
| FSC_BC*** | 27 | 24 | 9 | 0 | 57 | 49 | 166 |
| FSC_BO*** | 14 | 21 | 2 | 0 | 30 | 38 | 105 |
| FSC_CO*** | 28 | 11 | 3 | 2 | 54 | 47 | 145 |
| FSC_PC*** | 21 | 18 | 14 | 0 | 39 | 28 | 120 |
| FSC_PE*** | 19 | 9 | 2 | 2 | 59 | 42 | 133 |
| HEL* | 7 | 15 | 1 | 7 | 13 | 10 | 53 |
| ITTO* | 19 | 17 | 5 | 2 | 35 | 23 | 101 |
| LEI*** | 6 | 7 | 2 | 0 | 12 | 3 | 30 |
| LEP* | 15 | 7 | 3 | 4 | 16 | 5 | 50 |
| LUCID** | 54 | 12 | 25 | 12 | 60 | 30 | 193 |
| MCI*** | 19 | 9 | 5 | 2 | 59 | 49 | 143 |
| MON* | 10 | 7 | 7 | 4 | 24 | 10 | 62 |
| NRC** | 52 | 28 | 45 | 11 | 62 | 53 | 251 |
| SFI*** | 14 | 15 | 5 | 0 | 11 | 11 | 56 |
| TPA* | 5 | 7 | 4 | 2 | 30 | 9 | 57 |
| **Unique**** Indicators** | **203** | **123** | **96** | **35** | **251** | **158** | **866** |
| **Mean** | 20 | 13 | 9 | 3 | 33 | 25 | |
| **Std Dev** | 18 | 7 | 10 | 4 | 18 | 16 | |

See Table 3 for acronym definitions.
* International criteria and indicator initiative
** National criteria and indicator initiative
*** Third-party forest certification standard
**** The total number of indicators for each criteria and indicators standard do not sum to the total of the unique indicators since many of the standards share the same indicators

our analysis, the NRC and LUCID standards had the most indicators related to "soil and water," with the majority of the remaining standards having fewer then 10 indicators. The "productivity and health" criterion displayed the least variability between the standards. For the indicators related to "society," the FSC_BC, FSC_CO, FSC_PE, and MCI standards were found to have the greatest number, and the LEI, AIB, and LEP standards had the fewest.

### *Hierarchical Cluster Analysis*

The HCA grouped the 21 C&I standards into 11 similarity clusters (see Table 5). Figure 4 shows the proportion of indicators by criterion for each standard in each cluster. It is apparent why certain standards are grouped together. For example, the Montreal Process (MON) and CCFM standards (Cluster A), have a very similar distribution of indicators across the criteria; likewise, for Clusters F, G, H, and K. The CIFOR standard is alone in Cluster J. This is most likely because of the high proportion of biodiversity-related indicators in the standard. Similarly, the AFS standard had a greater proportion of forest productivity and health indicators when compared to the other standards. It is therefore alone in Cluster E. Some of the clusters comprise standards that do not appear to be similar. For example, in Cluster B the LEI standard appears to be different from the HEL and AIB standards. The number of indicators, combined with their distribution, is most likely influencing these groupings. Because of the apparent differences between standards in some of the clusters, we used ANOVA and Bonferroni tests to find where, if any, statistical differences and similarities existed. This provided a method for validating the results provided by the HCA.

### *ANOVA and Bonferroni Tests*

The ANOVA identified statistically significant differences between standards when all criteria were considered together, and when each criterion was tested separately ($P < 0.001$ for all tests). The Bonferroni

TABLE 5. Affiliated criteria and indicators (C&I) standards

| **C&I standards similarity clusters** | | | | | | | | | | |
|---|---|---|---|---|---|---|---|---|---|---|
| **A*** | **B** | **C** | **D** | **E** | **F*** | **G** | **H**** | **I** | **J** | **K** |
| MON | HEL | TPA | SFI | AFS | COM | FSC_BO | FSC_CO | FSC_BC | CIFOR | LUCID |
| CCFM | AIB | LEP | | | ITTO | FSC_PC | FSC_PE | | | NRC |
| | LEI | | | | | CERTF | MCI | | | |

See Table 3 for acronym definitions.
* C&I standards within group were statistically similar
** FSC_CO and MCI were statistically similar

FIGURE 4. Hierarchical cluster analysis (HCA) results for the criteria and indicators (C&I) standards.

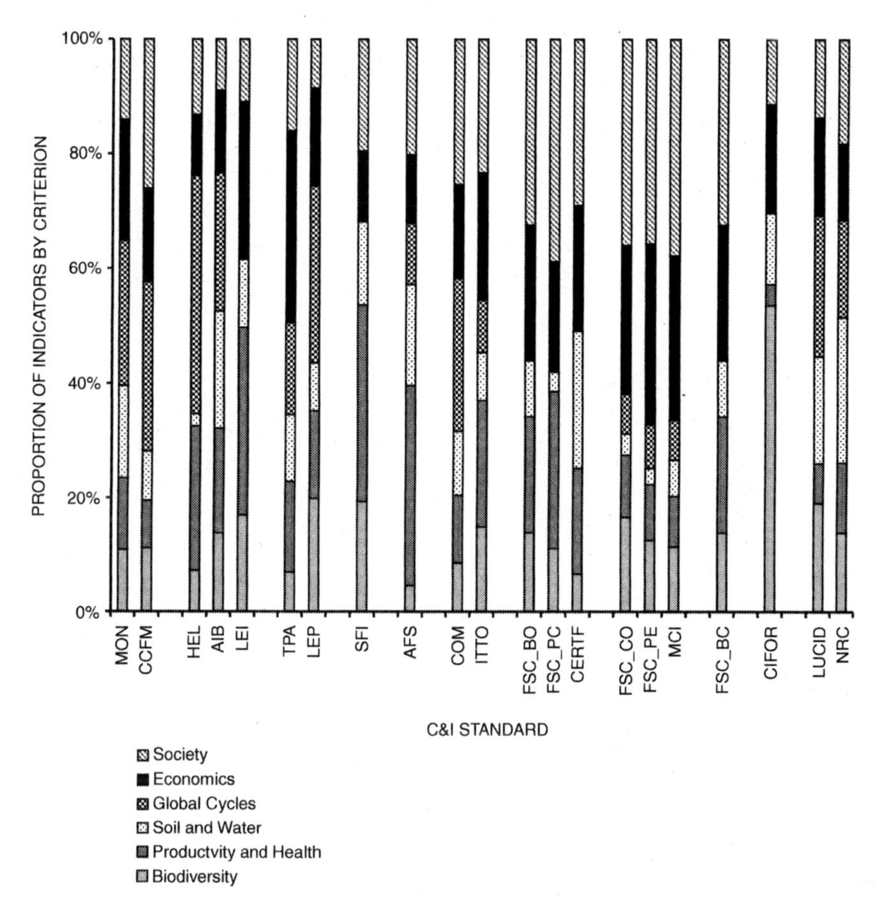


Figure includes the distribution of indicator proportions by SFM criterion. See Table 3 for acronym definitions.

tests identified which standards were similar and which ones were statistically different for all of the ANOVA tests (see Tables 6-12 and Figures 5 to 10). The results of each analysis are discussed separately along with a discussion on how the "groups of standards," also know as "clusters of similarity" (presented in Tables 6-12 and Figures 5-10), were generated from the Bonferroni tests.

*Interpreting the Bonferroni test results–creating "clusters of similarity."* Owing to the complex relationships that were identified by the

Bonferroni tests, we used both tables and figures to represent results from this analysis. The Bonferroni tests allowed standards to be assembled into groups based on homogeneity. These groups are analogous to the similarity clusters generated by the HCA. These groups are presented in Tables 6 through 12 and in Figures 5 through 10. If standards were not statistically different (i.e., variation between means did not exceed the Bonferroni critical distance), they were grouped together. Each group is then considered a separate "cluster of similarity." Standards that were significantly different from all other standards were considered to be a separate group. These unclustered standards are presented in Tables 6 through 12 under the heading "ungrouped standards."

*Interpreting the Bonferroni test results–creating aggregated clusters.* In some cases, the Bonferroni tests found that an association of one or more standards, for example *A*, was statistically similar to another association, *B*. In parallel, *B* was found to be statistically similar to another association, *C*. However, the Bonferroni tests detected a significant difference between *A* and *C*. In order to represent these complex relationships, groups of standards that exhibited this behaviour were assembled into aggregated groups with one or more standards identified as "connectors." These "connecting" standards formed a common link between associations of standards. These aggregated groups were considered as one "cluster of similarity." Figures 5 to 10 present the aggregated clusters of similarity. Within each aggregated "cluster of similarity," the standards that are connected by an arrow are not significantly different from each other, and standards that are not directly connected by an arrow were found to be significantly different from each other.

### *Comparison of All C&I Standards*

Based on the number of indicators related to each broad criterion, our analysis revealed that seven standards were significantly different (column labelled "ungrouped standards" in Table 6) and identified four groups of standards (columns labelled "Group 1," "Group 2," "Group 3," and "Group 4" in Table 6) that were similar. From this analysis we were able to identify 11 "clusters of similarity" (group 1, group 2, group 3, group 4, and the seven ungrouped standards in Table 6). The grouping of the standards into 11 "clusters of similarity" following the Bonferroni test was equal to the 11 "clusters of similarity" identified by the HCA (see Table 5). However, the Bonferroni test revealed different associations of standards than the clusters created by the HCA. Although there were differences in the makeup of the clusters, the Bonferroni

TABLE 6. Bonferroni test results for criteria and indicators (C&I) standards

| Group 1* | Group 2 | Group 3 | Group 4** | Ungrouped standards*** |
|---|---|---|---|---|
| ITTO | CIFOR | MCI | AIB | LEI |
| FSC_BO | FSC_PE | FSC_CO | LEP | AFS |
| COM | | | HEL | CERTF |
| | | | SFI | FSC_PC |
| | | | TPA | FSC_BC |
| | | | MON | LUCID |
| | | | CCFM | NRC |

Each group contains the standards that were not significantly different from each other, except for the column labelled "Ungrouped standards"
See Table 3 for acronym definitions.

* The standards grouped together in these columns (referred to as groups/clusters) are statistically different from all other standards but not statistically different from each other. This relationship holds for Tables 7 through 12.

** The Bonferroni test identified group 4 as an "aggregated cluster." This group has therefore been represented in Figure 5.

*** The standards in the column labelled "Ungrouped standards" are statistically different (i.e., variation between the means did not exceed the Bonferroni critical distance) from each other and all other standards. Each standard in this column represents a separate group/cluster. This relationship holds for Tables 7 through 12.

FIGURE 5. Criteria and indicators (C&I) standards: Group 4.

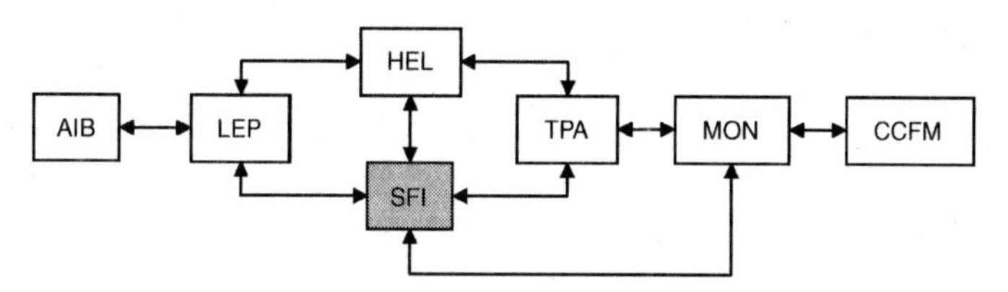


C&I standards that are connected by an arrow were not significantly different from each other. The highlighted standard allowed the standards to be aggregated by creating a common link. See Table 3 for acronym definitions.

This group (also referred to as a "cluster") is an aggregated cluster comprising standards that are not significantly different from one or more standards and which, in turn, are related to another standard or standards. Due to the multiplicity of links, the aggregation of the standards was the best way to illustrate similarities.

analysis did validate the HCA's grouping of the MON and CCFM standards (cluster A in Table 5 versus group 4 in Figure 5), the ITTO and Commonwealth of Australia (COM) standards (cluster F in Table 5 versus group 1 in Table 6) and the Malaysian Criteria and Indicators (MCI) and FSC Colombia (FSC_CO) standards (cluster H in Table 5 versus group 3 in Table 6). The Bonferroni test also validated the HCA's grouping of the AFS and FSC_BC as individual clusters (clusters E and I in Table 5 versus "ungrouped standards" in Table 6). Group 4 (Figure

5) is an aggregated grouping of standards, linked by the SFI standard. This aggregation shows that there are many similarities in the mean number of indicators across some of the higher-level initiatives. The process for comparing the Bonferroni test to the HCA, used in this section, should be repeated when interpreting the analysis done for each of the separate broad SFM criteria.

### *Broad SFM Criterion 1: Biological Diversity*

The analysis found that one standard was significantly different and identified five groups of standards that were statistically similar (see Table 7 and Figure 6). Within this criterion, six clusters of similarity were identified versus the 11 identified by the HCA. This analysis also validated the HCA's placement of the CIFOR standard by itself. The

TABLE 7. Bonferroni test results for "Biological diversity" criterion

| Group 1 | Group 2 | Group 3 | Group 4 | Group 5* | Ungrouped standards |
|---|---|---|---|---|---|
| TPA | ITTO | FSC_BC | NRC | MON | CIFOR |
| AFS | FSC_PE | FSC_CO | LUCID | AIB | |
| LEI | MCI | | | CCFM | |
| HEL | FSC_PC | | | COM | |
| CERTF | | | | FSC_BO | |
| | | | | SFI | |
| | | | | LEP | |

Each group contains the standards that were not significantly different from each other.
See Table 3 for acronym definitions.
*The Bonferroni test identified group 5 as an "aggregated cluster." This group has therefore been represented in Figure 6.

FIGURE 6. Biological diversity criterion: Group 5.

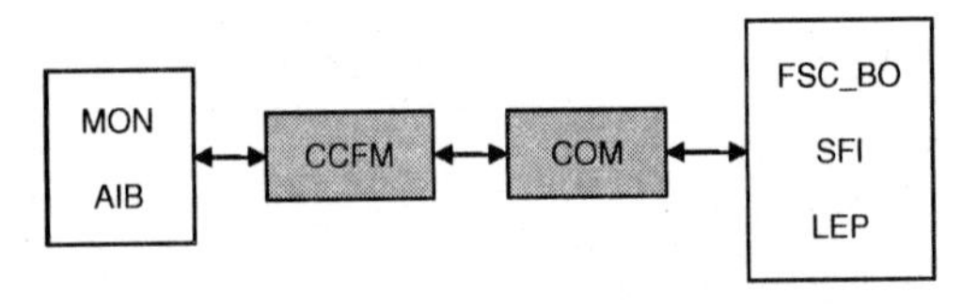


C&I standards that are connected by an arrow were not significantly different from each other. The highlighted standards allowed the standards to be aggregated. See Table 3 for acronym definitions.

CIFOR standard was the only standard that was significantly different in this test owing to its high proportion of biodiversity indicators. Group 5 (Figure 6) is an aggregated grouping of standards linked by the CCFM and COM standards. This aggregation shows that the Canadian and Australian national standards have a similar number of biodiversity indicators and shows that these national-level standards are congruent with some of the international standards. It also shows congruency between the MON, AIB, and LEP standards.

### *Broad SFM Criterion 2: Forest Productivity and Health*

The analysis found that nine standards were significantly different and identified five groups of standards that were similar (Table 8). Within this criterion, 14 clusters of similarity were identified versus the 11 identified by the HCA. This analysis also validated the HCA's placement of the AFS standard by itself. The AFS standard was significantly different in this test owing to its high proportion of productivity and health indicators. This analysis identified no aggregated groups; however, within the groups some interesting similarities were observed. The MON, TPA, LEP standards were found to be similar, suggesting congruency between these international standards. These standards were also related to the Indonesian LEI standard (local). Interestingly, the

TABLE 8. Bonferroni test results for "Forest Productivity and Health" criterion

| Group 1 | Group 2 | Group 3 | Group 4 | Group 5 | Ungrouped standards |
|---|---|---|---|---|---|
| MON | FSC_PE | COM | CERTF | HEL | CIFOR |
| TPA | MCI | FSC_CO | LUCID | SFI | CCFM |
| LEP | | | | | AIB |
| LEI | | | | | ITTO |
| | | | | | FSC_PC |
| | | | | | FSC_BO |
| | | | | | AFS |
| | | | | | FSC_BC |
| | | | | | NRC |

Each group contains the standards that were not significantly different from each other.
See Table 3 for acronym definitions.

FSC local standards were all found to be different from each other for this criterion and were not linked to any international-level standard based on the number of indicators.

### *Broad SFM Criterion 3: Soil and Water*

The analysis found that four standards were significantly different and identified two aggregated groups of similar standards (see Table 9 and Figures 7 and 8). Within this criterion, 6 clusters of similarity were identified versus the 11 identified by the HCA. The presence of two aggregated groups suggested that there was a high degree of parity between standards at international, national, and local scales. In group 1, the national level CIFOR and COM standards connected the MON and AIB international standards to the local-level AFS and FSC_BC standards. In group 2, the LEP and FSC_CO connected the local and national standards from across the Pacific Rim to the international HEL standard based on number of indicators.

### *Broad SFM Criterion 4: Global Cycles*

The analysis found that four standards were significantly different and identified four groups of standards that were similar (see Table 10). No aggregated groups were identified. Within this criterion, 8 clusters of similarity were identified versus the 11 identified by the HCA. The analysis identified that the majority of local-level standards had not yet incorporated indicators for global cycles that relate to national or international standards. Interestingly, the CIFOR standard is the only national-level standard that did not have any indicators related to global cycles. Group 2 shows that some local-level standards do link to an international standard, TPA, although this standard had the least number of indicators of all the international standards.

### *Broad SFM Criterion 5: Economics*

The analysis found that five standards were significantly different and identified four groups of standards that were similar (see Table 11 and Figures 9 and 10). Two of the four groups were aggregated. Within this criterion, 9 clusters of similarity were identified versus the 11 identified by the HCA. Many similarities were identified with local, national and international standards from different regions (e.g., groups 1 and 3). Group 4 identified that the local-level standards from Canada, USA,

TABLE 9. Bonferroni test results for "Soil and Water" criterion

| Group 1* | Group 2** | Ungrouped standards |
|---|---|---|
| MON | HEL | CERTF |
| AIB | FSC_BO | FSC_PC |
| COM | FSC_PE | LUCID |
| CIFOR | LEI | NRC |
| AFS | LEP | |
| FSC_BC | FSC_CO | |
| | TPA | |
| | CCFM | |
| | SFI | |
| | ITTO | |
| | MCI | |

These standards shared no common ground with any other standards.
See Table 3 for acronym definitions.
* The Bonferroni test identified group 1 as an "aggregated cluster." This group has therefore been represented in Figure 7.
** The Bonferroni test identified group 2 as an "aggregated cluster." This group has therefore been represented in Figure 8.

FIGURE 7. Soil and water criterion: Group 1.

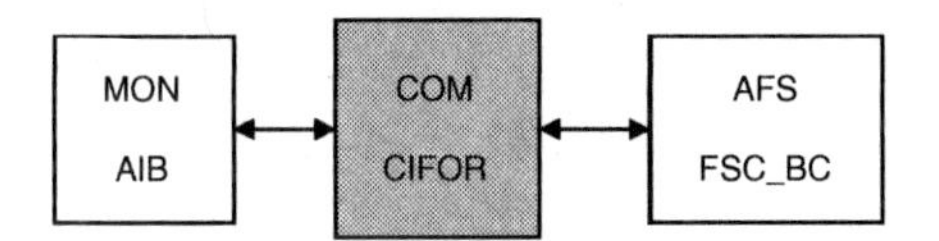


C&I standards that are connected by an arrow were not significantly different from each other. The highlighted standards allowed the standards to be aggregated. See Table 3 for acronym definitions.

FIGURE 8. Soil and water criterion: Group 2.

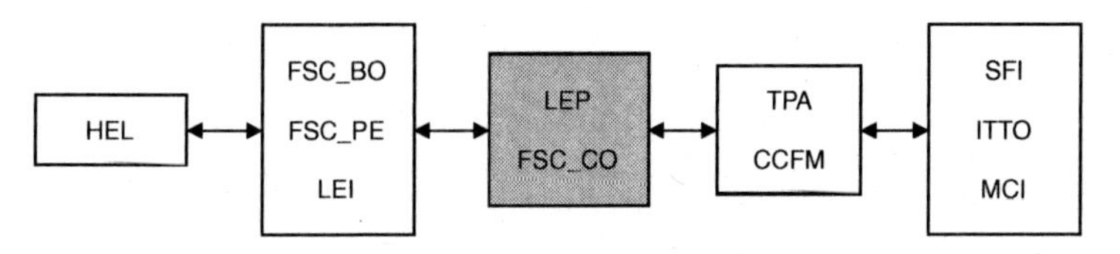


C&I standards that are connected by an arrow were not significantly different from each other. The highlighted standards allowed the standards to be aggregated. See Table 3 for acronym definitions.

TABLE 10. Bonferroni test results for "Global Cycles" criterion

| Group 1 | Group 2 | Group 3 | Group 4 | Ungrouped standards |
|---|---|---|---|---|
| CIFOR | TPA | MON | HEL | AIB |
| SFI | AFS | LEP | COM | CCFM |
| FSC_BC | FSC_CO | | | LUCID |
| FSC_BO | FSC_PE | | | NRC |
| FSC_PC | MCI | | | |
| CERTF | ITTO | | | |
| LEI | | | | |

Each group contains the standards that were not significantly different from each other. See Table 3 for acronym definitions.

TABLE 11. Bonferroni test results for "Economics" criterion

| Group 1 | Group 2 | Group 3* | Group 4** | Ungrouped standards |
|---|---|---|---|---|
| SFI | LEP | CERTF | FSC_BC | CCFM |
| LEI | AFS | FSC_BO | FSC_PE | MON |
| HEL | | TPA | MCI | ITTO |
| AIB | | COM | LUCID | FSC_PC |
| | | CIFOR | NRC | FSC_CO |

Each group contains the standards that were not significantly different from each other.
See Table 3 for acronym definition.
* The Bonferroni test identified group 3 as an "aggregated cluster." This group has therefore been represented in Figure 9.
** The Bonferroni test identified group 4 as an "aggregated cluster." This group has therefore been represented in Figure 10.

FIGURE 9. Economics criterion: Group 3.

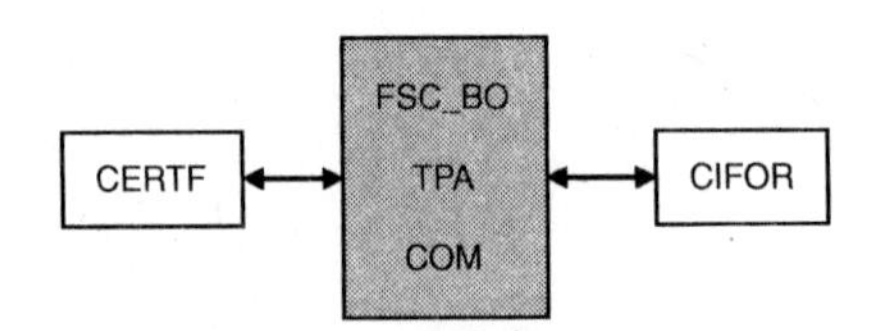


C&I standards that are connected by an arrow were not significantly different from each other. The highlighted standards allowed the standards to be aggregated. See Table 3 for acronym definitions.

Peru, and Malaysia were similar in terms of proportion of economic indicators. On closer inspection of the dataset, the Peruvian and Malaysian standards focused many of their economic indicators on monitoring corruption.

## *Broad SFM Criterion 6: Society*

The analysis found that 10 standards were significantly different and identified five groups of standards that were similar (see Table 12). No aggregated groups were identified. Within this criterion, 15 clusters of similarity were identified versus the 11 identified by the cluster analysis. The five groups of similarity show that there were congruencies

FIGURE 10. Economics criterion: Group 4.

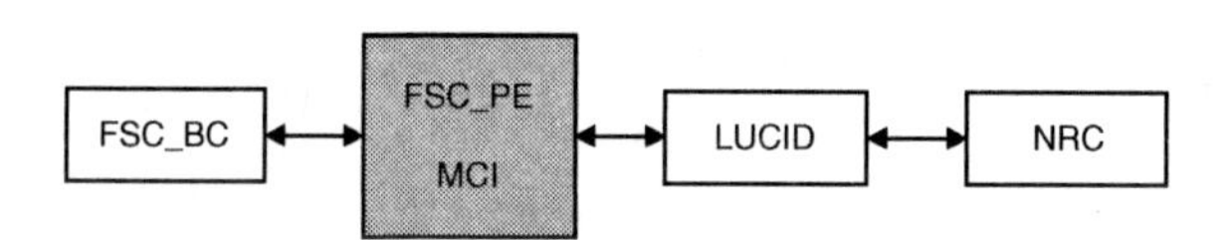


C&I standards that are connected by an arrow were not significantly different from each other. The highlighted standards allowed the standards to be aggregated. See Table 3 for acronym definitions.

TABLE 12. Bonferroni test results for "Society" criterion

| Group 1 | Group 2 | Group 3 | Group 4 | Group 5 | Ungrouped standards |
|---|---|---|---|---|---|
| LEP | HEL | ITTO | COM | FSC_BC | LEI |
| AIB | SFI | CERTF | LUCID | MCI | TPA |
| | MON | | | | CIFOR |
| | | | | | AFS |
| | | | | | CCFM |
| | | | | | FSC_PC |
| | | | | | FSC_BO |
| | | | | | FSC_PE |
| | | | | | FSC_CO |
| | | | | | NRC |

Each group contains the standards that were not significantly different from each other.
See Table 3 for acronym definitions.

between international standards (group 1), international and local standards (groups 2 and 3), national and local standards (group 4), and between local standards across the Pacific Rim (group 5). Interestingly, group 5 shows that a similarity existed between Malaysia's and British Columbia's local-level standards. This similarity is likely a result of their focus on indicators related to indigenous rights. This criterion revealed the most significant differences between standards.

### *Congruency Between C&I Standards in the Pacific Rim*

To determine the degree of overlap between standards, the number of significant results was calculated for each of the seven tests (by standard). Standards with a low number of significant responses indicate a high degree of overlap; for example, if a standard has zero significant responses then it overlaps with at least one other standard in each test. The results are summarized in Table 13. The HEL, LEP, COM, SFI, and MCI standards were found to share common ground with at least one other standard in all seven tests. The NRC and FSC_PC standards were found to be significantly different in five tests. With regard to the stan-

TABLE 13. Number of significant responses* by each criteria and indicators standard from all seven tests

| Standard | Number of significant responses | Standard | Number of significant responses |
|---|---|---|---|
| MON | 1 | FSC_BO | 3 |
| HEL | 0 | FSC_PC | 5 |
| ITTO | 2 | CERTF | 2 |
| TPA | 1 | AFS | 3 |
| LEP | 0 | LEI | 2 |
| AIB | 2 | LUCID | 3 |
| CIFOR | 3 | NRC | 5 |
| CCFM | 4 | FSC_CO | 2 |
| COM | 0 | FSC_PE | 1 |
| SFI | 0 | MCI | 0 |
| FSC_BC | 2 | | |

See Table 3 for acronym definitions.
* If a standard was not found to be significantly different (variation between the means did not exceed the Bonferroni critical distance) in any of the seven tests, then it was similar to at least one other standard in a test and will have zero significant responses.

dards being employed around the Pacific Rim, there appeared to be a substantial degree of congruency, with 5 out of the 21 standards overlapping with another standard in every test, and each standard overlapping with at least one other standard in at least one test. One of the most important findings is the degree of overlap that existed between the international standards and the lower level standards. The HEL, MON, ITTO, TPA, LEP, and AIB standards overlapped with the lower level standards in 5 out of the 7 tests. The national-level standards (CIFOR, CCFM, and COM) overlapped with both higher and lower level standards to varying degrees (see Table 13). Varying degrees of overlap were also observed between the local-level standards.

## *DISCUSSION*

A number of criteria and indicators initiatives have been adopted throughout the countries of the Pacific Rim. These have ranged from international processes to local initiatives such as forest certification. Although there is considerable variability in the issues facing forest policy makers in the countries of the Pacific Rim, it is often expected that criteria and indicators will reflect a level of comparability. Indeed, one of the driving forces behind the criteria and indicator initiatives is the need to promote comparable monitoring, information reporting, and "best" management practices. Although a number of comparative studies have been completed on the nature of forestry-related criteria and indicator initiatives, (see Hahn-Schilling et al., 1994; Lammerts van Bueren and Blom, 1996; Hornborg, 1999; Rickenbach, 1999; Nsenkyiere and Simula, 2000; CEPI, 2001; Ozinga, 2001; Meridian Institute, 2001; Pokorny and Adams, 2003; Holvoet and Muys, 2004; Hickey et al., 2006), none have focussed on the Pacific Rim region. This region is important to international forest policy because of its rich forest resources and high levels of trade in forest products.

It has been recognized that the data deemed relevant to SFM and adaptive management in one part of the world is often not relevant in others, yet the aim of global forestry data coordination and reporting remains (e.g., *Global Forest Resources Assessment* [FAO, 2001]; Montreal Process [MPTAC, 2003]; Hickey, 2004a). The existence of numerous C&I standards in the Pacific Rim suggests there is a lack of consensus on how to monitor SFM in the region. Conversely, the similarity of the standards, either in their entirety or in part, reflects a level of consensus on the underlying conceptualization of SFM in different jurisdictions. This observation is supported by Prabhu et al. (2002). The distribution and number of indicators does not allow us to infer which standards are the most effective. However, these data do give a coarse measure of (1) the main issues being faced within a particular jurisdiction, or (2) the perceptions of those in a position to develop C&I standards in different jurisdictions.

Bass (2002) noted that C&I standards are predisposed to "top down" control and often present "quick fix" solutions to complex problems. Nevertheless, it is important to consider that a set of indicators as a whole is greater than the sum of its parts (Prabhu et al., 2002). Our analysis revealed that the application of C&I standards across the Pacific Rim was nebulous. Although there were a lot of similarities between the standards, there were also a lot of differences. The high degree of overlap observed between the standards may mean that the international standards have been providing the foundations for many of the more region-focused standards (this is certainly true for the Australia Forestry Standard, which was based on the Montreal Process). The lower degree of overlap between local-level indicators most likely reflects differences in the wider operating environment of forestry in different jurisdictions (Hickey et al., 2005a). This is an important driver for the development of local-level indicators. Interestingly, when the local-level indicators from within the same country were examined, even less overlap was apparent. For example, FSC_BC (Canada) did not overlap with FSC_BO (Canada) on any broad SFM criterion in terms of distribution of indicators. In the Pacific Rim, the diverse nature of the C&I standards relevant at the local level of forest management can be seen to reflect the perceptions of those that are affected by SFM decisions and policies. This provides an insight into which indicators are deemed most relevant, given local conditions. These signals can be lost as we try to create international standards that will be more effective at promoting best management practices across larger and more diverse regions.

It is important to note that the countries in the Pacific Rim are at different stages of economic and social development. This has resulted in a number of different forest management paradigms operating within the region. Kimmins (1995) identified that forest management moves from exploitation to ecological forestry as foresters and researchers in a country develop a sense of environmental awareness. It can then shift to social forestry as the public becomes increasingly educated and concerned about the environment. The countries of the Pacific Rim can therefore be grouped into one of four broad management paradigms: (1) exploitation, (2) administrative, (3) ecological-based, or (4) social.

These paradigms were reflected in our results, with the standards from many of the developing countries focusing on indicators related to the mitigation of exploitative practices and corruption and to ensuring a sustainable harvesting yield (i.e., Peru, Columbia, Honduras, Malaysia, Indonesia, and Thailand). These results are supported by observations reported by the FAO in their *Global Forest Resources Assessment 2000*. The standards developed for use in many of the more industrialized countries were found to have a significant number of indicators that focussed on sustainable communities and the diversification of forest uses and products (i.e., Canada, Australia, and USA). These paradigms can provide a broad basis for identifying the important objectives for achieving SFM within a particular region or locale. The results are also useful for identifying and understanding the major challenges being faced by local forest managers and policy makers as they attempt to meet higher-level SFM objectives.

In Oceania and Central America, the major challenges being faced by policy makers include the conversion of forestland to agriculture and the exploitation of natural forests for fuel-wood and cattle production (see Tucker, 1999, 2000; Merry et al., 2002; Garcia-Romero et al., 2004). In both South and Central America, the continued exploitation of tropical forests (O'Neill et al., 2001) and the inability of governments to regulate the industry are significant SFM challenges (see FAO, 2000; Southgate et al., 2000). In Southeast Asia, illegal logging and forest degradation are major problems in many of the countries with well-developed forest management infrastructures (e.g., Indonesia [Dauvergne, 2001; Kato, 2005], Philippines [Tumaneng-Diete et al., 2005], and Thailand [Bhusal et al., 1998; Gallagher et al., 2000]). Shifting cultivation has also been identified as a major problem in the region. In Canada, the United States of America, Australia, and New Zealand, the SFM challenges are increasingly related to the participation of local communities and indigenous peoples in forest management. There are also difficult social issues related to post-materialist value systems and highly polarized opinions related to forest management.

The ebb and flow of economic and social transitions is a causal factor in the development of such a diverse set of C&I standards. What affects the perception of SFM at the international level can be both meaningful and meaningless at the local level, and this is reflected in the diversity of the C&I standards evaluated in this study. Although there are certainly international "expectations" regarding comparable forest-related information in different jurisdictions, as highlighted by the *Global Forest Resource Assessment* (FAO, 2001), our results suggest that the development of regular and comparable monitoring and information reporting programmes, especially at the local and subnational levels, continue as a challenge to policy makers. A situation where monitoring and information reporting is inconsistent or nonexistent has direct implications for the quality of national-level monitoring and international reporting. Similarly, the existence of standards that fail to reflect the wider societal issues affecting forest management in a particular jurisdiction will render national-level SFM-related reporting meaningless. Therefore, the challenge remains to maximize stakeholder confidence through processes designed to enhance transparency and facilitate international comparability, while still reflecting the operational reality of forest management within a particular jurisdiction (Hickey, 2004a). An important first step toward greater comparability is the use of a consistent set of SFM criteria at all scales of forest management. The subsequent indicators could then be tailored to suit the social, economic, and environmental conditions in each jurisdiction. In the case of the Pacific Rim region, using the seven criteria developed through the Montreal Process could serve as a valuable starting point for conceptualizing the main forest values associated with SFM monitoring and reporting across nations. This could also facilitate meaningful dialogue between stakeholders at all scales of forest management.

## *CONCLUSION*

Forest managers in the Pacific Rim face many different challenges for achieving sustainability. An important step in addressing these issues and moving toward sustainable forest management is improved monitoring and information reporting at the local, national, and international levels (Hickey, 2004b). Through a comparative analysis designed to identify similarities and differences in sustainable forest management criteria and indicators initiatives in the Pacific Rim region, broad ranges of similarities and differences have been revealed. The high degree of overlap observed between the international standards and the local-level standards suggests that finer scale C&I standards could be using the international standards as a guideline for their development. This makes the development of a C&I standard that increases the congruency between local, national, and international standards much more promising. The lower degree of overlap between local-level standards shows that there is still a need for local-level standards to deal with the social, economic, and environmental diversity that exists across the Pacific Rim. The development of an international standard that can bridge these differences may allow for increased transparency and reduce the cost of monitoring and information reporting for SFM. This is where the challenges lie for effective C&I development within regions that deviate considerably in their definitions of what is important for achieving SFM. When considered in the context of globalization, the research findings support international efforts to encourage comparability in SFM-related monitoring and information reporting. By harmonizing the monitoring and information reporting methods required by C&I standards and, where possible, encouraging a standardized approach to data collection and stakeholder coordination, international efforts to promote SFM will be fostered.

## REFERENCES

Archer, H., R. Kozak and D. Balsillie. 2005. The impact of forest certification labelling and advertising: an exploratory assessment of consumer purchase intent in Canada. *The Forestry Chronicle* 81(2):229-244.

Babbie, E. 2001. The practice of social research. 9th ed. Wadsworth, Belmont, CA. 498 pp.

Bass, S. 2002. Application of criteria and indicators to support sustainable forest management: some key issues. In: pp. 19-37. R.J. Raison, A.G. Brown and D.W. Flinn (eds.). Criteria and indicators for sustainable forest management. IUFRO Research Series 7. CABI Publishing, Oxford.

Bhusal, Y.R., G.B. Thapa and K.E. Weber. 1998. Thailand's disappearing forests: the challenge to tropical forest conservation. *International Journal of Environmental Pollution* 9(2-3):198-212.

Cashore, B., G. Auld and D. Newsom. 2003. Forest certification (eco-labeling) programs and their policy-making authority: explaining divergence among North American and European case studies. *Forest Policy and Economics* 5(3):225-247.

Clapp, R.A. 2001. Tree farming and forest conservation in Chile: Do replacement forests leave any originals behind? *Society and Natural Resources* 14(4):341-356.

Confederation of European Paper Industries [CEPI]. 2001. Comparative matrix of forest certification schemes. Brussels. Retrieved November 20, 2002 from http://www.cepi.org/files/Matrix01-171253A.pdf.

Contreras_Hermosilla, A. 2001. Illegal forest activities in the Asia Pacific Rim. Markets for Forest Conservation brief. Forest Trends, Washington, D.C. Retrieved June 23, 2005 from http://www.forest-trends.org/documents/publications/pri_illegallogging2.pdf

Creswell, J.W. 1997. Qualitative inquiry and research design: choosing among five traditions. Sage Publications, California, USA.

Dauvergne, P. 2001. Loggers and degradation in the Asia-Pacific: corporation and environmental management. Cambridge University Press. UK.

Environmental Systems Research Institute [ESRI]. 2004. ESRI ArcGIS 9. Environmental Systems Research Institute Ltd., Redlands, CA, USA.

Food and Agriculture Organization [FAO]. 2000. Global forest resources assessment 2000. FAO Forestry Papers 140. Rome, Italy.

Food and Agriculture Organization [FAO]. 2001. Global forest resources assessment 2000: main report. Food and Agriculture Organization of the United Nations, Rome, Italy.

Forest Research. 1999. Study of non-tariff measures in the forest products sector. APEC #99-CT-01.5. Prepared for Asia-Pacific Economic Cooperation, Singapore.

Gallagher, J., C. Wheeler, M. McDonough, and B. Namfa. 2000. Sustainable environmental education for a sustainable environment: lessons from Thailand for other nations. *Water, Air and Soil Pollution* 123(1-4):489-503.

Garcia-Romero, A., O. Oropeza-Orozco and L. Galicia-Sarmiento. 2004. Land-use systems and resilience of tropical rain forests in the Tehuantepec Isthmus, Mexico. *Environmental Management* 34(6):768-785.

Geist, H.J. and E.F. Lambin. 2001. What drives topical deforestation? Land-Use and Land-Cover Change (LUCC) Project Report Series No. 4. Louvain-la-Neuve, Belgium.

Genov, P.V. 1999. A review of the cranial characteristics of the wild boar (*Sus scrofa* Linnaeuss 1758) with systematic conclusions. *Mammal Review* 29(4):205-238.

Glaser, B.G. and A.L. Strauss. 1967. The discovery of grounded theory: strategies for qualitative research. Aldine Publishing Co., Chicago, IL.

Hahn-Schilling, B., J. Heuveldop and J. Palmer. 1994. A comparative study of evaluation systems for sustainable forest management (including principles, criteria and indicators) In: pp. 3-36. J. Heuveldop (ed.)., Assessment of sustainable tropical forest management (A contribution to the development of concept and procedure). Bundesforschungsanstalt fur Forest- und Holzwirtschaft, Hamburg.

Hickey, G.M. 2004a. Monitoring and information reporting for sustainable forest management in North America and Europe: Requirements, practices and perceptions. Ph.D. thesis. University of British Columbia, Vancouver, BC, Canada. 500 pp.

Hickey, G.M. 2004b. Regulatory approaches to monitoring sustainable forest management. *International Forestry Review* 6(2):89-98.

Hickey, G.M. and J.L. Innes. 2005. Monitoring sustainable forest management in different jurisdictions. *Environmental Monitoring and Assessment* 108(1-3):241-260.

Hickey, G.M., J.L. Innes, R.A. Kozak, G.Q. Bull and I. Vertinsky. 2005. Monitoring and information reporting for sustainable forest management: an international multiple case study analysis. *Forest Ecology and Management* 209(3):237-259.

Hickey, G.M., J.L. Innes, R.A. Kozak, G.Q. Bull and I. Vertinsky. 2006. Monitoring and information reporting for sustainable forest management: an inter-jurisdictional comparison of soft law standards. *Forest Policy and Economics* 9(4):297-315.

Holvoet, B. and B. Muys. 2004. Sustainable forest management worldwide: a comparative assessment of standards. *International Forestry Review* 6(2):99-122.

Hornborg, C. 1999. Comparison of forest certification schemes and standards in Finland, Sweden and Norway (Summary). Helsinki Department of Forest Ecology. 6 pp.

Innes, J.L. and G.M. Hickey. 2005. Certification of forest management and wood products. In: J.L. Innes, G.M. Hickey, and H.F. Hoen (eds.). Forestry and environmental change: socioeconomic and political dimensions. IUFRO Series 11. CABI Publishing, Oxford.

International Organization for Standardization [ISO]. 2004. The ISO Survey–2004. Retrieved October 20, 2005, from http://www.iso.org/iso/en/prods-services/otherpubs/pdf/survey2004.pdf

Kato, G. 2005. Forestry sector reform and distributional change of natural resource rent in Indonesia. *Developing Economies* 43(1):149-170.

Kimmins, J.P. 1995. Sustainable development in Canadian forestry in the face of changing paradigms. *The Forestry Chronicle* 71(1):33-40.

Lammerts van Bueren, E.M. and E.M. Blom. 1996. Hierarchical framework for the formulation of sustainable forest management standards. The Tropenbos Foundation, Wageningen, Netherlands.

Manly, B.F.J. 1994. Multivariate statistical methods: a primer. Chapman and Hall, London; New York.

Manly, B., L. Macdonald and D. Thomas. 1993. Resource selection by animals. Chapman and Hall, London.

Meridian Institute, 2001. Comparative analysis of the Forest Stewardship Council and Sustainable Forestry Initiative. Retrieved July 24, 2005 from http://www2.merid.org/comparison/.

Merry, F.D., P.E. Hildebrand, P. Pattie and D.R. Carter. 2002. An analysis of land conversion from sustainable forestry to pasture: a case study in the Bolivian lowlands. *Land Use Policy* 19(3):207-215.

Miles, M.B. and A.M. Huberman. 1984. Qualitative data analysis: a sourcebook of new methods. Sage Publications, Beverly Hills.

Montreal Process Technical Advisory Committee [MPTAC]. 2003. *Montreal Process: first forest overview report 2003.* Montreal Process. Retrieved February15, 2004 from www.mpci.org/rep-pub/2003/contents_e.html.

Myers, N., R.A. Mittermeier, C.G. Mittermeier, G.A.B. da Fonseca and J. Kent. 2000. Biodiversity hotspots for conservation priorities. *Nature* 403:853-858.

Nsenkyiere, E.O. and M. Simula. 2000. Comparative study on the auditing systems of sustainable forest management. International Tropical Timber Organization (ITTO), Yokohama, Japan. 83 pp.

O'Neill, G.A., I. Dawson, C. Sotelo-Montes, L. Guarino, M. Guariguata, D. Current and J.C. Weber. 2001. Strategies for genetic conservation of trees in the Peruvian Amazon. *Biodiversity and Conservation* 10(6):837-850.

Ozinga, S. 2001. Behind the logo: an environmental and social assessment of forest certification schemes. Fern, United Kingdom. Retrieved September 25, 2005 from http://www.fern.org/pubs/reports/behind/btlrep.pdf.

Pokorny, B. and M. Adams. 2003. What do criteria and indicators assess? An analysis of five C&I sets relevant for forest management in the Brazilian Amazon. *International Forest Review* 5(1):20-28.

Prabhu, R., H.J. Ruitenbeek, T.J.B. Boyle and C.J.P. Colfer. 2002. Between voodoo science and adaptive management: the role and research needs for indicators of sustainable forest management. In: pp. 39-66. R.J. Raison, A.G. Brown and D.W. Flinn (eds.). Criteria and indicators for sustainable forest management. IUFRO Research Series 7. CABI Publishing, Oxford.

Punj, F. and D.W. Stewart. 1983. Cluster analysis in marketing research: review and suggestions for application. *Journal of Marketing Research* 20:131-148.

Rametsteiner, E. 1999. Sustainable forest management certification: framework conditions, system designs and impact assessment. Ministerial Conference on the Protection of Forests in Europe, Liaison Unit, Vienna.

Rickenbach, M. 1999. Comparison of forest certification schemes of interest to USA forest owners. The Sustainable Forestry Partnership, Oregon State University, USA. Retrieved November 28, 2004 from http://sfp.cas.psu.edu/pdfs/Certification_matrix.pdf.

Southgate, D., P. Salazar-Canelos, C. Camacho-Saa and R. Stewart. 2000. Markets, institutions, and forestry: the consequences of timber trade liberalization in Ecuador. *World Development* 28(11):2005-2012.

Tabachnick, B.G. and L.S. Fidell. 2001. Using multivariate statistics: 4th Ed. Allyn and Bacon, Boston, MA, USA. 966 pp.

Thiele, R. 1995. Conserving tropical rain-forests in Indonesia–a qualitative assessment of alternative policies. *Journal of Agricultural Economics* 46(2):187-200.

Tucker, C.M. 1999. Private versus common property forests: forest conditions and tenure in a Honduran community. *Human Ecology* 27(2):201-230.

Tucker, C.M. 2000. Striving for sustainable forest management in Mexico and Honduras–the experience of two communities. *Mountain Research and Development* 20(2):116-117.

Tumaneng-Diete, T., I.S. Ferguson and D. MacLaren. 2005. Log export restrictions and trade policies in the Philippines: bane or blessing to sustainable forest management. *Forest Policy and Economics* 7(2):187-198.

Vincent, J.R. 1992. The tropical timber trade and sustainable development. *Science* 256(5064):1651-1655.

White A. and A. Martin. 2002. Who owns the world's forests? Forest tenure and public forests in transition. Forest Trends, Washington, D.C. ISBN 0-9713606-2-6. 30 pp.

World Trade Organization [WTO]. 1997. Environmental benefits of removing trade restrictions and distortions. Committee on Trade and Environment. Note by the Secretariat. WT/CTE/W/67. Geneva, Switzerland.

Zhang, Y. 2005. Multiple-use forestry vs. forestland-use specialization revisited. *Forest Policy and Economics* 7(2):143-156.

Zhang, Y.Q., J. Uusivuori and J. Kuuluvainen. 2000. Econometric analysis of the causes of forest land use changes in Hainan, China. *Canadian Journal of Forest Research* 30(12):1913-1921.

doi:10.1300/J091v24n02_08

# A European Network in Support of Sustainable Forest Management

Folke Andersson
Anders Mårell

**SUMMARY.** Today, sustainability is a political priority at the regional, national, and international levels. It is thought to be attained by political

---

Folke Andersson is Professor Emeritus in Terrestrial Ecosystem Science at the Forest Faculty, Swedish University of Agricultural Sciences, P.O. Box 7072, S-750 07 Uppsala, Sweden and Chairman of the COST Action E25 European Network for Long-Term Ecosystem and Landscape Research (ENFORS) (E-mail: folke.andersson@eom.slu.se).

Anders Mårell, Unité de Recherches Forestières Mediterranéennes UR 629, INRA Site Agropac, Domaine Saint Paul, F-84914 Avignon cedex 9, France (E-mail: anders.marell@sverige.nu; URL: www.enfors.org).

The ideas expressed in this paper are based on the work within COST Action E25 and are a result of collaboration between researchers, to whom the authors are grateful. The European Cooperation in the Field of Scientific and Technical Research (COST), Domain Forests and Forestry Products, provided financial support for networking. Financial support of the scientific coordination was provided by GIP ECOFOR (France), Brandenburg Technical University of Cottbus (Germany), Faculty of Forest Science, Swedish University of Agricultural Sciences, SLU (Sweden), Faculty of Forestry and Wood Technology, Mendel University of Agriculture and Forestry (the Czech Republic), Swiss Federal Institute for Forest, Snow and Landscape Research, WSL (Switzerland), Institute of Agro-Environmental and Forest Biology, CNR-IBAF (Italy), English Nature (United Kingdom), Lithuanian Forest Research Institute (Lithuania), and Cemagref (France). The first author received grants from the Swedish Ministry of Industry and the Research Council for Environment, Agricultural Sciences and Spatial Planning, FORMAS (Sweden).

[Haworth co-indexing entry note]: "A European Network in Support of Sustainable Forest Management." Andersson, Folke, and Anders Mårell. Co-published simultaneously in *Journal of Sustainable Forestry* (Haworth Food & Agricultural Products Press, an imprint of The Haworth Press, Inc.) Vol. 24, No. 2/3, 2007, pp. 279-293; and: *Sustainable Forestry Management and Wood Production in a Global Economy* (ed: Robert L. Deal, Rachel White, and Gary L. Benson) Haworth Food & Agricultural Products Press, an imprint of The Haworth Press, Inc., 2007, pp. 279-293. Single or multiple copies of this article are available for a fee from The Haworth Document Delivery Service [1-800-HAWORTH, 9:00 a.m. - 5:00 p.m. (EST). E-mail address: docdelivery@haworthpress.com].

Available online at http://jsf.haworthpress.com

doi:10.1300/J091v24n02_09

instruments and adequate management. The Ministerial Conference on the Protection of Forests in Europe (MCPFE) is the European forum and reference for sustainable forest management (SFM). The MCPFE has developed criteria and indicators for implementing SFM in Europe. However, there is a gap between policy makers, stakeholders, and researchers with regard to the implementation of the criteria and indicators. A research strategy is proposed by the COST Action E25–European Network for Long-Term Ecosystem and Landscape Research, whereby these gaps are identified, and means for bridging these gaps are suggested. For example, economic and social-cultural criteria and indicators are poorly developed or missing–there is a need for disciplinary and multidisciplinary research. Furthermore, the knowledge base of forests and forestry is well developed in Europe, but could be better communicated to parties involved. Decision-support systems and "landscape laboratories" are suggested as tools for bridging this gap. Furthermore, resources are scarce, so new approaches should be developed that take advantage of existing resources and experiences. Information systems for maintaining and developing long-term information, and increased European collaboration are needed to harmonize existing and future initiatives. doi:10.1300/J091v24n02_09 *[Article copies available for a fee from The Haworth Document Delivery Service: 1-800-HAWORTH. E-mail address: <docdelivery@haworthpress.com> Website: <http://www.HaworthPress.com>*


**KEYWORDS.** Biodiversity, ecology, economy, ecosystem and landscape, multidisciplinarity, social-cultural function, sustainable forest management

## *INTRODUCTION*

Today, sustainability is a political priority at the regional, national, and international levels. It is thought to be attained by political instruments and adequate management. A sound knowledge about the three pillars of sustainability (ecological, economic, and social-cultural values) is fundamental for reaching the goals of sustainable forest management (SFM). As sustainability is an evolving concept, which might be adequately defined at a theoretical level, scientists and managers have difficulty in going from theory to practice. In this science-policy context, the Ministerial Conference on the Protection of Forests in Europe (MCPFE) is the European forum and reference for European SFM. The

United Nations Forum on Forests (UNFF) and the Montreal process (Anon., 2000) are global equivalents.

The first MCPFE was held in 1990 in Strasbourg, France, where the need for an increased collaboration on the ecological knowledge base of Europe's forests was expressed (MCPFE, 2000). In particular, a commitment was made by the signatory states to establish a European network on forest ecosystem research (Resolution S6). In response to this political aim, EU-financed European research networks were established (MCPFE, 2003a): European Forest Ecosystem Research Network (EFERN), a pan-European network to promote coordination of forest ecosystem research and to improve communication among scientists, 1996-1999 (FAIR-CT95-0883); followed by European Network for Long-Term Forest Ecosystem and Landscape Research (ENFORS), a pan-European network for long-term research on sustainable forest management, 2001-2005 (COST Action E25).

This paper aims to give a background on ENFORS and to present recent results: (1) a network of facilities for ecosystem and landscape field research as a tool for SFM research, (2) a research strategy for future research and development on SFM, and (3) identification of important aspects related to the access of information and management of data that support science, policy, and management.

## *THE PAST–SFM AS AN EVOLVING CONCEPT*

The development of the sustainability concept in Europe began when a scientific approach to forest management was introduced to forestry education in Germany in the 18th century, and soon after in France. Initially it was defined as sustained yield of wood at the beginning of the 1900s, and it is thus a concept that has been practiced in forestry for more than 100 years. The sustainability concept has since evolved to mean sustained multiple use of all resources in the forest. Man's interaction with forests in Europe has a much longer history and complexity (Farrell et al., 2000). Kimmins (1992) described this development to have occurred in four stages:

1. Unregulated exploitation of local forests and clearing for agriculture and grazing.
2. Institution of legal and political mechanisms or religious taboos to regulate exploitation.

3. Development of an ecological approach to siliviculture and timber management of the biological resources of forests.
4. Social forestry, which recognizes the need to manage forests as a multifunctional resource in response to the diverse demands of modern society.

Along with the development of the concept of sustainability, the link between science and society has become closer. For example, politicians recognize that science plays an important role in today's knowledge-based society (Merkel, 1998). Consequently, scientists are nowadays expected to deliver practical solutions in return for funding, regardless of whether it concerns public or private funding. This could, however, be understood as a threat to the independence of science (Anon., 2001). Although the interface between science and policy is recognized to be important, communication between policy makers and scientists seems not to be that straightforward (Bradshaw and Borchers, 2000). Other parties also need to be considered–the stakeholders such as forest owners and the public in general. Stakeholder participation in forest management has increased lately and will become more and more important (Petheram et al., 2004). The lack of (1) institutional structures that correspond to the demand of knowledge transfer, (2) a common language to communicate this knowledge, and (3) methodologies that take into consideration the different ways of working among scientists, policy-makers, forest managers, landscape planners, and the general public (e.g., through nongovernmental organizations), are major obstacles for an effective communication at the science-policy interface and successful stakeholder participation.

## *THE PRESENT–BUILDING ON RESULTS FROM TWO SUCCESSIVE RESEARCH NETWORKS*

### *EFERN, European Forest Ecosystem Research Network, 1996-1999*

The European Forest Ecosystem Research Network operated during a period when the SFM concept was rapidly evolving (stages 3 and 4, see above). This paradigm shift is reflected in the book *Pathways to the Wise Management of Forests in Europe* (Führer et al., 2000), where the history of the utilization of Europe's forests along with the development of research is discussed. Fundamental ecological concepts are dealt with as well as the present understanding of the forest ecosystem and its processes. Based on these principles, regional problems for different parts of Europe were analyzed.

It was concluded that the conventional stand or ecosystem approach, used in Europe, is no longer adequate. A new dimension was needed in order to handle multifunctionality and all three dimensions of SFM (ecological, economic, and social-cultural)–the landscape dimension. Andersson et al. (2000) suggested a new concept "ecosystem and landscape forestry," which aimed to take into consideration not only the different natural science aspects, but also the economic, social, and cultural aspects. "Landscape-based forestry planning" was considered a means for balancing the different interests expressed by the stakeholders, in using the forest resource.

The recent evolution of the concept of sustainability (as described above) is reflected in the resolutions of MCPFE; the first resolutions concerned mainly ecological issues, and economic and social-cultural issues have been included with an increasing number in later resolutions (MCPFE, 2003c). For example, the second conference dealt with the definition of SFM (Resolution H1, Helsinki 1993), the third made the definition operational by defining criteria and indicators (Resolution L2, Lisbon 1998), and the fourth emphasized the multifunctional role of forestry (Resolution V1, Vienna 2003).

Based on the conclusions made by EFERN and the changes of the sustainability concept as reflected in the MCPFE process, the members of EFERN suggested a continued European network on forest ecosystem and landscape research (MCPFE, 2003a). The aim was to use existing resources, but with a new and broader scope (Mårell et al., 2003).

### *ENFORS, European Network for Long-Term Forest Ecosystem and Landscape Research, 2001-2005*

The European Network for Long-Term Forest Ecosystem and Landscape Research is an activity (Action E25) within the European Cooperation in Science and Technical Research (COST). Presently, it has 27 member countries with participants representing 100 research organizations. The vision of ENFORS is to become a European meeting point and platform for scientists and stakeholders by using a network of 90 field facilities (Figure 1). In the following, we treat briefly the three major ENFORS components: (1) the field facilities or "landscape laboratories," (2) a research strategy, and (3) management of information.

FIGURE 1. Map of location of ENFORS field facilities–virtual landscape laboratories–that aim at developing landscape approaches in support to integrated forest resource management. Background: Forest cover (FAO, 1999).

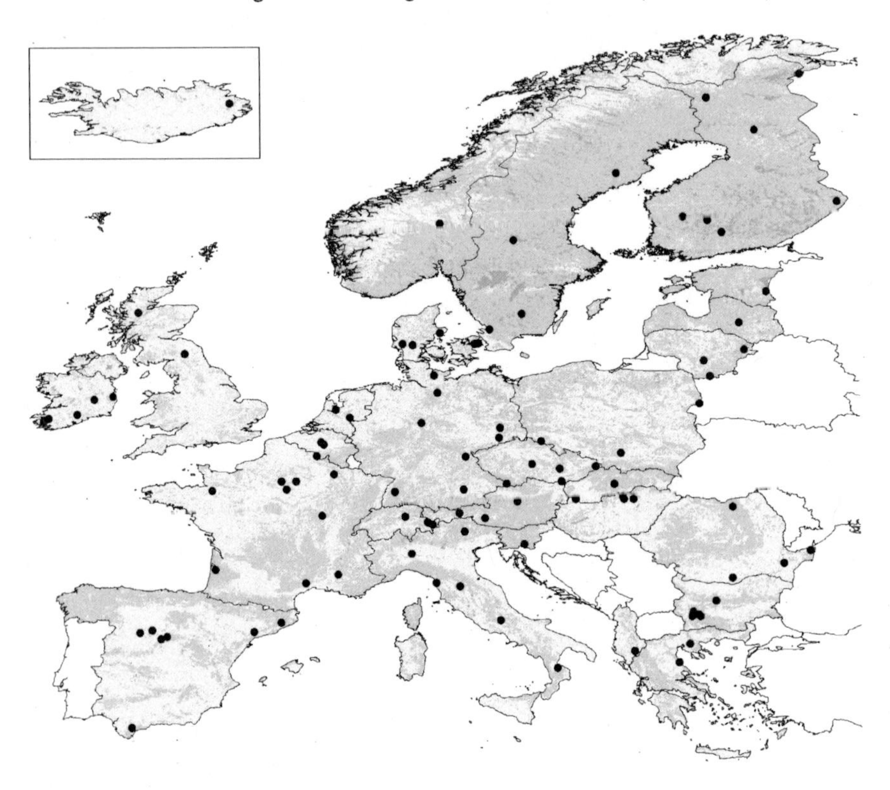

*ENFORS field facilities or landscape laboratories: a tool for SFM research.* Experimental manipulations of environmental factors at the landscape level (experimental treatments that cover thousands of square meters or more) are few (e.g., Holt et al., 1995). There are three major reasons for the lack of landscape experiments: (1) the complexity of the environment is generally increasing when going from the ecosystem to the landscape level, and experiments at high spatial scale thus become more difficult to control; (2) experimental manipulations at the landscape level need to take into consideration ethical dilemmas that are increasingly difficult to justify in front of a growing group of stakeholders (e.g., landowners, public); and (3) landscape experiments are economically very costly. However, our knowledge about how naturally dynamic systems such as forests function tells us that research is needed not only at the ecosystem but also at the landscape level in order to understand high-level processes such as those involved when forests respond to disturbances or long-term environmental changes (Schnitzler and Borlea, 1998; Andersson et al., 2000; Antrop, 2005). Despite the difficulties described above, examples of landscape experiments in forest research do exist (e.g., Likens et al., 1977; Rosén, 1982; Tyrväinen et al., 2006). However, most "landscape-level experiments" (e.g., Lindenmayer et al., 1999; Martin and Baltzinger, 2002) lack full control of treatments, including absence of randomization. These landscape-level experiments are rather to be considered as "quasi-experiments" (Campbell and Stanley, 1963). Thus, the most feasible way to carry out forest research at the landscape level is to use quasi-experiments and observational studies as a complement to true experimental manipulations.

Although experimental manipulations are difficult to achieve at the landscape level, a suite of landscape-scale case studies can be viewed as a "labscape" (Kohler, 2002) or "landscape laboratory" (Angelstam and Törnblom, 2004). The U.S. Long-Term Ecological Research (LTER) sites, a collaborative effort for investigating ecological processes over long temporal and broad spatial scales (Hobbie et al., 2003), the Canadian Model Forests (Sinclair and Smith, 1999; Besseau et al., 2002), the United Nations Educational, Scientific, and Cultural Organization (UNESCO) Man and Biosphere reserves (Malcolm, 2002), and the ENFORS field facilities (Mårell and Leitgeb, 2005a, 2005b) are examples of such networks of case studies, which address problems at the ecosystem as well as the landscape level. Whereas the U.S. LTER sites are mainly science-oriented, the focus of the Canadian Model Forests and the UNESCO Man and Biosphere reserves are more applied, and relate to the implementation of science, policy, and legal frameworks. The idea behind the ENFORS field facilities is to combine fundamental research with applied approaches (Mårell and Leitgeb, 2005b).

Landscape laboratories, such as the ENFORS field facilities, can become essential tools for developing leading-edge research on sustainable resource management. With a common strategy for research activities, they could contribute to the implementation of a sustained use of forest resources in real landscapes through a set of representative case studies. By linking survey and monitoring with observational and experimental research, landscape laboratories provide arenas for integrated and multidisciplinary research approaches as well as a meeting place for interactions between science and society.

*A research strategy toward integrated forest resource management.* A principle such as SFM can be understood as defining ecological, economic, and social-cultural criteria (McDonald and Lane, 2004), which in turn are assessed by combining indicators with performance targets (Angelstam et al., 2004). When equal emphases are put on all three elements of sustainability, SFM can be considered to be truly integrated resource management. By integrated we mean that the elements of sustainability are combined to form a whole. From a research perspective, this demands a holistic approach, and ENFORS has proposed a strategy for research and development on the topic of SFM that correspond to these needs (Andersson et al., 2005). The strategy is holistic in the sense that (1) future research is considered in the context of society, and (2) all three elements of sustainability are analyzed from the perspective of the main functions of the forests (Table 1). It is composed of two elements (Figure 2):

- The *P-D-P-M-A-element* aims to bridge the gap between policy and research through the elaboration of *P*redictions with development of scenarios, syntheses of existing facts that lead to *D*ecision-support systems, which serve as a base for establishing *P*olicies. The policy decisions should lead to accepted *M*anagement activities, which need to be followed by *A*ssessment activities.
- The *R-S-M-element* aims to develop and maintain the knowledge base through experimental and observational *R*esearch on the understanding of ecological, economic and social-cultural forest functions in real landscapes, and based on proper knowledge of the forest and its surrounding environment through *S*urveys and of ongoing changes and trends observed through *M*onitoring.

The starting point for implementing the idea to analyze, assess, and communicate the elements of sustainability to policy makers, managers, and stakeholders, is to consolidate a truly integrated and multidiciplinary research agenda. By using the SFM criteria as a base, the needs for improved knowledge and efficient tools for management and planning can be identified and improved. At the same time, we argue that the development of truly integrated and multidisciplinary research should not be at the expense of advances within disciplines concerning biodiversity, economic, protective, productive, and social-cultural functions of the forests. We identified the driving forces and underlying processes for these functions, and in doing so, we give examples of research needs at a disciplinary as well as multidisciplinary level for im-

TABLE 1. Definition of activities within respective elements of the ENFORS (European Network for Long-Term Forest Ecosystem and Landscape Research, COST Action E25) research approach

| Activity | Definition |
|---|---|
| Experimental Research | • International collaboration on particular topics related to research themes on SFM.<br>• Forum for scientific exchange. |
| Survey | • Common policy for sharing and accessing survey data. |
| Monitoring | • Common policy for sharing and accessing monitoring data (mutual collaboration with existing monitoring programs and networks).<br>• Recommendations for standardized and harmonized monitoring programs (complementing existing national and international programs). |
| Prediction | • Predictive models and scenarios. |
| Decision support | • Decision-support systems.<br>• Information systems. |
| Policy | • Identify the gaps between policy making, research, and application and provide means for bridging these gaps. |
| Management | • Management recommendations.<br>• Expertise. |
| Assessment | • Assessment tools.<br>• Network of case studies/model forest landscapes demonstrating regional best practices. |

proving the knowledge base (Figure 3) (Andersson et al., 2005). This also means improvements of current criteria and indicators, and development of performance targets.

*Management and retrieval of information.* A survey by ENFORS (Mårell et al., 2004) has revealed a rich asset of experiments, case studies, research projects, watershed studies, and monitoring activities in Europe, with their associated long-term data sets (Mårell and Leitgeb, 2005a; Web-based meta-database accessible at www.enfors.org). The earliest European forest experiments date back to the second half of the 19th century, and several national forest inventories have nearly 100 years of data collection. However, this asset with long-term data is now jeopardized because of insufficient financial support for its mainte-

FIGURE 2. Main elements in the ENFORS (European Network for long-term Forest Ecosystem and Landscape Research, COST Action E25) proposal for future research and development on sustainable forest management.

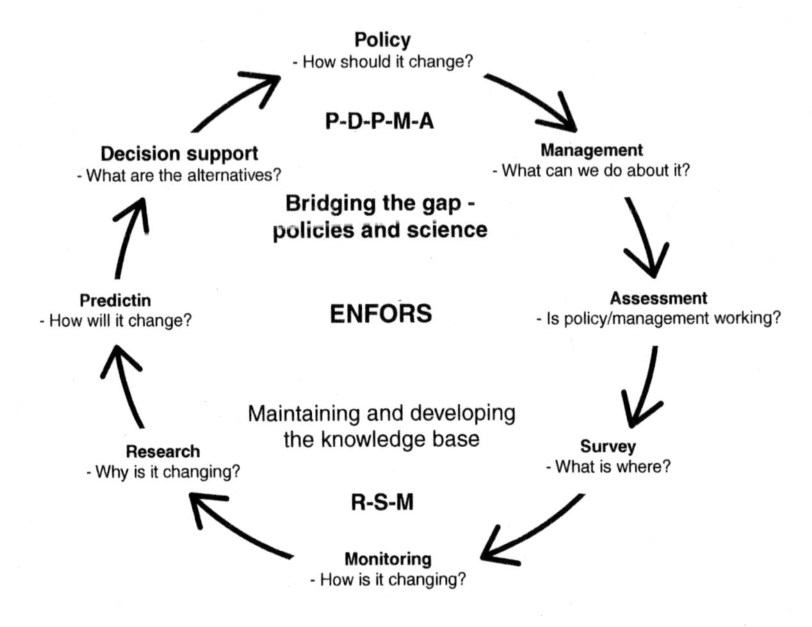


nance, proper documentation, and management. This is a major concern because (1) access to long-term data is vital to our ability to detect changes where long-term processes are involved (e.g., global change) (Parr et al., 2003); (2) long-term data are necessary for making predictions of the future (including model validation) (Freckleton, 2004); and (3) sustainable resource management (including its political framework) is a highly information-demanding activity where access to data at multiple spatial and temporal scales is needed for its assessment (Sverdrup and Stjernquist, 2002; Angelstam et al., 2004).

The loss of information and access to data are recognized problems in environmental sciences (Michener et al., 1997; National Research Council, 1997). One way to overcome these difficulties is to introduce guidelines for proper documentation in terms of meta-data, as well as the development of information systems for data management and exchange. This needs to be communicated and implemented at a large scale by the forest scientific community as within other environmental sciences (Michener et al., 1997). However, careful considerations in designing these information systems must be taken, as the requirements

FIGURE 3. Main functions of the forests as interpreted from the criteria and indicators elaborated by the Ministerial Conference on the Protection of Forests in Europe (MCPFE, 2003b) as well as their driving forces and processes indicating research needs for developing the knowledge base on forests and forested landscapes.

| Functions | Socio-Cultural | Economic | Biodiversity | Protective | Productive |
|---|---|---|---|---|---|
| Driving forces | Human needs | Demand | Management<br>Landscape fragmentation<br>Disturbance | Chemical-physical environment<br>Environmental change | Demand for forest products |
| Major processes | Equity & justice<br>Governance<br>Cultural identity | Globalisation<br>Economic growth | Population dynamics<br>Succession | Water cycling*<br>Erosion<br>Organic matter formation | C-cycling*<br>Element cycling* |
| Sub-processes | Policy-making<br>Law<br>Citizen power<br>Discursive practices | Subsidy<br>Taxation<br>Technology | Self-regulation<br>Fungal/bacteria symbiosis<br>Migration<br>Extinction<br>Food-chains | Water flow and budgets<br>Mineralisation*<br>Decomposition*<br>Weathering*<br>Uptake*<br>Leaching* | Primary production* |
| Knowledge base | Participation/<br>Qualitative research<br>Political science | Economics<br>Science & technology of production | Species requirements<br>Ecosystem functioning* | Forest health & vitality<br>Resilience | Inventories<br>Silviculture<br>Forest growth<br>Modeling |

*Processes of ecosystem functioning.

on information and data systems differ between those designed for scientific or decision-making purposes (Vos et al., 2000). Effective information and data handling systems designed for scientific exchange as well as for decision-making at the political and management levels are required if criteria and indicators are to become a successful tool for implementing SFM (Haklay, 2003; Joyce, 2003). In the near future, care, management, and retrieval of information can become a major bottleneck in our efforts toward sustainable use of natural resources.

## *CONCLUSIONS FOR THE FUTURE*

We believe that the landscape is the appropriate level for implementing a principle such as SFM. We argue that the landscape is a suitable arena for interactions between scientists, policy makers, managers, and forest users. This level provides the necessary elements for analyzing the effects of ecological and economic as well as social-cultural processes at lower and higher spatial levels. The European Network for

Long-Term Forest Ecosystem and Landscape Research has created the basis for a European network of "landscape laboratories," and we suggest that these facilities could be a means of implementing SFM at regional levels through a closer collaboration between scientists, policy makers, and other stakeholders. However, this cannot be done without a common framework. The ENFORS research strategy, where gaps of knowledge and future research needs have been identified, will serve this purpose.

For example, the process to improve and develop new criteria and indicators for SFM in Europe must continue. In particular, criteria and indicators for economic and social-cultural aspects are missing or poorly developed. In this perspective, the use of qualitative indicators is problematic. Our research methodologies and decision systems need to be adapted accordingly. With regard to global change, prediction models and scenarios need to be further developed to better correspond to the needs required for risk assessment analyses. Here we foresee an increased collaboration between disciplines. It is thought that the ENFORS research strategy is best implemented through a bottom-up approach where researchers specify the suggested research needs into projects with clear and testable hypotheses, and with collaboration at national or international levels.

Furthermore, we consider that most of the necessary basic knowledge exists, but that this knowledge (1) is poorly communicated to decision makers, (2) could be more intensively used through increased collaboration across disciplines, and (3) is partly lost because of the lack of proper management of long-term information. Decision-support systems (DSS) is an example of an important tool for overcoming these problems. Members of ENFORS are involved in projects that aim at creating DSS at the national level, and an initiative for exchange and collaboration at the European level is under development. Furthermore, projects to create and improve information systems for handling long-term data are steadily increasing in many European countries. However, the efficiency and value of these national systems would dramatically increase if there were common standards and coordination at the European or global level.

## REFERENCES

Andersson, F., P. Angelstam, K.-H. Feger, H. Hasenauer, A. Mårell, U. Schneider, and P. Tabbush (eds.). 2005. A research strategy for sustainable forest management in Europe. Technical Report 5, Working Groups 2 and 3, COST Action E25, ECOFOR, Paris, France. 166 pp.

Andersson, F., K.-H. Feger, R.F. Hüttl, N. Kräuchi, L. Mattson, O. Sallnäs, and K. Sjöberg. 2000. Forest ecosystem research–priorities for Europe. *Forest Ecology and Management* 132:111-119.

Angelstam, P., S. Boutin, F. Schmiegelow, M.-A. Villard, P. Drapeau, G. Host, J. Innes, G. Isachenko, T. Kuuluvainen, M. Mönkkönen, J. Niemelä, G. Niemi, J.-M. Roberge, J. Spence, and D. Stone. 2004. Targets for boreal forest biodiversity conservation–a rationale for marcoecological research and adaptive management. *Ecological Bulletins* 51:487-509.

Angelstam, P. and J. Törnblom. 2004. Maintaining forest biodiversity in actual landscapes–European gradients in history and governance systems as a "landscape lab." In: pp. 299-313. M. Marchetti (Ed.). Monitoring and indicators of forest biodiversity in Europe–from ideas to operationality. EFI Proceedings 51, Joensuu, Finland.

Anon. 2000. The Montreal Process. Year 2000 progress report. Progress and innovation in implementing criteria and indicators for the conservation and sustainable management of temperate and boreal forests. The Montréal Process and Liaison Office, Ottowa, Canada.

Anon. 2001. Independent scientists an endangered species. ISIS report, September 4. Retrieved November 15, 2005 from http://www.i-sis.org.uk/.

Antrop, M. 2005. Why landscapes of the past are important for the future. *Landscape and Urban Planning* 70:21-34.

Besseau, P. Dansou, K. and F. Johnson. 2002. The International Model Forest Network (IMFN): elements of success. *Forestry Chronicle* 78:648-654.

Bradshaw, G.A. and J.G. Borchers. 2000. Uncertainty as information: narrowing the science-policy gap. *Conservation Ecology* 4(1):7. Retrieved November 15, 2005 from http://www.consecol.org/vol14/iss1/art7/.

Campbell, D.T. and J.C. Stanley. 1963. Experimental and quasi-experimental designs for research. Houghton Mifflin Co., Boston.

Farrell, E.P., E. Führer, D. Ryan, F. Andersson, R. Hüttl, and P. Piussi. 2000. European forest ecosystems: building the future on the legacy of the past. *Forest Ecology and Management* 132:5-20.

Food and Agriculture Organization of the United Nations [FAO]. 1999. FRA 2000. Global forest cover map. Forest Resource Assessment Programme, Working Paper 19, Food and Agriculture Organization of the United Nations, Rome, Italy. Retrieved November 10, 2005, from http://www.fao.org/forestry/fo/fra/index.jsp.

Freckleton, R.P. 2004. The problems of prediction and scale in applied ecology: the example of fire as a management tool. *Journal of Applied Ecology* 41:599-603.

Führer, E., F. Andersson, and E.P. Farrell (eds.). 2000. Pathways to the wise management of forests in Europe. *Forest Ecology and Management* 132:1-119.

Haklay, M.E. 2003. Public access to environmental information: past, present and future. *Computers, Environment and Urban Systems* 27:163-180.

Hobbie, J.E., S.R. Carpenter, N.B. Grimm, J.R. Gosz, and T.R. Seastedt. 2003. The US Long-Term Ecological Research Program. *BioScience* 53:21-32.

Holt, R.D., G.R. Robinson, and M.S. Gaines. 1995. Vegetation dynamics in an experimentally fragmented landscape. *Ecology* 76(5):1610-1624.

Joyce, L.A. 2003. Improving the flow of scientific information across the interface of forest science and policy. *Forest Policy and Economics* 5:339-347.

Kimmins, H. 1992. Balancing act–environmental issues in forestry. UBC Press, Vancouver.

Kohler, R.E. 2002. Landscapes and labscapes. Exploring the lab-field border in biology. The University of Chicago Press, Chicago.

Likens, G.E. Bormann, F.H. Pierce, R.S. Eaton, J.S. and N.M. Johnson. 1977. Biochemistry of a forested ecosystem. Springer-Verlag, Berlin.

Lindenmayer, D.B. Cunningham, R.B. and M.L. Pope. 1999. A large-scale "experiment" to examine the effects of landscape context and habitat fragmentation on mammals. *Biological Conservation* 88:387-403.

Malcolm, H. (Ed.). 2002. Biosphere reserves. Special places for people and nature. United Nations Educational, Scientific, and Cultural Organization, Paris.

Mårell, A., O. Laroussinie, N. Kräuchi, G. Matteucci, F. Andersson, and E. Leitgeb. 2003. Scientific issues related to sustainable forest management in an ecosystem and landscape perspective. Technical Report 1, Working Group 1, COST Action E25, Ecosystèmes Forestiers (ECOFOR), Paris, France. 64 pp.

Mårell, A. and E. Leitgeb (eds.). 2005a. European long-term research for sustainable forestry: experimental and monitoring assets at the ecosystem and landscape level. Part 1: Country reports. Technical Report 3, Working Group 1, COST Action E25, Ecosystèmes Forestiers (ECOFOR), Paris, France, 316 pp.

Mårell, A. and E. Leitgeb (eds.). 2005b. European long-term research for sustainable forestry: experimental and monitoring assets at the ecosystem and landscape level. Part 2: ENFORS field facilities. Technical Report 4, Working Group 1, COST Action E25, Ecosystèmes Forestiers (ECOFOR), Paris, France. 80 pp.

Mårell, A., E. Leitgeb, O. Laroussinie, N. Kräuchi, G. Matteucci, and F. Andersson. 2004. Guidelines for national inventories of field research facilities. Technical Report 2, Working Group 1, COST Action E25. GIP Ecosystèmes Forestiers (ECOFOR), Paris. 44 pp.

Martin, J.-L. and C. Baltzinger. 2002. Interaction among deer browsing, hunting, and tree generation. *Canadian Journal of Forest Research* 32:1254-1264.

McDonald, G.T. and M.B. Lane. 2004. Converging global indicators for sustainable forest management. *Forest Policy and Economics* 6:63-70.

Merkel, A. 1998. The role of science in sustainable development. *Science* 281(5375): 336-337.

Michener, W.K., J.W. Brunt, J.J. Helly, T.B. Kirchner, and S.G. Stafford. 1997. Nongeospatial metadata for the ecological sciences. *Ecological Applications* 7(1): 330-342.

Ministerial Conference on the Protection of Forests in Europe [MCPFE]. 2000. General declarations and resolutions adopted at the Ministerial Conferences on the Protection of Forests in Europe: Strasbourg 1990, Helsinki 1993, Lisbon 1998. Liaison Unit, Vienna, Austria.

Ministerial Conference on the Protection of Forests in Europe [MCPFE]. 2003a. Implementation of MCPFE commitments. National and pan-European activities 1998-2003. Liaison Unit, Vienna, Austria.

Ministerial Conference on the Protection of Forests in Europe [MCPFE]. 2003b. Improved pan-European indicators for sustainable forest management as adopted by the MCPFE Expert Level meeting. Liaison Unit, Vienna, Austria.

Ministerial Conference on the Protection of Forests in Europe [MCPFE]. 2003c. MCPFE work programme. Pan-European follow-up of the fourth Ministerial Conference on the Protection of Forests in Europe. Adopted at the MCPFE Expert Level meeting. Liaison Unit, Vienna, Austria.

National Research Council. 1997. Bits of power: issues in global access to scientific data. National Academy Press, Washington, DC.

Parr, T.W., A.R.J. Sier, R.W. Barrarbee, A. Mackay, and J. Burgess. 2003. Detecting environmental change: science and society–perspectives on long-term research and monitoring in the 21st century. *The Science of the Total Environment* 310:1-8.

Petheram, R.J., P. Stephen, and D. Gilmour. 2004. Collaborative forest management: a review. *Australian Forestry* 67:137-146.

Rosén, K. 1982. Supply, loss and distribution of nutrients in three coniferous forest watersheds in central Sweden. Ph.D. Dissertation. Rapporter i Skogsekologi och skoglig marklära, Rapport 41, Swedish University of Agricultural Sciences, Uppsala, Sweden.

Schnitzler, A. and F. Borlea. 1998. Lessons from natural forests as keys for sustainable management and improvement of naturalness in managed broadleaved forests. *Forest Ecology and Management* 109:293-303.

Sinclair, A.J. and D.L. Smith. 1999. The model forest program in Canada: Building consensus on sustainable forest management? *Society and Natural Resources* 12:121-138.

Sverdrup, H. and I. Stjernquist (eds.). 2002. Developing principles and models for sustainable forestry in Sweden. In the series: Managing forest ecosystems. Kluwer Academic Publishers, Dordrecht, The Netherlands.

Tyrväinen, L., R. Gustavsson, C. Konijnendijk, and Å. Ode. 2006. Visualization and landscape laboratories in planning, design and management of urban woodlands. *Forest Policy and Economics* 8:811-823.

Vos, P., E. Meelis, and W.J. Ter Keurs. 2000. A framework for the design of ecological monitoring programs as a tool for environmental and nature management. *Environmental Monitoring and Assessment* 61:317-344.

doi:10.1300/J091v24n02_09

# Index

*Note:* Page numbers followed by the letter "t" designate tables; numbers followed by the letter "f" designate figures.


Adams, D.M., 77
AFS. *See* Australian Forestry Standard (AFS)
AIC. *See* Akaike's information criterion (AIC)
Akaike's information criterion (AIC), 73
Alig, R.J., 78,209
Amazon Cooperation Treaty, 253
Andersson, F.P., xx,279,283
Andrean-Patagonia Forestry Research and Advisory Centre (CIEFAP), 88
ANOVA, in monitoring of SFM in Pacific Rim region, 260-262, 263t-265t,267t-269t
APEC forum. *See* Asia-Pacific Economic Cooperation (APEC) forum
Asia-Pacific Economic Cooperation (APEC) forum, 251
Association of Forest Communities in the Peten (ACOFOP), 198
Australia, viewpoint on integrated wood production within sustainable forest management, 19-40. *See also* Wood production, sustainable forest management and, integration of, Australian viewpoint
Australian Forestry Standard (AFS), 253

Baeza, A.P., 123
Barbour, R.J., 77
Bass, S., 272
Bayesian information criterion (BIC), 74
Benson, G.L., xxi
Berlik, M.M., 232
Best, C., 11
Bhopal-India Process, 253
BIC. *See* Bayesian information criterion (BIC)
Bigsby, H., 85
Bonferroni critical distance, 254
Bonferroni tests, in monitoring of SFM in Pacific Rim region, 260-262, 263t-265t,267t-269t
Breunig, K., 231
Broadleaf Forest Development Project, 195
Brundtland, G.H., 143

CACOFOP. *See* Association of Forest Communities in the Peten (ACOFOP)
Calcer, M., 25
Campbell, S.M., 241
Canadian Council of Forest Ministers (CCFM), 253


Available online at http://jsf.haworthpress.com

Canadian International Development Agency, 195
Canadian Model Forests, 285
Carabelli, E., xx,85
CATIE. *See* Tropical Agricultural Research and Higher Education Center (CATIE)
CATIE-based Center for the Competitiveness of Ecoenterprises (CeCoEco), 199,200
CCF. *See* Continuous cover forestry (CCF)
CCFM. *See* Canadian Council of Forest Ministers (CCFM)
CeCoEco. *See* CATIE-based Center for the Competitiveness of Ecoenterprises (CeCoEco)
Center for International Forestry Research (CIFOR), 89
Central America
  forest resources in, 190,191t
  sustainable forestry in, barriers to, 189-207
    fragmented support for community-based forest management, 200-201
    gaps in available knowledge for SFM, 204-205
    introduction to, 190-191,191t
    lack of business management skills required for rural enterprise development, 197-200,197t
    lack of enabling environment that favors community-based SFM, 202-204
    lack of legal access to forests, 191-194
    lack of technical capacity to carry out forest management operations, 194-196,195t
Centre for International Forest Research (CIFOR), 253,258, 260
Certification, in wood production, 33
Chacko, K.C., 109
Chemonics, 198
Chilean Magellan Region, *Nothofagus betuloids* production forests in
  prior knowledge about, 125-128
    ecological aspects, 126
    forest species record, 125
    geographical distribution, 25-126
    stand volume and tree growth, 127-128
    vegetation, 126-127
  structural and biometric characterization of, 123-140
    study of
      discussion of, 135-137
      introduction to, 124-125
      materials and methods in, 128-131,128f
      results of, 131-135,131t-135t
      sites of, 128-129,128f
      stand description in, 129-131
Chubut Forest Service, 88
C&I schemes. *See* Criteria and indicator (C&I) schemes
CIEFAP. *See* Andrean-Patagonia Forestry Research and Advisory Centre (CIEFAP)
Ciesla, W.M., 170
CIFOR. *See* Centre for International Forest Research (CIFOR)
Clapp, R.A., 251
Climate change, effects on sustainability of forests, 175
CLTAP. *See* Convention on Long-Range Transboundary Air Pollution (CLTAP)
COATLAHL. *See* Regional Agroforestry Cooperative, Colon, Atlantida, Honduras Ltd. (COATLAHL)
CODEFORSA. *See* Forest Development Commission of San Carlos (CODEFORSA)

Codes of Practice, in wood production, 27-29
COHDEFOR. *See* Honduran Forest Development Corporation (COHDEFOR)
Commission of the European Communities, 185
Common(s), forests as set of, 8-11
Commonwealth of Australia, 253
Commonwealth Parliament, 21
Commonwealth-State National Competition Policy, 31
Commonwealth-State National Forest Policy Statement, 20-22,23
Conacher, A., 25,33
Conacher, J., 25,33
CONAP. *See* National Protected Areas Council (CONAP)
Constitution, 21,34
Continuous cover forestry (CCF), 50,52,53
Convention on Long-Range Transboundary Air Pollution (CLPTA), of UNECE, 173-174,181
COST. *See* European Cooperation in Science and Technical Research (COST)
COST Action E25–European Network for Long-Term Ecosystem and Landscape Research, 280
Cost-effectiveness forest certification, 236-237
Criteria and indicator(s) (C&I)s, in measuring sustainability, in Tierra del Fuego, Argentina, 88-89,90f,94-95,96t-103t
Criteria and indicator (C&I) schemes, 85
Critical Threshold Value (CTV), 93
CTV. *See* Critical Threshold Value (CTV)
Cullen, R., 85
Damery, D.T., 229,233
Dargavel, J., 25,32
Davey, S., 22
Deal, R.L., xx,xxi,141
Department of International Development of the United Kingdom, 195
Department of Prime Minister and Cabinet, 22
Donaubauer, E., 170
Donoso, C., 137
Douglas-fir region, wood product proportions in, size-, age-, and time-related changes in, 59-83
  future directions in, 74-79,75f,76f
  introduction to, 60-64
  study of
    analysis of, 67-69
    data from, 64-67,66t,67t,68f
    discussion of, 72-79,75f,76f
    materials and methods in, 64-69, 66t,67t,68f
    results of, 69-72,70t,71f,72f
Dovers, S., 28

Earth Summit on Sustainable Development, 175
Ecological Site Classification system, 54
Ecological Sustainable Production (ESP) team, 64
*EcoNegocios Forestales–Forest EcoBusiness*, 199
Economic values, in wood production, 32
Ecosystem, managing within, 157-159
EFERN. *See* European Forest Ecosystem Research Network (EFERN)
EMEP, 174,182
EMS. *See* Environmental Management System (EMS)

ENFORS. *See* European Network for Long-Term Forest Ecosystem and Landscape Research (ENFORS)
Environmental change, sustainability of European forests and, 165-187. *See also* European forest(s), sustainability of, environmental change and
Environmental Management System (EMS), 253
EU. *See* European Union (EU)
EU COST program, 180
Eucalypt plantations, tropical, productivity of, increasing and sustaining over multiple rotations in, 109-121. *See also* Tropical Eucalypt plantations, productivity of, increasing and sustaining over multiple rotations in
European Community, 182
European Cooperation in Science and Technical Research (COST), 283
European forest(s), biodiversity assessment in, monitoring of SFM in, 182-184,183t
European Forest Ecosystem Research Network (EFERN), 281-283
European Forestry Strategy, 185
European forests, sustainability of
  environmental change and, 165-187
    climate change, 175
    critical loads and ecosystem approaches in forestry, 181-182
    described, 169-175,172t
    EU monitoring scheme and national forces inventories, 179-180,180f,181f
    forest area and, 176-179, 178f-179t
    forest decline, 169-170
    fungal pathogens, 171-173,172t
    future issues in, 184-185
    insect pathogens, 170-171
    introduction to, 166-169,167f
    pollution, 173-174
  monitoring and evaluation of, 175-184,178f,179t,180f,181f, 183t
European network, in support of SFM, 279-293. *See also* Ministerial Conference on the Protection of Forests in Europe (MCPFE)
European Network for Long-Term Forest Ecosystem and Landscape Research (ENFORS), 281,283-289, 284f,287t,288f,289t
European SFM, 280
European Union (EU), 168
European Union (EU) Forestry Strategy, 168

FAO. *See* Food and Agriculture Organization (FAO)
FCS. *See* Forestry Commission Scotland (FCS)
Ferguson, I.S., xix,19,28
Food and Agriculture Organization (FAO), 247,249,273
FORESCOM, 198
Forest(s)
  area of, effects on sustainability of forests, 176-179,178f-179t
  in Central America, lack of legal access to, 191-194
  decline of, effects on sustainability of forests, 169-170
  European, sustainability of, environmental change and, 165-187. *See also* European forest(s), sustainability of, environmental change and
  native, wood production in, future of, 35-37

Scottish, silviculture of, at time of change. *See also* Silviculture, of Scottish forests, at time of change
as set of commons, 8-11
sustainability of, environmental threats to, 169-175,172t
in U.S.
ownership of, 211-214,212f
value of, 214-216
Forest and Rangeland Renewable Resource Planning Act timber assessment, 74
Forest certification, cost-effectiveness, 236-237
Forest Development Commission of San Carlos (CODEFORSA), 195
Forest Practices Code of British Columbia, 62
Forest resources, in Central America, 190,191t
Forest Stewardship Council (FSC), 194,236,253
C&I scheme of, 88
Forest-GALES, 54
Forestland conditions, in U.S., changes in, 219-222
Forestry
continuous cover, 50
sustainable. *See* Sustainable forestry
Forestry Commission, 44
Forestry Commission Scotland (FCS), 49
Forestry Stewardship Council (FSC), 43
Freer-Smith, P., xx, 165
FSC. *See* Forestry Stewardship Council (FSC)
Fungal pathogens, effects on sustainability of forests, 171-173,172t

Galloway, G.E., xx,189
GDP. *See* U.S. Gross Domestic Product (GDP)
Glentress Forest, 52
Global Forest Resource Assessment, of United Nations Food and Agriculture Organization, 177
*Global Forest Resources Assessment 2000*, 273
Grennfelt, P., 181
Grove, T.S., 109

Hardin, G., 9
Haynes, R.W., xix,1,74,78
HCA. *See* Hierarchical cluster analysis (HCA)
Helliwell, D.R., 52
Helms, J.A., 144
Helsinki Ministerial Conference (1993), 43
Herath, G., 25
Hickey, G.M., xx,245,246
Hierarchical cluster analysis (HCA), 254,260,260t,261f
Holvoet, B., 255
Honduran Forest Development Corporation (COHDEFOR), 195
Horwitz, P., 25
Hunter, M.L., Jr., 158

Indonesian Ecolabelling Institute (LEI), 253,258
Insect(s), effects on sustainability of forests, 170-171
INTA. *See* National Institute of Agricultural Research (INTA)
Intercropping, legume, in study of increasing and sustaining over multiple rotations in Tropical Eucalypt plantations productivity, 114-115,115t, 116t
Intergovernmental Panel on Climate Change (IPCC), 175

International Cooperative Programme on Assessment and Monitoring of Air Polutants Effects on Forests, 174
International Organization for Standardization (ISO), 253
International Organization for Standardization (ISO) 14001 Environmental Management System framework, 27
International Tropical Timber Organization (ITTO), 253
International Union of Forest Research Organizations (IUFRO)
  2005 World Congress of, 222
  World Congress of, xix,xx
IPCC. *See* Intergovernmental Panel on Climate Change (IPCC)
ISO. *See* International Organization for Standardization (ISO)
ITTO. *See* International Tropical Timber Organization (ITTO)

Jones, G., 33

Kerala, India, Tropical Eucalypt plantations in, productivity of, increasing and sustaining over multiple rotations, 109-121. *See also* Tropical Eucalypt plantations, productivity of, increasing and sustaining over multiple rotations
Kimmins, J.P., 272,281
Kirkpatrick, J.B., 25
Kyoto Protocol process, 225

Lalela, E., 127
Land use, U.S. view on changes affecting SFM, 209-227. *See also* United States (U.S.), view on changes in land use and land values affecting SFM
Land values, U.S. view on changes affecting SFM, 209-227. *See also* United States (U.S.), view on changes in land use and land values affecting SFM
Landowner-driven SFM, value-added processing and, in Massachusetts, U.S., 229-243. *See also* Massachusetts Woodlands Cooperative, LLC (MWC)
  cooperation as solution, 234-235
  cost-effectiveness forest certification, 236-237
  future of, 240-241
  introduction to, 230-233,232t
  landowner goals, 243-244,244t
  operations and marketing, 237-240
Latta, G.S., 77
LEED architectural standards. *See* Low-energy and environmental design (LEED) architectural standards
Legume intercropping, in study of increasing and sustaining over multiple rotations in Tropical Eucalypt plantations productivity, 114-115,115t, 116t
LEI. *See* Indonesian Ecolabelling Institute (LEI)
Lepaterique Process for Central America, 253
Limited Liability Corporation (LLC), 235
Lindenmayer, D., 28
Lisbon Ministerial Conference (1998)
LLC. *See* Limited Liability Corporation (LLC)
Low-energy and environmental design (LEED) architectural standards, 239
Lower Mississippi Riverine Forest, 158

LTER. *See* U.S. Long-Term Ecological Research (LTER)

Madariaga, G.C., xx,123
Magellan Region, Chile, *Nothofagus betuloids* production forests in, structural and biometric characterization of, 123-140. *See also* Chilean Magellan Region, *Nothofagus betuloids* production forests in, structural and biometric characterization of
Management, plans of, in wood production, 29-30
Marell, A., xx,279
Mason, W.L., xix,41
Massachusetts, U.S., landowner-driven SFM and value-added processing in, 229-243. *See also* Landowner-driven SFM, value-added processing and, in Massachusetts, U.S.
Massachusetts Woodlands Cooperative, LLC (MWC), 229-243. *See also* Landowner-driven SFM, value-added processing and, in Massachusetts, U.S.
  cost-effectiveness forest certification, 236-237
  formation of, 235-236
  future of, 240-241
  membership growth, 240-241
  mission of, objectives of, 235
  operations and marketing of, 237-240
Maya Biosphere Reserve (MBR), 192
MBR. *See* Maya Biosphere Reserve (MBR)
McDonald, J., 32
McKinnell, F.H., 26
MCPFC Pan-European criteria, 176
MCPFE. *See* Ministerial Conference on the Protection of Forests in Europe (MCPFE)
Mendham, D.S., 109
Mercer, D., 25,33
Ministerial Conference on the Protection of Forests in Europe (MCPFE), 168,176,177,182,185,279-293
  building on results from successive research networks, 282-289, 284f,287t,288f,289f
  EFERN and, 282-283
  ENFORS and, 283-289,284f,287t, 288f,289t
  future directions for, 289-290
  introduction to, 280-281
Mobbs, C., 25
Monserud, R.A., xix,59
Montreal Process, 144-145,162,253, 260,274,281
  C&I of, 88
Multilateral Investment Fund of the Interamerican Development Bank, 200
Musselwhite, G., 25
Muys, B., 255
MWC. *See* Massachusetts Woodlands Cooperative, LLC (MWC)

National Conservation Reserve System, 34
National Forest and Rangeland Renewable Resources Planning Act (RPA), 219
National Forest Inventory, 22
National Forest Policy, 19
National Forest Policy Agreement, 25
National Forest Policy Statement, 20-22,23
National Forest System, 63
National Geographic Society, 149
National Institute of Agricultural Research (INTA), 89

National Protected Areas Council (CONAP), 192
National System of Protected Wild Areas of the State (SNASPE), 126
Native forest(s), wood production in, future of, 35-37
Natural Forest Integrated Management Project, 195
Natural Resource Conservation Service (NRCS), 217
Natural Resources Canada (NRC), 258
Navarro, R.C., 123
Network for the Management of Broadleaf Tropical Forests in Honduras (REMBLAH), 198, 200-201
Nilsson, J., 181
1993 Helsinki Ministerial Conference, 43
1998 Lisbon Ministerial Conference, 43
1998 UK Forestry Standard, 43
NIPF. *See* Nonindustrial private forest (NIPF)
Nitschke, C.R., xx,245
Nonindustrial private forest (NIPF), 211,217,230
Northwest Forest Plan, 14,62
Northwest Research Station, 64
*Nothofagus betuloids* production forests, in Chilean Magellan Region, structural and biometric characterization of, 123-140. *See also* Chilean Magellan Region, *Nothofagus betuloids* production forests in, structural and biometric characterization of
NRC. *See* Natural Resources Canada (NRC)
NRCS. *See* Natural Resource Conservation Service (NRCS)

O'Connell, A.M., 109
OLAFO. *See* Project for the Sustainable Development of Central America (OLAFO)
Oliver, C.D., xx,141,144

Pacific Northwest Research Station, of United States Forest Service, 160
Pacific Rim region
- described, 245,247f
- SFM in
  - background of, 246-253, 248t-249t,250f,252t
  - initiatives in, 252-253
  - monitoring of, 245-278
    - assumptions, 257-258
    - discussion of, 271-274
    - forest ecology and management, 246-250,248t-249t, 250f
    - forest products and trade, 250-252,252t
    - introduction to, 246,247f
    - jurisdiction and documentation, 255,255t,256f
    - limitations in, 257-258
    - methodology in, 253-258, 255t,256f,257f
    - results of, 258-271,259t, 260t,261f,263f, 263t-265t,264f, 267f-269f,267t-270t
      - ANOVA in, 260-262, 263t-265t,267t-269t
      - biological diversity in, 264-265,264f,264t
      - Bonferroni tests in, 260-262,263t-265t, 267t-269t
      - comparison of C&I standards in, 262-264,263f,263t

congruency between C&I standards, 270-271, 270t
constant comparison and content analysis, 258-260,259t
economics in, 266,268f, 268t,269,269f
forest productivity and health in, 265-266, 265t
global cycles in, 266,268t
HCA, 260,260t,261f
society and, 269-270,269t
soil and water in, 266, 267f,267t
Pandalai, R.C., 109
"Pan-European Criteria" (PEC), 43
Pathogen(s), fungal, effects on sustainability of forests, 171-173,172t
*Pathways to the Wise Management of Forests in Europe*, 282
PEC. *See* "Pan-European Criteria" (PEC)
Peri, P., 85
Plantation profitability, timber prices and, 48-50,49t
Pollution, effects on sustainability of forests, 173-174
Prabhu, R., 91,253,272
Prestemon, J.P., 68
Price(s), of timber, plantation profitability and, 48-50,49t
Private land management, challenges facing, 11-13,13
Production forests, *Nothofagus betuloids*
of Chilean Magellan Region, structural and biometric characterization of, 123-140
of Chilean Magellan Region, structural and biometric characterization of. *See also* Chilean Magellan Region, *Nothofagus betuloids* production forests in, structural and biometric characterization of
*Profiles from Working Woodlands*, 241
Project for the Sustainable Development of Central America (OLAFO), of CATIE, 192
Public Land Use Commission, 33
Public participation, in wood production, 32-33

Rainforest Alliance, 237
Regional Agroforestry Cooperative, Colon, Atlantida, Honduras Ltd. (COATLAHL), 194,199
Regional Forest Agreement, 25,26
REMBLAH. *See* Network for the Management of Broadleaf Tropical Forests in Honduras (REMBLAH)
Rio Declaration of Forest Principles, 176
Rising prices, as integrating concern about wood production and sustainable forest management, 3-8,5f,6f,7t
Rural Business–Cooperative Services Grant, of USDA, 237

Saaty, T.L., 92
Sankaran, K.V., 109,114
Scottish forest(s), silviculture of, at time of change, 41-57. *See also* Silviculture, of Scottish forests, at time of change
Scottish Forestry Strategy, 43,51-52
Scottish Parliament, 43
Seymour, R.S., 158
SFI. *See* Sustainable Forestry Initiative (SFI)
SFM. *See* Sustainable Forest Management (SFM)
Silviculture, of Scottish forests
current plantation management, 45-48,47t
at time of change, 41-57

alternative practices, 50-53,51f
introduction to, 42-45
plantation profitability, 48-50,49t
timber prices, 48-50,49t
Slash manipulation, in study of increasing and sustaining over multiple rotations in Tropical Eucalypt plantations productivity, 114-115,115t, 116t
Small and medium forest enterprises (SMFE), 197
Smartwood, 237
SMFE. *See* Small and medium forest enterprises (SMFE)
SNASPE. *See* National System of Protected Wild Areas of the State (SNASPE)
Social values, in wood production, 32
Society of American Foresters, 144
Southgate, D., 251
Statement of Forest Principles, 176
Stewart, J., 33
Stoian, D., xx,189
Sustainable Agriculture Research and Education program, of USDA, 241
Sustainable Forest Management (SFM), 43
changes in land use and land values effects on, U.S. view on, 209-227. *See also* United States (U.S.), view on changes in land use and land values affecting SFM
European network in support of, 279-293. *See also* Ministerial Conference on the Protection of Forests in Europe (MCPFE)
as evolving concept, 281-282
landowner-driven, in Massachusetts, U.S., 229-243. *See also* Landowner-driven SFM, value-added processing and, in Massachusetts, U.S.
MCPFE in, 279-293. *See also* Ministerial Conference on the Protection of Forests in Europe (MCPFE)
monitoring of, biodiversity assessment in European forests and, monitoring of, 182-184, 183t
in Pacific Rim region, monitoring of, 245-278. *See also* Pacific Rim region, SFM in, monitoring of
in Tierra del Fuego, Argentina, measuring of. *See also* Tierra del Fuego, Argentina, measuring SFM in
wood production and, in United States, integrating concerns about, 1-18. *See also* Wood production, SFM and, in U.S., integrating concerns about
Sustainable forest management (SFM), in Tierra del Fuego, Argentina, measuring of, 85-108
Sustainable forestry
in Central America, barriers to, 189-207. *See also* Central America, sustainable forestry in, barriers to
defining of, 142-155,145f, 146t-148t,151f,152f
in different spatial scales, 141-163
working definition of, 141-163
spatial scales in, study of
integrated management areas in, 157, 158-159
intensive tree plantation in, 159
managing within ecosystem, 157-159
protected areas in, 158
regional implementation in, 159-161
within-country analyses in, 155-157,156f,157f

Sustainable Forestry Initiative (SFI), 253
Sustainable management, one tenure-one use vs., 33-34
Sustainable Wood Production Initiative (SWPI), 160
Sustainable yield, in wood production, 30-31
Sverdrup, 180
Swiss Agency for Development and Cooperation, 195
SWPI. *See* Sustainable Wood Production Initiative (SWPI)

Tasmanian Government, 36
Thiele, R., 251
Tierra del Fuego, Argentina, measuring SFM in, 85-108
  C&I in, 88-89,90f
  introduction to, 86-88
  study of
    C&I in, 94-95
      identification of, 90-92
      importance of, 92-93
      weights and individual sustainability scores for, 94,96t-103t,97
    C&I scores in, as indicators of degree of sustainability, 93
    methodology in, 90-94
    overall sustainability score calculation in, 96t-103t,97,99
    results of, 94-104,96t-103t,104f
    surveys in, 93-94
    using sustainability scores to improve sustainability in, 99,102-104,104f,105t
Timber, prices of, plantation profitability and, 48-50,49t
Toyne, P., 34
*Tragedy of the Commons*, 9
Tropical Agricultural Research and Higher Education Center (CATIE), 192,195,196,200
  OLAFO of, 192
Tropical Eucalypt plantations, productivity of, increasing and sustaining over multiple rotations in, 109-121
  introduction to, 110-111
  study of, 111-121
    discussion of, 117-119
    experimental design in, 112-114, 113t
    methods in, 111-114,112t,113t
    results of, 114-117,115t,116t, 117f
    sites of, 111-114,112t,113t
    slash manipulation and legume intercropping in, 114-115, 115t,116t
    statistical analyses in, 114
    tree growth measurements in, 114
    weeding and fertilizer effects in, 116-117,117f
Tumaneng-Diete, T., 251
Turner, B.J., 30
2005 World Congress, of International Union of Forest Research Organizations, 222

UK Forestry Standard (1998)
UKWAS. *See* United Kingdom Woodland Assurance Standard (UKWAS)
UNCED. *See* United Nations Conference on Environment and Development (UNCED)
UNECE. *See* United Nations Economic Commission for Europe (UNECE)
UNESCO. *See* United Nations Educational, Scientific, and Cultural Organization (UNESCO) Man and Biosphere Reserves

UNFF. *See* United Nations Forum on Forests (UNFF)
United Kingdom Woodland Assurance Standard (UKWAS), 43,44,52
United Nations Conference on Environment and Development (UNCED), 144,168,185
United Nations Convention of Long-Range Transboundary Air Pollution, 177
United Nations Economic Commission for Europe (UNECE)
  CLTAP convention of, 173-174
  Timber Committee of, 177
United Nations Educational, Scientific, and Cultural Organization (UNESCO) Man and Biosphere Reserves, 285
United Nations Food and Agriculture Organization (FAO), 149
  European Forestry Committee of, 177
  Global Forest Resource Assessment of, 177
United Nations Forum on Forests (UNFF), 185,281
United States (U.S.)
  Massachusetts, landowner-driven SFM and value-added processing in, 229-243. *See also* Landowner-driven SFM, value-added processing and, in Massachusetts, U.S.
  view on changes in land use and land values affecting SFM, 209-227
    forestland conditions, 219-222
    forests in rural-urban continuum, 218
    introduction to, 210
    1953-2002, 216-219
    ownership of forests, 211-214, 212f
    spatial changes, 218-219
    state of land, 222-225
    total forest area, 217-218
    value of forests, 214-216
  wood production and SFM in, integrating concerns about, 1-18. *See also* Wood production, SFM and, in U.S., integrating concerns about
United States Forest Service, Pacific Northwest Research Station of, 160
University of Massachusetts–Amherst, 236
U.S. Agency for International Development (USAID), 195,198
U.S. Department of Agriculture (USDA)
  Rural Business–Cooperative Services Grant of, 237
  Sustainable Agriculture Research and Education program of, 241
U.S. Department of Agriculture (USDA) Forest Service, 64,219,236
U.S. Gross Domestic Product (GDP), 215
U.S. Long-Term Ecological Research (LTER), 285
*U.S. National Report on Sustainable Forests*, 223
USAID. *See* U.S. Agency for International Development (USAID)
USDA. *See* U.S. Department of Agriculture (USDA)

Value(s)
  economic, in wood production, 32
  of forests, in U.S., 214-216
  land, U.S. view on changes affecting SFM, 209-227. *See also* United States (U.S.), view on changes in land use and land values affecting SFM

social, in wood production, 32
Value-added processing, landowner-driven SFM and, in Massachusetts, U.S., 229-243. *See also* Landowner-driven SFM, value-added processing and, in Massachusetts, U.S.
Vanclay, J.K., 30
Venn Diagram approach, 144,145f
Verrugio, G.C., 123

Warren, D.D., 77
White, R., xxi
Wilson, E.R., 52
Wood
nonwood uses of, funding of, 31
pricing of, in wood production, 31
Wood product proportions, size-, age-, and time-related changes in, in Douglas-fir region, 59-83. *See also* Douglas-fir region, wood product proportions in, size-, age-, and time-related changes in
Wood production
in native forests, future of, 35-37
broadening debate, 35-36
relationships between commonwealth and states, 36-37
utilising framework, 36
objectives of, 26-33
certification, 33
Codes of Practice, 27-29
plans of management, 29-30
pricing of wood and funding of nonwood uses, 31
public participation, 32-33
social and economic values, 32
sustainable yield, 30-31
tradeoffs with other values, 31-32
SFM and
integration of, Australian viewpoint, 19-40
conservation objectives, 25-26
future challenges facing, 24-33, 35-37
introduction to, 20-21
one tenure-one use vs. sustainable management, 33-34
outcomes of, 24-33
regional forest agreements, 22-24,23f
wood production objectives, 26-33. *See also* Wood production, objectives of
in U.S., integrating concerns about, 1-18
changing propensity to regulation, 13-14
contribution of rising prices, 3-8,5f,6f,7t
forests as set of commons, 8-11
introduction to, 2-3
private land management challenges, 11-13,13
World Meteorological Organisation, 175
World Trade Organization (WTO), 251
World Wildlife Fund, 149
WTO. *See* World Trade Organization (WTO)

Yale Global Institute of Sustainable Forestry, 150
Yield, sustainable, in wood production, 30-31

Zhou, X., xix,59,74